Biofuels and Bioenergy

Processes and Technologies

GREEN CHEMISTRY AND CHEMICAL ENGINEERING

Series Editor: Sunggyu Lee

Ohio University, Athens, Ohio, USA

Proton Exchange Membrane Fuel Cells: Contamination and Mitigation Strategies
Hui Li, Shanna Knights, Zheng Shi, John W. Van Zee, and Jiujun Zhang

Proton Exchange Membrane Fuel Cells: Materials Properties and Performance
David P. Wilkinson, Jiujun Zhang, Rob Hui, Jeffrey Fergus, and Xianguo Li

Solid Oxide Fuel Cells: Materials Properties and Performance
Jeffrey Fergus, Rob Hui, Xianguo Li, David P. Wilkinson, and Jiujun Zhang

Efficiency and Sustainability in the Energy and Chemical Industries: Scientific Principles and Case Studies, Second Edition
Krishnan Sankaranarayanan, Jakob de Swaan Arons, and Hedzer van der Kooi

Nuclear Hydrogen Production Handbook
Xing L. Yan and Ryutaro Hino

Magneto Luminous Chemical Vapor Deposition
Hirotsugu Yasuda

Carbon-Neutral Fuels and Energy Carriers
Nazim Z. Muradov and T. Nejat Veziroğlu

Oxide Semiconductors for Solar Energy Conversion: Titanium Dioxide
Janusz Nowotny

Lithium-Ion Batteries: Advanced Materials and Technologies
Xianxia Yuan, Hansan Liu, and Jiujun Zhang

Process Integration for Resource Conservation
Dominic C. Y. Foo

Chemicals from Biomass: Integrating Bioprocesses into Chemical Production Complexes for Sustainable Development
Debalina Sengupta and Ralph W. Pike

Hydrogen Safety
Fotis Rigas and Paul Amyotte

Biofuels and Bioenergy: Processes and Technologies
Sunggyu Lee and Y. T. Shah

GREEN CHEMISTRY AND CHEMICAL ENGINEERING

Biofuels and Bioenergy

Processes and Technologies

Sunggyu Lee and Y. T. Shah

CRC Press
Taylor & Francis Group
Boca Raton London New York

CRC Press is an imprint of the
Taylor & Francis Group, an **informa** business

Reprint 2018

CRC Press
Taylor & Francis Group
6000 Broken Sound Parkway NW, Suite 300
Boca Raton, FL 33487-2742

ISBN: 978-1-138-56941-6 (hbk)

Digitally Printed at Replika Press Pvt. Ltd.

Library of Congress Cataloging-in-Publication Data

Lee, Sunggyu.
Biofuels and bioenergy : processes and technologies / Sunggyu Lee, Y.T. Shah.
p. cm. -- (Green chemistry and chemical engineering)
Includes bibliographical references and index.
ISBN 978-1-4200-8955-4 (hardback)
1. Biomass energy. I. Shah, Yatish T. II. Title.

TP339.L44 2012
662'.88--dc23 2012021353

Visit the Taylor & Francis Web site at
http://www.taylorandfrancis.com

and the CRC Press Web site at
http://www.crcpress.com

Contents

Authors

Sunggyu Lee is Russ Ohio Research Scholar in Coal Syngas Utilization and Professor of Chemical and Biomolecular Engineering at Ohio University, where he is also the Director of the Sustainable Energy and Advanced Materials (SEAM) Laboratory. Dr. Lee holds 32 U.S. patents and has authored over 450 journal articles and conference papers. He has authored and edited 15 books in the fields of chemical, energy, and materials processing. His research specialty areas include alternative fuels, syngas conversion, supercritical fluid technology, chemical process engineering and reactor design, and polymer synthesis and processing.

Dr. Y.T. Shah received his Bachelor degree in Chemical Engineering at University of Michigan and Master and Sc.D degrees in Chemical Engineering at M.I.T. During more than 40 years of academic career, he has served as department head of Chemical Engineering, Dean of Engineering and Provost in more than five different institutions. Currently he is a professor of engineering at Norfolk State University, Virginia, USA. Dr. Shah's research interests are in chemical reaction and reactor engineering, particularly as applied to energy and environmental topics. He is an author of three books, about 50 refereed chapters or review articles, and more than 200 refereed journal publications. He has been a consultant to numerous industrial and governmental organizations, particularly in the area of synthetic fuels.

Series Preface

Green Chemistry and Chemical Engineering

A Book Series by CRC Press/Taylor & Francis

The subjects and disciplines of chemistry and chemical engineering have encountered a new landmark in the way of thinking about, developing, and designing chemical products and processes. This revolutionary philosophy, termed *green chemistry and chemical engineering*, focuses on the designs of products and processes that are conducive to reducing or eliminating the use or generation of hazardous or potentially hazardous substances. In dealing with such substances, there may be some overlaps and interrelationships between environmental chemistry and green chemistry. Although environmental chemistry is the chemistry of the natural environment and the pollutant chemicals in nature, green chemistry proactively aims to reduce and prevent pollution at its very source. In essence, the philosophies of green chemistry and chemical engineering tend to focus more on industrial application and practice rather than academic principles and phenomenological science. However, as both a chemistry and chemical engineering philosophy, green chemistry and chemical engineering derive from and build upon organic chemistry, inorganic chemistry, polymer chemistry, fuel chemistry, biochemistry, analytical chemistry, physical chemistry, environmental chemistry, thermodynamics, chemical reaction engineering, transport phenomena, chemical process design, separation technology, automatic process control, and more. In short, green chemistry and chemical engineering is the rigorous use of chemistry and chemical engineering for pollution prevention and environmental protection.

The Pollution Prevention Act of 1990 in the United States established a national policy to prevent or reduce pollution at its source whenever feasible. And adhering to the spirit of this policy, the Environmental Protection Agency (EPA) launched its Green Chemistry Program in order to promote innovative chemical technologies that reduce or eliminate the use or generation of hazardous substances in the design, manufacture, and use of chemical products. Global efforts in green chemistry and chemical engineering have recently gained a substantial amount of support from the international community of science, engineering, academia, industry, and governments in all phases and aspects.

Some of the successful examples and key technological developments include the use of supercritical carbon dioxide as a green solvent in separation technologies, application of supercritical water oxidation for destruction of harmful substances, process integration with carbon dioxide sequestration steps, solvent-free synthesis of chemicals and polymeric materials, exploitation of biologically degradable materials, use of aqueous hydrogen peroxide for efficient oxidation, development of hydrogen proton exchange membrane (PEM) fuel cells for a variety of power generation needs, advanced biofuel production, devulcanization of spent tire rubber, avoidance of the use of chemicals and processes causing generation of volatile organic compounds (VOCs), replacement of traditional petrochemical processes by micro-organism-based bioengineering processes, replacement of chlorofluorocarbons (CFCs) with nonhazardous alternatives, advances in the design of energy-efficient processes, use of clean, alternative, and renewable energy sources in manufacturing, and much more. This list, even though it is only a partial compilation, is undoubtedly growing exponentially.

This book series on Green Chemistry and Chemical Engineering by CRC Press/Taylor & Francis is designed to meet the new challenges of the twenty-first century in the chemistry and chemical engineering disciplines by publishing books and monographs based upon cutting-edge research and development to effect reducing adverse impacts upon the environment by chemical enterprise. And in achieving this, the series will detail the development of alternative sustainable technologies that will minimize the hazard and maximize the efficiency of any chemical choice. The series aims at delivering to readers in academia and industry an authoritative information source in the field of green chemistry and chemical engineering. The publisher and its series editor are fully aware of the rapidly evolving nature of the subject and its long-lasting impact upon the quality of human life in both the present and future. As such, the team is committed to making this series the most comprehensive and accurate literary source in the field of green chemistry and chemical engineering.

Sunggyu Lee

Preface

Humans have a long history of using a wide variety of biomass resources as sources of energy and fuel. The discovery and use of fossil energy, represented largely by coal, natural gas, and petroleum, have drastically reduced the utilization of biomass fuels. Technologies for generating electricity using biomass, producing bioliquid fuels, and powering motor vehicles using bioalcohols and blended gasolines have been developed and practiced since the early twentieth century. Up until recently, however, development interest in biofuels had lessened due to the availability of relatively inexpensive fossil energy resources as well as the handling and transportation convenience of these conventional fuel sources.

Due to the strong growth of global transportation fuel demand, sharply escalating worldwide fossil energy prices, fear over the dwindling supply of petroleum and natural gas for the near future, and credible evidence linking global warming and climate change issues with the emission of greenhouse gases, global interest and R&D efforts in renewable alternative fuels have become intense and fiercely competitive, targeting both short- and long-term solutions to alternative energy needs. Although there are a number of options and routes for energy sustainability and independence via renewable alternative energy, bioenergy and biofuels certainly possess outstanding potential to provide solutions and relief to many of the immediate, intermediate, and long-term societal needs of clean energy and their associated challenges. Bioenergy and biofuels are quite broadly defined, including fuels derived from biological resources such as agricultural and forestry products, biomass and biomass-derived energy, fuels and fuel precursors secondarily derived from the primary biofuels and their by-products, fuels and energy derived from biological activities and processes, biodiesel from microalgae, and much more.

Accordingly, the resources for biofuels are ultimately renewable and totally independent of fossil energy availability and its distribution pattern. Biofuels are still, in some sense, similar to fossil fuels in that: (a) both are carbon energy sources, (b) direct combustion generates carbon dioxide emission, (c) both can be used as gas, liquid, and solid fuels, (d) both can be used for heat and power generation, and (e) both can be used as transportation fuels for conventional and futuristic motors. However, biofuels are drastically different from fossil energy in that biofuel resources are renewable, biomass is distributed worldwide, biomass typically contains much less sulfur, biofuels can also originate from nonmanufacturing and nonmining industrial sectors such as agriculture and forestry, and carbon dioxide generated by combustion of biofuels essentially originated from carbon dioxide removed from the atmosphere by plants through photosynthesis.

Because plant material in nature utilizes sunlight and carbon dioxide to produce energy and hydrocarbons, the combustion of biofuel derived from biomass by itself does not contribute to a net increase of carbon dioxide in the atmosphere. When biofuel is processed in conjunction with other carbon-neutral energies such as solar, wind, and hydrothermal energy, biofuels could be made available as a nearly carbon-neutral fuel. Furthermore, if biomass feedstock is chosen from abundantly available but nonfood crops whose growth does not require arable land, both sustainability and renewability would be warranted. Biofuel process R&D has directly benefited from the technological and scientific advances in C_1-chemistry, coal science and technology, petroleum science and technology, natural gas science and engineering, hydrocarbon processing, environmental science and engineering, separation science and technology, heat and mass transfer, catalysis and reactor engineering, materials science and engineering, biochemistry and biotechnology, agriculture, food science, and more. For example, the processes for fast pyrolysis, gasification, and liquefaction of biomass are quite similar to those developed and adopted for coal and oil shale processing. Gas cleaning and upgrading of bioliquid fuels also share common routes in conventional fuel processing with, in particular, those of natural gas and petroleum, with modifications and adaptations. Ingenious energy integration schemes are also receiving technological enhancements and know-how support from the well-practiced petroleum, petrochemical, and power generation industries. Biodiesel and bioethanol technologies have also benefited from the lessons-learned approaches of internal combustion engine developers and fuel engineers. Needless to say, the bioethanol industry is a direct and successful example of advances made in biotechnology, separation science and technology, agriculture, and animal food science. As such, the subject of biofuels cannot be treated as a new and independent stand-alone discipline, but rather as a multidisciplinary subject that has its foundation in C_1-chemistry, petroleum and coal science, natural gas engineering, environmental science and technology, process engineering and design, separation science and technology, biotechnology and biological engineering, heat and mass transfer, reactor engineering, catalysis and enzymology, energy management, public policies, and more. Therefore, this book is written as a comprehensive source book on the subject area covering all of the aforementioned subjects of relevance from the standpoints of fuel process engineers and fuel scientists as well as from the viewpoints of energy technologists.

This book is intended to provide the most comprehensive background in the science and technology of biofuels and bioenergy, including the most up-to-date and in-depth coverage of definitions and classifications of biofuels and related matters, characterization and analysis of biofuels, primary processing technologies of biofuel resources, secondary processing technologies of biofuels and biofuel precursors, upgrading of crude biofuels, issues involving the ethanol economy and the hydrogen economy, chemistry of process conversion, process engineering and design of biofuel production and associated environmental technologies, combined cycle processes, coprocessing of biomass with other fuels, energy balances and energy efficiencies, reactor designs and process configurations, energy materials and process equipment, commercial biofuel processes and significant practices, energy integration strategies and schemes, flowsheet analysis, relation to and integration with other conventional fossil fuel processes, by-product utilization, process economics of biofuel technology, environmental and ecological impacts and benefits, sustainability issues, governmental regulations and policies, global bioenergy trend and outlook, and much more.

Chapter 1 focuses on the introductory subjects of the book including the definition of biofuels, global energy outlook in reference to bioenergy and biofuels, issues of sustainability and carbon neutrality, generalized view of biomass feedstock and its availability, brief overview of technological trends of biofuel and bioenergy processes, and a general discussion of environment and ecology in relation to the utilization of bioenergy and biofuels.

Chapter 2 provides scientific and technological details of crop oils, algae fuels, and biodiesel. Vegetable oils, or plant oils, and their utilization as straight biofuels and biofuel feedstock are discussed in depth, including characterization, extraction, purification, and chemical conversion to biodiesel via transesterification. Also presented are algae biofuel and its technological details involving harvesting, extraction, and chemical conversion with detailed descriptions of the proposed or practiced technologies and their flowsheets.

Chapter 3 deals with the science and technology of producing bioethanol from starch crops, specifically corn. The fuel properties of ethanol as an oxygenated fuel for the internal combustion engine and as a blend fuel for conventional gasoline are presented. Particulars of corn ethanol process technologies, both dry milling and wet milling, are explained in necessary detail. Socioeconomic issues of "food versus oil" for corn ethanol production as well as technoeconomic issues of net energy balance of corn ethanol production are also discussed based on the literature data and recently published information. Also mentioned is the beneficial and profitable utilization of process by-products and coproducts in a variety of industrial sectors.

Chapter 4 is devoted to the production of ethanol from lignocellulose. An historical perspective of alcohol fermentation is presented along with its technological evolution and trend. Essential scientific and technological details of cellulose, hemicellulose, and lignin are provided and a variety of enzymes that are found and developed for processing these components are also discussed. Processing steps and options for lignocellulosic alcohol production involving prehydrolysis and pretreatment, hydrolysis and enzymatic treatment, and fermentation are elucidated. Particular attention is also paid to the fermentation of xylose, C_5-sugar, and to the co-fermentation of xylose with glucose, C_6-sugar, using genetically engineered micro-organisms. Also included in the chapter is the beneficial utilization of lignin in areas other than cellulosic ethanol production.

Chapter 5 is mainly focused on the thermochemical conversion of biomass into an assortment of gaseous products (biomass syngas), bioliquid (bio-oil), and solid fuel (biochar). Details of process technologies developed or proposed for fast pyrolysis of biomass as well as for gasification of biomass are presented with explanations of the merits and potential limitations associated with the specific technological steps. Process parameters, operating conditions, product spectra and properties, and pertinent reactor design issues are also discussed. Process options for further upgrading of crude products of thermochemical intermediates are also examined.

Chapter 6 provides an in-depth overview of the conversion of wastes to biofuels, bioproducts, and bioenergy. More specifically, the chapter deals with strategies for waste management, methods for waste preparation and pretreatment, an extensive list of waste-to-energy conversion technologies, socioeconomic and environmental issues of waste conversion, and the future of the waste management and conversion industry. Conversion technologies covered in the chapter include incineration, gasification, pyrolysis, plasma technology, supercritical processing, transesterification, anaerobic digestion, fermentation, and product upgrading.

Chapter 7 discusses various thermochemical technologies for processing a mixed feedstock such as a mixture of coal and biomass. The discussions are divided into two principal parts: (a) the technologies that are mainly generating gases such as combustion, gasification, and high-severity pyrolysis or plasma technology, and (b) those that are for generating liquids such as low-severity pyrolysis, liquefaction, and supercritical process. The chapter also reviews essential topics of reactor configurations and associated technologies, handling of product streams, process configurations, and examples of commercial processes, while discussing various technoeconomic benefits of the technology including early expansion of bioenergy utilization, mitigated environmental concerns, and enhanced thermal and economical efficiencies.

Even though this book covers in-depth knowledge on the subject of interdisciplinary biofuels and bioenergy, it is written for readers with college-level backgrounds in chemistry, biology, physics, and engineering. This book can be used ideally as a textbook for a three-credit-hour semester course in bioenergy or biofuels. If used as a textbook, two weeks each may be allocated for each chapter of this book, with one final week opted for open-ended discussions on regionally important topics or contemporary or arising issues related to the subject area. This book can also be adopted as a textbook or reference book for upper-level undergraduate or graduate-level courses in the fields of energy and fuels, renewable energy, alternative fuels, and fuel science and engineering.

This book will also serve as an excellent desk reference book for professionals who are engaged in renewable bioenergy- and biofuel-related industries as well as environmental engineering fields. Furthermore, this book will serve as a single source encompassing the most comprehensive and up-to-date scientific and technological information for researchers in the fields of biofuels as well as alternative and renewable energies.

Finally, the authors wish to acknowledge and thank their families, former and current graduate students, friends, and colleagues for their support and assistance in completing this book project.

1

Introduction to Biofuels and Bioenergy

1.1 Definition

Bioenergy is energy derived or obtained from any fuel that is derived or originated from biomass which includes recently living organisms and their metabolic by-products. Similarly, biofuels are defined as fuels made from biomass resources, or their processing and conversion derivatives [1]. *Biomass* is defined as all plant and animal matter on the Earth's surface. Therefore, harvesting biomass such as crops, trees, or dung and using it to generate heat, electricity, or motion, is bioenergy [2]. Biomass is a very broad term that is used to describe materials of recent biological origin that can be used as an energy source or for their chemical ingredients. According to this definition, biomass includes crops, trees, algae, and other plants as well as agricultural and forest residues. It also includes many other materials that are regarded as waste by most people, including food and beverage manufacturing effluents, sludges, manures, industrial organic by-products, and organic fraction of household waste [2]. The word "recent" in the defining statements of biomass is of significance, because it eliminates any logical ground for fossil fuels to be considered as such.

Biomass has a number of different end uses such as heating (thermal energy), power generation (electrical energy), and transportation fuels. The term *bioenergy* is usually used for biomass energy systems that produce heat or electricity, whereas the term *biofuels* is typically used for liquid fuels for transportation [2]. For example, biofuels include corn ethanol, cellulosic ethanol, biodiesel, algae diesel, biomass-derived methanol, biomass-derived Fischer–Tropsch fuels, and more.

Historically speaking, biomass is the oldest fuel known to humans in all regions of the world. In the current world, biomass is a clean and renewable fuel source that can produce heat, power, and transportation fuels. In the future world, biomass will be a sustainable energy source whose utilization will have little or minimal impact on the environment and climate change.

Biomass may be considered as a form of stored solar energy that is captured through the process of photosynthesis in growing plants. Utilizing biomass as a biofuel or bioenergy source means that carbon dioxide (which was captured from the air by growing plants) is released back to the air when the biofuel is eventually combusted. Therefore, the system based on biomass energy is carbon neutral, or at least close to being carbon neutral. The term *carbon neutral* means removing as much carbon dioxide from the atmosphere as we put in, thus leaving a net zero impact on the atmospheric carbon dioxide amount.

1.2 Global Energy Outlook

The global energy consumption outlook forecast by the U.S. Energy Information Administration (EIA) is shown in Figure 1.1, which shows two sets of data series for the OECD (Organization for Economic Cooperation and Development) and the non-OECD nations [3]. As shown in the chart, much of the increase in energy consumption is predicted for the non-OECD nations due to their strong long-term economic growth and continuous industrialization. Although about 45% of cumulative increase in the annual energy consumption is predicted from 1990 to 2035 for the OECD countries, more than a threefold increase is expected for the same period for the non-OECD countries. For the entire world, annual energy consumption for Year 2035 is predicted to be 770 quadrillion BTU, whereas that for year 1990 was recorded as 354 quadrillion BTU. A quadrillion BTU is 1×10^{15} BTU and is also called a *quad*. Assuming that the average heating value (HV) of gasoline is 125,000 BTU/gal, a quad of energy is equivalent to an aggregated heating value of approximately 8 billion U.S. gallons of gasoline.

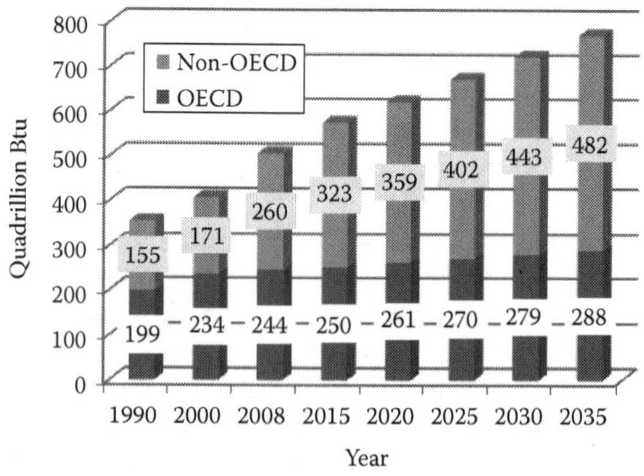

FIGURE 1.1
World energy consumption for 1990–2035. (From U.S. Energy Information Administration (EIA). 2011. *International Energy Outlook 2011*. Tech. Rep. DOE/EIA-0484, September 19.)

The marketed energy consumption of modern society is closely tied to its economic strength. It is also affected by the efforts of energy conservation and energy efficiency enhancement by various sectors. Petroleum prices on the global market have been largely dictated by the market's supply and demand dynamics. Sharply increasing oil prices in recent years have been due to a lack of sufficient supply to meet the growing demand in the marketplace, which was frequently intensified by regional supply disruptions caused by political unrest. High energy costs also add strong causes to inflation and adversely affect the global economy. Furthermore, it is difficult to predict or accurately assess technological advances in future energy technologies with respect to the technoeconomic constraints of future societies. Therefore, an energy outlook forecast is by no means an easy task. The U.S. EIA presented five different cases/scenarios: viz. (1) reference case, (2) high oil price case, (3) low oil price case, (4) traditional high oil price, and (5) traditional low oil price [4]. The reference case scenario is based on the baseline world economic growth at 3.5%/year for 2008–2015, 3.3%/year for 2015–2035, and the world light sweet crude oil prices growing to $125 per barrel by 2035 (2009 dollars). On the other hand, the high oil price case, a more pessimistic case scenario, is assuming that the world light sweet crude oil price grows to $200 per barrel by 2035 (2009 dollars) and the low oil price case, a more optimistic case scenario, is based on the assumption that the light sweet crude oil prices are going to be $50 per barrel (2009 dollars). The reference case scenario by the U.S. Energy Information Administration (EIA)'s forecast data for world energy consumption by fuel is presented in Table 1.1 [4].

As shown in Table 1.1, the lowest annual growth rate is estimated for liquid fuels or liquids among all fuel types. A very strong demand for transportation vehicles is expected particularly for non-OECD nations, however, a more serious and rigorous global effort in vehicle fuel efficiency enhancement is also to be practiced, thus offsetting the effect from the increase in the number of vehicles. Even with the modest 1% increase, a large portion of the increase is expected to come from the increased consumption of biofuels as transportation fuels.

Table 1.2 shows the projections for world consumption of hydroelectricity and other renewable energy for the OECD and the non-OECD countries. As shown in Tables 1.1 and 1.2, the annual growth rate for this category is substantially higher than the conventional fossil energy sources of petroleum, natural gas, and coal. However, it should be noted that common biofuels for transportation such as bioethanol and biodiesel are not included in this category of "Other," but in the category of "Liquids."

The category of "Liquids" in the EIA's reporting, as also shown in Table 1.1, includes petroleum and other liquid fuels. Bioethanol and biodiesel, for example, are included in this category as other liquid fuels. Therefore, this category of "Liquids" includes both renewable and nonrenewable liquid fuels as well as conventional and unconventional supplies. Table 1.3 shows the world liquid fuel consumption by region for 1990–2035 [5]. The

TABLE 1.1

World Total Energy Consumption by Fuel: Reference Case Scenario

	2005		2010		2015		2020		2025		2030		2035		Annual Growth Rate (%)
	quads	%	quads	%	quads	%	quads	%	quads	%	quads	%	quads	%	(2008–2035)
World															
Liquids	170.8	18.1	173.2	16.6	187.2	16.3	195.8	15.8	207	15.4	216.6	15.0	225.2	14.6	1.00
Natural Gas	105	13.6	116.7	13.4	127.3	13.3	138	13.2	149.4	13.2	162.3	13.2	174.7	13.3	1.60
Coal	122.3	18.4	149.4	19.8	157.3	18.9	164.6	18.2	179.7	18.2	194.7	18.3	209.1	18.3	1.50
Nuclear	27.5	5.1	27.6	4.6	33.2	4.9	38.9	5.3	43.7	5.4	47.4	5.5	51.2	5.5	2.40
Other	45.4	8.8	55.2	9.6	68.5	10.7	82.2	11.7	91.7	12.0	100.6	12.2	109.5	12.5	2.90
Total	471.1	100	522	100	573.5	100	619.5	100	671.5	100	721.5	100	769.8	100	1.60

Source: U.S. Energy Information Administration (EIA). 2011. *International energy outlook* 2011: *World energy consumption by region and fuel*. Tech. Rep. DOE/EIA-0484, September 19.

Note: (a) 1 quad = 1×10^{15} BTU. (b) The annual growth rate is taken for the period of 2008–2035. (c) "Other" in the table represents the hydroelectricity and other renewable energy.

TABLE 1.2

World Consumption of Hydroelectricity and Other Renewable Energy: Reference Case

	2005	2006	2007	2008	2010	2015	2020	2025	2030	2035
OECD	20	20.4	20.7	22.1	23.7	29.3	33.6	37.1	39.4	41.4
Non-OECD	25.5	26.8	27.8	29.2	31.5	39.3	48.6	54.6	61.2	68.1
World Total	45.4	47.1	48.5	51.3	55.2	68.5	82.2	91.7	100.6	109.5

Source: U.S. Energy Information Administration (EIA). 2011. *International Energy Outlook: World Consumption of Hydroelectricty and Other Renewable Energy by Region, Reference Case.* Tech. Rep. DOE/EIA-0484, September 19.

Note: The values are in the units of quadrillion BTU.

worldwide consumption of petroleum and other liquid fuels in 1990 was 67 million barrels a day, 85.7 million in 2008, 83.9 million in 2009, and 86.0 million in 2010. A decrease in 2009 was due to the global recession. The total liquids consumption for the world is projected to increase to 112.2 million barrels a day (225 quadrillion BTU for the year) in 2035, which is an increase of 26.5 million barrels a day compared to 2008. Of the total increase from 2008 to 2035, 17.2 million barrels a day, or about 75% of the increase, is expected to come from non-OECD nations. This demand growth of liquids is driven by the projected world GDP growth of 3.6%/year for 2008–2020 and 3.2%/year for 2020–2035. This is also based on the aforementioned reference case scenario, where the world oil price rises to $125 per barrel (2009 dollars) by 2035. This also implies that in the long term, despite the high oil price assumed for the reference case, the liquids consumption will still increase steadily. In order to satisfy this increase in global liquids consumption in the reference case, liquids production has to increase by 26.6 million barrels per day from 2008 to 2035. Even with the expected increase in the world liquids consumption, the growth in demand for liquids in the OECD nations is expected to slow due to a number of factors including: (a) governmental policies, (b) efforts of increasing the fuel efficiencies of motor vehicles, and (c) various incentives. In Japan and OECD Europe, the consumption of liquids is predicted to decline by average annual rates of 0.4%/year and 0.2%/year, respectively [5].

The increased portion of the global liquids demand of 26.6 million barrels per day from 2008 to 2035 will have to come from both conventional supplies (such as crude oil and lease condensate, natural gas plant liquids, and refinery gain) and unconventional supplies (such as biofuels, oil sands, extra-heavy oil, coal-to-liquids (CtL), gas-to-liquids (GtL), and shale oil) [6]. Sustained high oil prices will provide incentives for the unconventional supplies by making them more competitive. Unconventional liquids production is predicted to be increasing at a rate of about 5%/year for 2008–2035, according to the reference case. In all five oil price cases of projection models by the EIA, Canadian bitumen (oil sands) production is an important factor, making up more than 40% of the total non-OPEC unconventional liquids production, ranging from

TABLE 1.3

Projected World Liquids Consumptions by Region: Reference Case

Region/Country	2008	2009	2010	2015	2020	2025	2030	2035	AGR%
OECD									
OECD Americas	**24**	**23**	**24**	**25**	**25**	**26**	**26**	**27**	**0.40**
United States	20	19	19	20	21	21	21	22	0.40
Canada	2	2	2	2	2	2	2	2	0.20
Mexico/Chile	2	2	2	3	3	3	3	3	0.70
OECD Europe	**16**	**15**	**14**	**14**	**15**	**15**	**15**	**15**	**−0.20**
OECD Asia	**8**	**8**	**8**	**8**	**8**	**8**	**8**	**8**	**0.10**
Japan	5	4	4	4	5	5	5	4	−0.40
South Korea	2	2	2	2	2	2	3	3	0.70
Australia/New Zealand	1	1	1	1	1	1	1	1	0.50
Total OECD	**48**	**45**	**46**	**47**	**48**	**49**	**50**	**50**	**0.20**
Non-OECD									
Non-OECD Europe & Eurasia	**5**	**5**	**5**	**5**	**5**	**5**	**5**	**6**	**0.40**
Russia	3	3	3	3	3	3	3	3	0.10
Other	2	2	2	2	2	2	3	3	0.80
Non-OECD Asia	**17**	**18**	**19**	**23**	**26**	**30**	**33**	**34**	**2.60**
China	8	8	9	12	14	16	16	17	2.90
India	3	3	3	4	5	6	7	8	3.50
Other	6	6	6	7	8	9	9	10	1.70
Middle East	**7**	**7**	**7**	**8**	**8**	**8**	**9**	**10**	**1.40**
Africa	**3**	**3**	**3**	**3**	**3**	**4**	**4**	**4**	**0.90**
Central & South America	**6**	**6**	**6**	**7**	**7**	**7**	**8**	**8**	**1.40**
Brazil	2	2	3	3	3	3	4	4	1.70
Other	3	3	3	4	4	4	4	4	1.10
Total Non-OECD	**38**	**38**	**40**	**46**	**49**	**54**	**58**	**62**	**1.90**
Total World	**86**	**84**	**86**	**93**	**98**	**103**	**108**	**112**	**1.00**

Source: Energy Information Administration (EIA). 2011. *International Energy Outlook* 2011: *Liquid Fuels.* Tech. Rep. DOE/EIA-0484, September 19.

Note: AGR = Annual growth rate in %/year.

3.1 million barrels/day (low oil and traditional low oil case scenarios) to 6.5 million barrels/day (high and traditional high oil price case scenarios).

Biofuels, which are the main topical area of this book, come under this unconventional liquids category in the EIA projections. The world biofuel production in 2010 exceeded 105 billion liters (28 billion U.S. gallons or 667 million barrels), which was a remarkable 17% increase from the 2009 production. The liquid biofuels, mostly made up of bioethanol and biodiesel, accounted for about 2.7% of the world's transportation fuel in 2010 [7].

As a long-term projection, the global biofuels production in the reference case scenario of IEO2011 is projected to increase from 1.5 million barrels/day in 2008 to 4.7 million barrels/day in 2035, at an average annual growth rate of 4.3% per year. The largest increase in biofuels production and consumption is expected to be in the United States, whose annual biofuels production grows from 0.7 million barrels/day in 2008 to 2.2 million barrels/day in 2035 in the reference case scenario. Another very strong growth in biofuels production is expected to take place in Brazil, a traditionally strong biofuels nation, whose annual production will increase from 0.5 million in 2008 to 1.7 million barrels/day in 2035, based on the reference case scenario. To achieve these goals, many biofuel-using nations set mandates for the amount of biofuels used and provide tax credits or incentives for biofuel producers. For example, the United States mandates 36 billion gallons of biofuels by 2022 under the Energy Independence and Security Act of 2007 (EISA of 2007), which is explained in more detail in Chapters 3 through 5.

The combined total of the two nations accounts for about 84% of the world increase in biofuels production. It has to be noted here that the biofuel projections by the EIA have received significant changes and adjustments in their predictions from the IEO2009 reference case to the IEO 2010 and 2011 reference cases. The revised projection of 2010 and 2011 shows a 40% lower projected biofuels production in 2030, compared to the IEO2009 reference case. This rather significant adjustment was made based on several compounded factors including:

- Some recent studies suggest that biofuels may not be as effective in reducing greenhouse gas (GHG) emissions as previously thought. As a result, many countries such as Germany [8] have relaxed or postponed renewal of their mandates.
- The global economic recession of 2009 has dampened investment in biofuels development.
- Some of the timetables originally set for technological exploitation of new and enhanced biofuels technologies on commercial scales have been pushed back.

Biofuels will become more competitive with conventional petroleum products over time, as the oil price remains high or continues to rise and new technologies are continuously introduced and enhanced. A strong relationship between future biofuels production and the future petroleum price trend is explicitly reflected in the IEO's projections. The IEO2011's low oil price case scenario predicts the total global biofuels production of 3.5 million barrels/day in 2035, whereas the IEO2011's traditional high oil price case scenario predicts it at 6.2 million barrels/day in 2035[6].

It should also be noted that the future years' projections of biofuels production will have to be continuously adjusted and modified based on evolving market conditions and changing technoeconomic constraints, which include: (a) oil price trend, (b) economic strengths of major markets,

(c) global politics, (d) R&D progress of new technologies and commercialization efforts, (e) changing views and concerns of environmental and sustainability issues with regard to biofuels utilization, (f) governmental mandates and incentives, and so on.

1.3 Sustainability

A principal reason for the use of biofuels and bioenergy is its renewable nature. Utilizing biomass as a biofuel feedstock means that carbon dioxide in the atmosphere that was absorbed and transformed via a plant's photosynthesis is released back into the atmosphere upon combustion of the biofuel. By considering the starting feedstock of the biomass and the end product of biofuels only, without considering any ancillary carbon energy input during the conversion process, the system can be said to be *carbon neutral*. Carbon neutrality means achieving net zero carbon emission, thus leaving a net zero carbon footprint. Furthermore, by maintaining a balance between plant growth and biomass use, the energy system is both renewable and sustainable. Sustainability of a feedstock is defined by availability of the feedstock, very positive and beneficial impact on GHG emissions, and no negative impact on biodiversity and land use [9].

There is a growing consensus that carbon dioxide emission and its accumulation in the atmosphere is a major culprit of global climate change via human interference with natural cycles of greenhouse gases. Two of the more direct causes for carbon dioxide accumulation in the modern era have been recognized as combustion of fossil fuels and land-use change, in particular, deforestation. The use of biofuels undoubtedly reduces the use of fossil fuels and helps restore the needed balance between the carbon dioxide uptake and release.

Agriculture is also expected to continue to evolve and adapt to new technologies and changing circumstances. Biotechnology is advancing agriculture by making available genetically altered varieties of corn and soybeans as well. According to the National Corn Growers Association, biotech hybrids accounted for 40% of the total planted acreage of the United States in 2004 [10]. Crop yields are very important because they directly affect the amount of residue generated as well as the amount of land needed to meet food, feed, and other demands. A joint study sponsored by the U.S. Department of Energy and the U.S. Department of Agriculture offers three scenarios in its report (2005) [10]:

- Scenario 1: Current sustainable availability of biomass feedstocks from agricultural lands

- Scenario 2: Biomass availability through a combination of technology changes focused on conventional crops only
- Scenario 3: Biomass availability through technology changes in both conventional crops and new perennial crops together with significant land usage change

Current availability is the baseline case that summarizes sustainable biomass resources under current crop yields, tillage practices (20–40% no-till for major crops), residue collection technology (~40% recovery potential), grains to bioethanol and biodiesel production, and use of secondary and tertiary residues [10]. Summing up, the total amount of biomass currently available in the United States as of 2005 for bioenergy and bioproducts was about 194 million dry tons annually. This was about 16% of the 1.2 billion dry tons of plant materials produced on agricultural land of the United States. The single largest source of this biomass potential in 2005 was corn residues or corn stover totaling close to 75 million dry tons [10,11]. Considering that the U.S. corn production was 282 million metric tons in 2005 and 333 million metric tons in 2010, the biomass potential from corn residues alone would have totaled more than 85 million dry tons in 2010. On the other hand, the total biomass derived from forestlands in the United States was estimated to be about 142 million dry tons in 2005 [10]. Therefore, from the standpoints of resource availability and sustainability, intensive R&D efforts focused on efficient conversion of corn residues into biofuels as well as cost-effective conversion of cellulose into ethanol are rationally grounded and well justified.

According to a more recent study conducted and reported in the *U.S. Billion-Ton Update* [11], for the baseline scenario, projected consumption of currently used resources, the forest residues and wastes, the agricultural residues and waste, and energy crops show a total of 1,094 million dry tons in 2030. Under the baseline assumptions, up to 22 million acres of cropland and 41 million acres of pastureland shift into energy crops by 2030 at a simulated farmgate price of $60 per dry ton [11]. This study also shows that the total biomass potential in the United States for the currently used and potential forest and agricultural biomass at $60 per dry ton or less, under the baseline scenario, sharply increases from 473 million in 2012, to 676 million in 2017, to 914 million in 2022, and to 1094 million dry tons in 2030.

1.4 Biomass Feedstocks

Even though biomass includes all plants and plant-derived materials, all plant matter is not equal as biomass feedstock for bioenergy and biofuels production. The success of the biofuels and bioenergy industry depends on the quality and quantity of biomass available as well as the ability to utilize it cost-efficiently to produce biofuels and bioenergy [12]. Various factors are taken into consideration for suitability determination and choice of biomass for the biofuels program and they include:

- Sustainable feedstock production
- Arable land requirement
- Feedstock logistics
- Regional strength
- Food crops or not
- Grains or nongrains
- Feedstock properties and compositions
- Pretreatment cost
- Availability of efficient conversion/transformation technology
- Feedstock cost
- Capital investment and operating cost involved
- Environmental benefits
- Desirable biofuel products and their values

A variety of biomass feedstock can be converted into alternative transportation fuels. Currently, a dominant majority of biofuels produced in the United States are corn ethanol. Although corn is an excellent source of starch which can be easily converted into fuel ethanol and is also the most abundantly produced U.S. crop, the corn feedstock also has concerns of being a food crop and demanding high feedstock cost. More R&D is being focused on the development of cellulosic feedstock such as corn stover, switchgrass, and woody cellulose. Figure 1.2 shows resource-based biorefinery pathways, as presented in the Biomass Program's website, U.S. DOE, Office of Energy Efficiency and Renewable Energy [12]. The Biomass Feedstock Platform has three phases of program foci in its R&D Platform and they are

- *In the immediate and near term,* it will focus on the sustainable production, collection, and use of readily available low-cost agricultural residues and industrial wastes.
- *In the near to mid-term,* it will address additional agricultural and forestry residues and a potentially few dedicated energy crops.
- *In the longer term,* it will involve the development and use of both herbaceous and woody dedicated energy crops.

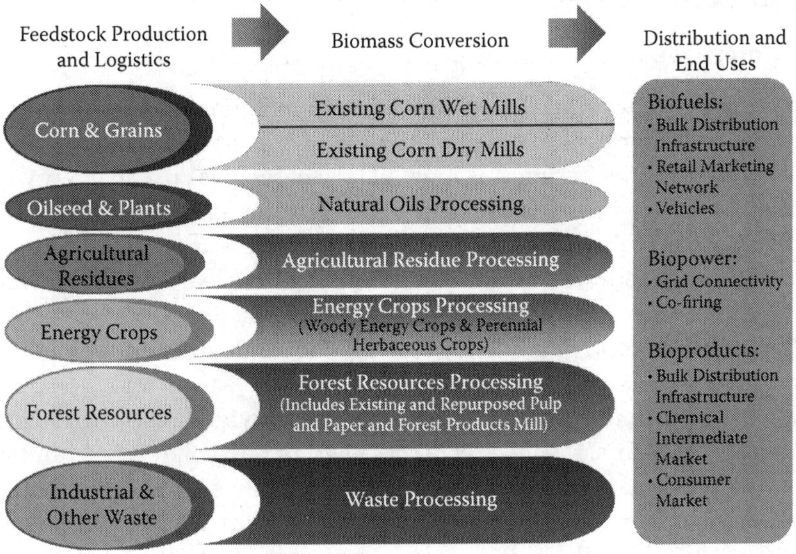

FIGURE 1.2
Resource-based biorefinery pathways by U.S. DOE, EERE. (From U.S. Department of Energy, 2011. *Energy Efficiency & Renewable Energy, Biomass Program. Biofuels, Biopower, and Bioproducts: Integrated Biorefineries.* http://www1.eere.energ y.gov/biomass/in dex.html.)

Biofuels are classified into two broad categories, namely first-generation and second-generation biofuels, based on the kinds of feedstock as well as the types of process technologies applied or applicable. *First-generation* (1G) *biofuels* refer to fuels that have been derived from biological sources such as starch, sugar, animal fats, and vegetable oil. These conventional biofuels are produced from oil crops, sugar crops, and cereal crops, using established technology. Some of the most typical types of first-generation biofuels include vegetable oils, conventional biodiesel, bioalcohols, biogas, and biosyngas. On the other hand, *second-generation* (2G) *biofuels* are derived from cellulosic materials (lignocellulosic feedstock). These raw materials for 2G biofuels may result in more biofuel per unit area of agricultural land and also require less chemical and energy input for production and harvesting, which in turn results in a higher net energy yield [9]. As such, raw materials for 2G biofuels may be considered more sustainable than those for 1G biofuels and do not create the issues of "food versus fuel," either. Many second-generation biofuels are under active research and development and include cellulosic ethanol, algae fuel, biohydrogen, biomethanol, bio-dimethylether (bio-DME), Fischer–Tropsch diesel, 2,5-dimethyl furan (DMF), mixed alcohols, and wood diesel. The second-generation biofuels are also referred to as advanced sustainable biofuels.

1.5 Processes and Technologies

Biomass is the oldest fuel known to and used by humans in every conceivable aspect of their lives. Biomass is of a widely diverse kind and is also the most versatile. Considering this long history of biomass use in energy generation, its development as a modern energy source has been slow-paced, mainly due to the rapidly gained public acceptance and popularity of other competing fuel sources such as fossil fuels and nuclear energy. However, biomass has clear advantages over fossil fuels in terms of renewability, sustainability, carbon neutrality, political independence, and regional economic benefits.

Biomass occurs in many different forms with widely varying chemical compositions. Biomass can be readily converted into solid, liquid, and gaseous fuel products. For example, wood is typically chopped into chunks, chipped for easier handling, or pelletized for pumping. Biomass can be pyrolyzed into bio-oil liquid fuel as a major product along with the byproducts including solid biochar and gaseous fuel. Or, biomass can be gasified to generate synthesis gas which is rich in hydrogen. Biomass can also be burned to generate heat (such as hot water or steam) or to produce electricity (or power). Thus, biomass combustion can be used in a combined heat and power (CHP) mode. Alternately, some biomass can be coprocessed with other conventional fossil fuels such as lignite coal to generate power. However, some biomass is too wet to be efficiently burned and instead is fermented to produce liquid fuel such as bioethanol. In addition, some biomass such as oil crops and microalgae can be processed into biodiesel via both physical and chemical treatments. Furthermore, bio-oil and biosyngas can be upgraded for subsequent chemical and fuel processing as well as for other diverse end uses.

Different forms and shapes, diverse origins, varying feedstock composition, and substance-specific properties of biomasses inevitably require different types, levels, and sequences of process treatments before they can be used effectively and cleanly in modern technological society. Typical process treatments in biomass conversion and utilization technologies are discussed in the next sections.

1.5.1 Feedstock Preparation and Pretreatments

Pretreatment processing usually takes place in the prescreening of feedstock, where unsuitable materials are screened out. Typical pretreatment operations include drying, devolatilization, size reduction and particle size classification, torrefaction, pelletizing, and mixing.

1.5.2 Chemical and Biochemical Reactions

Highly oxygenated and high molecular-weighted biomass organic materials need to be converted to lighter molecular-weighted and simpler chemicals via a variety of targeted and accompanying chemical reactions, depending upon the desired outcomes. Such reactions include complete combustion, partial oxidation, pyrolysis, gasification, reformation, hydrocracking, liquefaction, hydrolysis, enzymatic hydrolysis, fermentation, and more. The reactions listed here are not single chemical reactions and each is rather a class of chemical reactions. For example, gasification includes steam gasification, hydrogasification, carbon dioxide gasification, partial oxidation, and a combination of these. Conventional process technologies are more applicable to first-generation biofuels, whereas more advanced process technologies are being developed for second-generation biofuels.

1.5.3 Heat Transfer Enhancement and Management

Efficient control of heat transfer to and from biomass in every step of the process treatments becomes crucially important. In particular, depending upon the desired heat transfer characteristics of biomass conversion technology, a wide variety of process designs and reactor configurations have been proposed and demonstrated. Some of the examples in the reactor design include bubbling fluidized bed, circulating fluidized bed, entrained flow, ablative tube, rotating cone, vacuum, twin-screw, and others. Examples of a heat transfer medium utilized in such processes include hot sand which is usually recirculated in the process system as a direct-contact heat transfer medium as well as a means for thermal energy preservation.

1.5.4 Downstream Processing of Raw or Intermediate Biofuel Products

Crude products of biomass conversion technology include both principal products and by-products. Principal products of biomass processing include crude bio-oil from fast pyrolysis, biochar from slow pyrolysis, biomass syngas from gasification, biodiesel from transesterification, crude ethanol beer from corn ethanol, straight vegetable oils from oil crops, biogas from anaerobic digestion, bioenergy and biofuels from waste and refuse, and more. These principal products can be used for direct end uses after purification, as raw feedstock for further upgrading and transformation into alternative fuels that supplant other conventional fuels and chemicals, or as captive energy sources for the fuel conversion and power generation technology and its ancillary processing. By-products include biochar from fast pyrolysis, bio-oil from gasification, dry distillers' grains and corn syrups from the corn ethanol industry, crude glycerin from the transesterification process, and more. Profitable by-product portfolios for specific biomass technologies not only enhance the profitability of the biofuels industry, but also contribute to the sustainability of the technological society via promoted utilization of renewable materials.

1.5.5 Energy Integration and Energy Efficiency Enhancement

A key to the overall success of the biomass program depends upon its energy efficiency of the overall technology implementation. The net energy value (NEV) of corn ethanol technology is often a matter of concern and a subject of debate, for example. In all of the biomass processes, ingenious energy integration schemes are adopted, to list a few, drying of biomass feed using hot effluent stream, combustion of biomass residues for reheating the heat transfer medium such as sand, generation and utilization of process steam, efficient process alignment for combined heat and power, autothermal reformation, and more.

1.5.6 Product Purification and Separation

A significant cost factor comes from product purification of a biofuel product. Even if a highly selective chemical transformation process is adopted, the nonuniform and heterogeneous nature of biomass feedstock invariably produces products of a broad spectrum of chemical species, targeted versus nontargeted, or desired versus not desired. The unwanted species in the product composition include source-specific and treatment-specific ingredients that need to be separated out from the principal products. Examples include ethanol purification from crude ethanol beer, separation of methanol and salt from crude glycerin that was produced by the transesterification process, removal of trace minerals from biosyngas and bio-oil for upgrading, char removal from bio-oil, denitrification of bio-oil, removal of carbon dioxide from biosyngas, and so on.

1.6 Environment and Ecology

Sound management and efficient utilization of biomass can provide substantial benefits in terms of increased biodiversity, local amenity, and even rehabilitation of land and water courses [13]. Examples include income-generating management of native woodlands, growing energy crops such as short rotation coppice (SRC) that has biodiversity and soil conservation benefits, utilization of low-fertility and abandoned lands, returns and increase of certain declining animal populations, and energy independence of local or rural communities, among others. When biomass energy is exploited properly, the renewability of biomass via photosynthesis absorbing carbon dioxide as a pivotal species helps reduce the carbon footprint of the energy generation and consumption cycle associated with human activities, thereby positively contributing to greenhouse gas management in both direct and indirect ways.

Furthermore, the utilization of biomass helps contribute to income generation and increase for rural and agricultural communities, while reducing dependence on nonrenewable fossil fuels and helping establish a sustainable future.

References

1. Bioenergy Feedstock Information Network (BFIN). 2011. *Glossary of Useful Bioenergy Feedstock Terms.* [https://bioenerg y.ornl.gov/main.aspx].
2. International Energy Agency (IEA) Bioenergy. 2011. *What Is Biomass?* [www.aboutbioenergy.info/definition.html].
3. U.S. Energy Information Administration (EIA). 2011. *International Energy Outlook 2011*. Tech. Rep. DOE/EIA-0484, September 19.
4. U.S. Energy Information Administration (EIA). 2011. *International Energy Outlook* 2011*: World Energy Consumption by Region and Fuel.* Tech. Rep. DOE/EIA-0484, September 19.
5. U.S. Energy Information Administration (EIA). 2011. *International Energy Outlook: World Consumption of Hydroelectricty and Other Renewable Energy by Region, Reference Case.* Tech. Rep. DOE/EIA-0484, September 19.
6. U.S. Energy Information Administration (EIA). 2011. *International Energy Outlook* 2011: *Liquid Fuels.* Tech. Rep. DOE/EIA-0484, September 19.
7. Next Big Future. 2011. Global biofuels production increased 17% in 2010. [http://nextbigfuture.com/2011/08/global-biofuels-production-increased-17.html].
8. Licht, F.O. 2009. Germany: Government to lower 2009 biofuels quota. *World Ethanol Biofuels Rep.* [website www.agra-net.com]. 7(3): 55.
9. European Biofuels Technology Platform. 2011. Biofuel production. [http://www.biofuelstp.eu/fuelproduction.html].
10. Perlack, R.D., Wright, L.L., Turhollow, A.F., Graham, R.L., Stokes B.J., and Erbach, D.C. 2005. *Biomass as Feedstock for a Bioenergy and Bioproducts Industry: The Technical Feasibility of a Billion-Ton Annual Supply*. Tech. Rep. DOE/GO-102005-2135 and ORNL/TM-2005/66, April.
11. Perlack, R.D. and Stokes, B.J. 2011. *U.S. Billion-Ton Update: Biomass Supply for a Bienergy and Bioproducts Industry*. Oak Ridge National Laboratory, Tech. Rep. ORNL/TM-2011/224, August.
12. U.S. Department of Energy, 2011. *Energy Efficiency & Renewable Energy, Biomass Program. Biofuels, Biopower, and Bioproducts: Integrated Biorefineries*. http://www1.eere.energy.gov/biomass/index.html.
13. International Energy Agency (IEA) Bioenergy. 2005. *What are the environmental benefits of using biomass?* http://www.task29.net/assets/files/broshures/Brochure2005.pdf. Accessed December 2011.

2

Crop Oils, Biodiesel, and Algae Fuels

First-generation biofuels refer to the fuels that have been derived from biological sources such as starch, sugar, animal fats, and vegetable oils, whereas *second generation biofuels* are derived from lignocellulosic crops. The oils under the category of first generation biofuels are obtained using conventional production techniques. Some of the most typical types of first generation biofuels include vegetable oil, conventional biodiesel, bioalcohols, biogas, and synthesis gas (biosyngas).

2.1 Vegetable Oils

2.1.1 Background

Vegetable oils are lipid materials derived from plants and are in liquid phase at room temperature. Vegetable oils are mostly composed of triglycerides whose molecular structures are tri-esters of fatty acids based on glycerin backbones. Vegetable fats are classified basically in the same group as vegetable oils, except that vegetable fats are solid at room temperature and are made of higher molecular weight materials. The distinction between vegetable oils and fats may intuitively appear to be in their melting point differences. However, both vegetable oils and fats are actually mixtures of similarly structured triglyceride molecules, shown in Figure 2.1, as opposed to unimolecular substances; as such, a precise definition of melting point is impractical for them. Instead, melting point ranges are used to characterize vegetable oils and fats.

Vegetable oils have long been used by humans in a variety of essential end uses including food ingredients, cooking oil, heating, lighting, medicinal treatment, and lubrication. Some vegetable oils are directly edible, but many others are not. Examples of edible oils include sesame oil, corn oil, coconut oil, palm oil, sunflower oil, olive oil, peanut oil, rice bran oil, and soybean oil, whereas examples of inedible vegetable oils include linseed oil, tung oil, and castor oil which are used in lubricants, solvents, stains, paints, cosmetics, pharmaceuticals, and other industrial purposes. Basically, all edible oils and fats can be described as triglycerides where the acyl group (also known as alkanoyl group, RCO–) is a fatty acid moiety, which is from an aliphatic

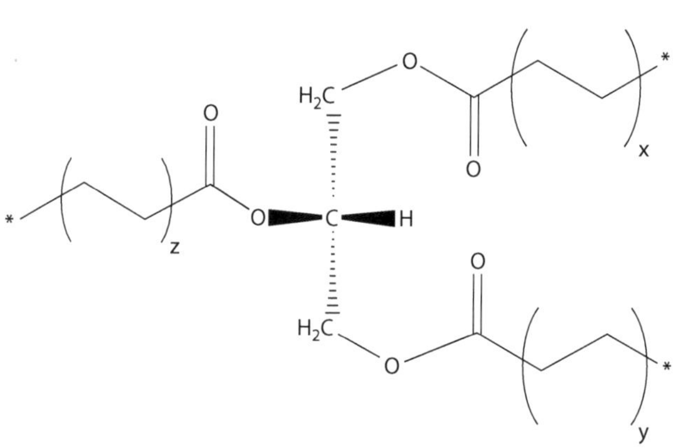

FIGURE 2.1
Molecular structure of triglycerides.

carboxylic acid with an even number of carbon atoms that may contain one or more double bonds [1, 2]. Edible oils and fats always contain several different fatty acids, and inasmuch as almost all triglycerides in an oil or fat also contain various fatty acid moieties, the actual number of unique triglycerides in an oil or fat can be quite large. As such, a range or a distribution of melting points is used to characterize the oil and fat rather than a single unique value for its melting point. Because the triglyceride composition controls the physical and chemical properties of the oil or fat, it is this composition that may have to be modified to alter the physical and chemical properties in order to meet the specific application requirements [1, 2]. Converting vegetable oil into conventional biodiesel via well-publicized and established transesterification reaction is a good example of this viewpoint.

Unsaturated vegetable oil molecules contain a number of unsaturated C=C double bonds in their molecular structures which can be hydrogenated by a relatively simple catalytic hydrogenation process. Hydrogenation of unsaturated vegetable oils can be achieved by bubbling or sparging the oil in the presence of a hydrogenation catalyst with hydrogen at high temperature and pressure. Precious metals such as platinum, palladium, rhodium, and ruthenium make highly active catalysts. However, the catalyst most commonly used is a powdered nickel catalyst (such as Raney nickel or Urushibara nickel) for economical reasons, as in most hydrogenation reactions. The nickel-based catalysts are less active and require higher pressures such as 60–70 atm. The reactor type used for such an operation is a three-phase fluidized bed reactor. As the hydrogenation reaction converts unsaturated vegetable oils toward saturated vegetable oils, both partially and fully saturated oils, the viscosity and melting point of the resultant oil increase. Although hydrogenation of vegetable oils can alter the oils' physical properties and textures, the results of hydrogenation of vegetable oils are not necessarily beneficial, especially to human health. Partial hydrogenation of vegetable oils results in the formation of a large amount of *trans*-fat in the resultant oil mixture, which is considered very unhealthy in edible oils [3]. *Trans*-fat is a common name for unsaturated fat with *trans*-isomer fatty acids and can be either mono- or poly-unsaturated in its structure. As well publicized, the negative health-related consequences and concerns of *trans*-fat consumption go far beyond cardiovascular risk to humans.

2.1.2 Production and Use of Vegetable Oils

Vegetable oils have long been used by humans throughout the entire world for a variety of applications, including traditional uses such as cooking, food ingredients, medicinal ingredients, lubrication, and heating and lighting as well as more recent uses such as motor oil, lubricants, drying oils, alternative diesel fuel, raw materials for biodiesel, and so on.

In order to distinguish vegetable oil for fuel applications from biodiesel, some people refer to it as *waste vegetable oil* (*WVO*) if it is originated from discarded sources such as used restaurant grease, and as *straight vegetable oil* (*SVO*) or *pure plant oil* (*PPO*) if it has not undergone any chemical treatment or reaction such as transesterification, just as the name implies.

According to the USDA [4, 5], the total world production of major vegetable oils in 2008–2009 includes: 43.19 million metric tons of palm oil, 36.26 of soybean oil, 20.22 of rapeseed oil, 11.46 of sunflower seed oil, 5.15 of peanut oil, 5.10 of palm kernel oil, and 4.72 of cottonseed oil. The world consumption of vegetable oils has been steadily increasing at a rate of 4.5% annual growth and their trade volumes of import and export have also been steadily increasing. See Table 2.1.

2.1.3 Extraction of Vegetable Oils

Vegetable oil can be extracted from oil seeds by several different methods, including solvent extraction, expeller method, and supercritical extraction. To maximize extraction efficiency, oil seeds are usually "crushed," thus decreasing particle size and lowering the mass transfer resistances of oil removal. The traditional method for extraction is using an expeller machine, of which there are two variations: a screw type and a ram type. Solvent extraction is a more recent method, and usually utilizes a petroleum-based product such as hexane. Expeller methods are typically used for edible products such as nutrients and food oils, whereas solvent extraction is for more modern oil applications. Supercritical extraction is used for extraction of value-added nutrient and medicinal ingredients from oil seeds and is still in the development stage. Typical oil extraction efficiency by these methods is shown in Table 2.2. However, it should be noted that the actual efficiency depends on the specific type of oil seeds, specific design of the machinery used, specific extraction conditions employed, the moisture level of raw materials, and different operational parameters.

TABLE 2.1

Global Production/Consumption of Vegetable Oils[a]

Oils	2008/09 World Consumption (Million Metric Tons)[a]	Remarks
Palm Oil	43.19	The most widely produced tropical oil. Also used for biofuel manufacture.
Soybean Oil	36.26	An oil extracted from the seeds of soybean. Accounts for about half of worldwide edible oil production. Also used as a drying oil (for printing inks and oil paints).
Rapeseed Oil	20.22	One of the most commonly used cooking oils; Canola is one of two cultivars of rapeseed.
Sunflower Seed Oil	11.46	A common cooking oil for frying. Used in cosmetic formulations as an emollient. Also used for biodiesel manufacture.
Peanut Oil	5.15	A mild-flavored common cooking oil for frying (French fries and chicken). Has a high smoke point.
Palm Kernel Oil	5.10	An edible plant oil derived from the kernel of the African palm tree.
Cottonseed Oil	4.72	A major cooking oil. A naturally hydrogenated oil used as a stable frying oil and for food processing.
Coconut Oil	3.64	Extracted from the kernel or meat of matured coconut from the coconut palm.
Olive Oil	2.97	Widely used in cooking, cosmetics, and soaps. Has a high smoke point. Used as a liquid fuel for traditional oil lamps.
Total	**132.70**	This total reflects a steady increase from 111.47 (2004–2005), 118.49 (2005–2006), 121.33 (2006–2007), and 127.86 (2007–2008).

Source: [a]*Oilseeds: World Markets and Trade*, Foreign Agricultural Service, USDA, and *Economic Research Service. Oil Crops Yearbook*. U.S. Department of Agriculture.

Note: The worldwide production data for 2008–2009 are forecast values.

2.1.4 Composition of Vegetable Oils

Vegetable oils may or may not be edible, as explained in Section 2.1.1. Even though most efforts in producing biodiesel have been using edible oils, biofuel and, in particular, biodiesel can also be produced using inedible oils [6].

The most frequently used method of characterizing the chemical structure of a vegetable oil is the fatty acid composition, which shows the distribution of different kinds of fatty acids, the carbon numbers of ingredient fatty acids, location of double bonds in the fatty acid molecular structure, ratio of saturated versus unsaturated fatty acids, and more. Table 2.3 shows fatty

TABLE 2.2

Typical Oil Extraction Efficiency of Various Extraction Methods

Extraction Method	Percentage Extracted (%)
Expeller Method	34–37
Solvent Extraction	40–43
Supercritical Fluid Extraction	35–80

acid compositions of common edible oils in terms of percent by weight of total fatty acids.

The specific gravity of most edible vegetable oil ranges between 0.91 and 0.93, depending upon the kind of vegetable oil, specific composition of the oil, purity, and other factors. The specific gravity of castor oil, an inedible oil, is approximately 0.957–0.961.

2.1.5 Use of Vegetable Oil as Alternative Diesel Fuel

The diesel engine is named after the original developer of the engine, Rudolf Diesel, who initially attempted to run the engine with coal dust and later redesigned the engine to run with vegetable oil. Several similar efforts of the early 1900s were reported. Since then, the R&D efforts of using straight vegetable oils as a diesel substitute or alternative diesel fuel have continued or received public interest in response to escalating and fluctuating petroleum prices.

If properly modified, most diesel engines on automobiles can be run on vegetable oils, that is, SVO or PPO. Principal modifications involve reduction of viscosity and surface tension of SVO before injection of the fuel. High viscosity and surface tension of vegetable oils, if not altered, can cause poor atomization of the fuel, which results in incomplete combustion. Incomplete combustion causes coking or carbonization, which produces polycyclic aromatic hydrocarbons (PAHs), soot, or coke. The reduction of viscosity and surface tension can be achieved via preheating of the fuel before its injection, which can be accomplished using the waste heat of the engine or the automobile's electric power. In such a case, an additional tank of normal diesel (i.e., petrodiesel or biodiesel) would still be needed in addition to the main SVO tank for smooth engine operation. The cold engine is started with normal diesel and once the engine gets warmed up, the fuel is switched from normal diesel fuel to vegetable oil which is preheated using a heat exchanger. In a very cold climate, this limitation in cold start is potentially a significant hurdle to overcome.

Alternately, an innovative blend chemistry may be devised for property modification of vegetable oils for their direct use in diesel engines. Vegetable oil can be mixed with other fuels or chemicals such as kerosene, diesel, and gasoline, thereby reducing the viscosity (in particular, kinematic viscosity) and surface tension of the blended fuel. The blending chemical/fuel typically

TABLE 2.3

Fatty Acid Compositions of Common Edible Oils[a]

Oil	Unsat./Sat Ratio	Saturated					Monounsaturated	Polyunsaturated	
		Capric Acid C10:0	Lauric Acid C12:0	Myristic Acid C14:0	Palmitic Acid C16:0	Stearic Acid C18:0	Oleic Acid C18:1	Linoleic Acid (ω6) C18:2	Alpha Linolenic Acid (ω3) C18:3
Almond Oil	9.7	–	–	–	7	2	69	17	–
Canola Oil	15.7	–	–	–	4	2	62	22	10
Cocoa Butter	0.6	–	–	–	25	38	32	3	–
Cod Liver Oil	2.9	–	–	8	17	-	22	5	–
Coconut Oil	0.1	6	47	18	9	3	6	2	–
Corn Oil (Maize Oil)	6.7	–	–	–	11	2	28	58	1
Cottonseed Oil	2.8	–	–	1	22	3	19	54	1
Flaxseed Oil	9.0	–	–	–	3	7	21	16	53
Grape seed Oil	7.3	–	–	–	8	4	15	73	–
Olive Oil	4.6	–	–	–	13	3	71	10	1
Palm Oil	1.0	–	–	1	45	4	40	10	–
Palm Olein	1.3	–	–	1	37	4	46	11	–
Palm Kernel Oil	0.2	4	48	16	8	3	15	2	–
Peanut Oil	4.0	–	–	–	11	2	48	32	–
Safflower Oil[b]	10.1	–	–	–	7	2	13	78	–
Sesame Oil	6.6	–	–	–	9	4	41	45	–

Shea Nut	1.1	–	1	–	4	39	44	5	–
Soybean Oil	5.7	–	–	–	11	4	24	54	7
Sunflower Oil[b]	7.3	–	–	–	7	5	19	68	1
Walnut Oil	5.3	–	–	–	11	5	28	51	5

Source: http://www.scientificpsychic.com/fitness/fattyacids1.html; http://www.connectworld.net/whc/images/chart.pdf; http://curezone.com/foods/fatspercent.asp

Note: Percentages may not add to 100% due to rounding and other constituents not listed in the table. Where percentages vary, average values are used.

[a] By weight of total fatty acids.

[b] Not high-oleic variety.

is chosen from lower molecular weight hydrocarbons. This type of blending is often referred to as "cutting," "diluting," or "cosolvent mixing." However, there are some concerns regarding blending, which are largely based on higher rates of wear and tear in conventional fuel pumps and piston rings when using such blends. Advances in automobile component design including fuel injectors, cooling system, and glow plugs as well as in materials of construction are expected to be made.

As an example of recent advances, automobiles powered by indirect injection engines equipped with in-line injection pumps are capable of running on pure SVO in most moderate climates, except during cold winter temperatures. Some of these vehicles are equipped with a coolant-heated fuel filter, which functions as a fuel preheater that helps reduce the viscosity of the SVO.

2.1.6 Use of Vegetable Oil in Direct Heating

Higher heating values (HHVs) of vegetable oils range between 39,000–48,000 kJ/kg, or 16,770–20,650 BTU/lb, depending upon the kind of vegetable oil. The HHV of vegetable oil is higher than that of anthracite, spent tire rubber, or wood. Table 2.4 shows a comparison of higher calorific values of common fuels and energy sources [7].

The higher heating value is also known as the gross calorific value, higher calorific value, gross energy, or gross heat. The HHV of a fuel is defined as the amount of heat released per unit mass (initially at 25°C) once it is combusted and the products have returned to a temperature of 25°C. This means that the HHV value is the total heat recoverable, including the energy contained in water vapor released due to combustion reaction. In other words, the higher (or gross) heating value is the gross calorific value (gross CV) when all products of combustion are cooled back to the precombustion temperature, water vapor formed during combustion is also condensed, and necessary corrections have been made.

On the other hand, the lower heating value (LHV) is also known as net calorific value, or net CV. The LHV of a fuel is defined as the amount of heat released due to combustion of a unit mass of fuel (initially at 25°C or another reference state) and returning the temperature of the combustion products to 150°C. As such, the energy contained in water vapor released during combustion is not wholly included in the LHV. A major portion of the energy amount excluded from the HHV, in obtaining the LHV, is the latent heat of vaporization of water.

As shown in Table 2.4, the heating value of vegetable oil is quite high, higher than other naturally derived fuels such as coal and wood. Vegetable oils have long been used for cooking, lighting, and heating throughout the world. Residential furnaces and boilers that are designed to burn heating oil No. 2 can be modified to burn vegetable oils, including filtered waste vegetable oil. The required modification is based on a similar ground as

TABLE 2.4

Higher Heating Values (HHVs) of Common Fuels

Fuel	Higher Heating Value (Gross Calorific Value—GCV)		Ref.
	kJ/kg	BTU/lb	
Acetone	29,000	12,470	National Institute of Standards and Technology (NIST) [7]
Acetylene	49,900	21,460	National Institute of Standards and Technology (NIST) [7]
Anthracite coal	32,500–34,000	14,000–14,500	Speight, 1994 [8]
Biodiesel	40,168	17,280	National Biodiesel Board (NBB), 2011 [9]
Bituminous coal	17,000–23,250	7,300–10,000	Speight, 1994 [8]
Bituminous coal (dmmf basis)	24,000-35,000	10,320-15,050	Speight, 1994 [8]
Butane	48,590	20,900	National Institute of Standards and Technology (NIST) [7]
Carbon	32,800	14,100	Walker, Rusinko, and Austin, 1959 [10]
Charcoal	29,600	12,800	Speight, 1994 [8]
Coke	28,000–31,000	12,000–13,500	Speight, 1994 [8]
Diesel (Petro-)	44,800	19,300	Lee, Speight, and Loyalka, 2007 [11]
Ethane	51,900	22,400	National Institute of Standards and Technology (NIST) [7]
Ethanol	29,700	12,800	Lee, Speight, and Loyalka,, 2007 [11]
Dimethylether	31,680	13,625	Lee, Speight, and Loyalka,, 2007 [11]
Gasoline	47,300	20,400	Lee, Speight, and Loyalka, 2007 [11]
Glycerin	19,000	8,170	National Institute of Standards and Technology (NIST) [7]
Hydrogen	141,790	61,000	National Institute of Standards and Technology (NIST) [7]
Lignite coal	16,200	7,000	Speight, 1994 [8], Lee, Speight, and Loyalka, 2007 [11]
Methane	55,530	23,884	National Institute of Standards and Technology (NIST) [7]
Methanol	22,700	9,800	Lee, 1990 [12]
Natural gas	52,225	22,453	Lee, 1997 [13]
Paper wastes (mixed)[a]	13,955–15,543	6,002–6,682	Ucuncu and Vesilind, 1991 [14]
Peat	13,800–20,500	5,500–8,800	Lee, Speight, and Loyalka, 2007 [11]
Peat (damp)	6,000	2,500	Lee, Speight, and Loyalka, 2007 [11]
Petroleum crude	43,000	18,490	Lee, Speight, and Loyalka, 2007 [11]
Propane	50,350	21,660	National Institute of Standards and Technology (NIST) [7]

(Continued)

TABLE 2.4 (CONTINUED)

Higher Heating Values (HHVs) of Common Fuels

Fuel	Higher Heating Value (Gross Calorific Value—GCV) kJ/kg	BTU/lb	Ref.
Sulfur	9,200	3,960	National Institute of Standards and Technology (NIST) [7]
Tar	36,000	15,480	Speight, 1994 [8]
Spent Tire	37,200	16,000	Lee, Speight, and Loyalka, 2007 [11]
Turpentine	44,000	18,920	National Institute of Standards and Technology (NIST) [7]
Vegetable oil	39,000–48,000	16,770–20,650	Lee, Speight, and Loyalka, 2007 [11]
Wood (dry)	14,400–17,400	6,200–7,500	Lee, Speight, and Loyalka, 2007 [11]

[a] dmmf = dry mineral matter free basis.

that for diesel engines for SVO, in which the viscosity of vegetable oil is reduced by appropriate preheating. Although this method results in substantial cost savings, it has not been popularly practiced in the United States.

2.1.7 Use of Vegetable Oil for Combined Heat and Power (CHP)

A number of manufacturers offer compressed ignition engine generators optimized to run on SVOs in which the waste engine heat is recovered. As with all types of generators, including diesel and jet fuel generators, the operational issues involve (1) fuel specificity; (2) extreme climate operation including hot, cold, wet, and dusty; (3) trouble-free operational longevity; (4) maintenance frequency and burden; and (5) overall energy efficiency, among others.

2.1.8 Use of Vegetable Oil for Biodiesel Manufacture

Triglycerides, which are the principal ingredients of vegetable oils and algae oils, are chemically converted into biodiesel via catalytic transesterification reaction under mild reaction conditions. The effects of transesterification on selected fuel properties have been examined by Bello, Magaji, and Agge [15]. Their results showed that the density reduces by 7–9%, and the cetane number (CN) increases by 60–78% by transesterification. They also found that transesterification overall tends to make the properties of vegetable oils close to those of petrodiesel [15]. Major parts of ensuing sections of this chapter are devoted to the processing and manufacture of biodiesel and its precursor feedstocks.

2.2 Algae Oil Extraction of Straight Vegetable Oil

2.2.1 Introduction

Despite the ever-escalating price for petroleum products and rapidly growing concerns regarding carbon dioxide emissions, the world still remains heavily dependent upon fossil fuels. The *2011 International Energy Outlook* [16] predicts by its Reference Case Scenario that the total world consumption of marketed energy will increase by roughly 42% by 2035 from 2010 with an increase in liquid fuels consumption of 30% by 2035 from 2010 in the transportation sector. This increase in demand, however, cannot be met by petroleum alone.

Many commodities, such as biodiesel and bioplastics, are made from the triglycerides found in vegetable oils, and other petrochemicals can also be derived or synthesized using processing by-products including glycerin. Algae, specifically microalgae, are a promising source of oil because, compared to other crops, they have fast growth rates, potential for higher yield rates, and the ability to grow in a wide range of conditions [17]. The yield of oil per unit area of algae is at least seven times greater than that of palm oil, the second highest yielding crop [18]. Another benefit of using algae is that they overcome the "food versus fuel" issue of other vegetables and grains because algae is not a universal food crop, and it does not take arable land from other crops. Algae grown on 9.5 million acres, compared to the 450 million acres used for other crops, could provide enough biodiesel to replace all petroleum transportation fuels in the United States [18]. Based on the currently available technologies, harvesting of algae and the extraction of oil is technologically challenging and energy intensive [17]. There are several different methods proposed and developed for extracting oil, an important step in making biofuels as well as bioplastics. Unlike straight vegetable oils (SVOs) and conventional biodiesels based on crop oils, algae fuels are classified as second-generation biofuels.

The lipid containing oil from algae must be separated from the proteins, carbohydrates, and nucleic acids. The steps for extracting the oil involve breaking the cell wall, separating the oil from the remaining biomass, and purifying the oil [19]. Oil can be extracted from algae by either mechanical or chemical methods. The three well-known methods for extraction of oil from algae are expeller pressing (or oil pressing), subcritical solvent extraction, and supercritical extraction [17]. Other methods include enzymatic extraction, osmotic shock, and ultrasonic-assisted extraction [18]. Expeller pressing and ultrasonic-assisted extraction are mechanical processes, and subcritical and supercritical extraction methods are chemical processes. Each of the methods has its advantages and drawbacks. Extraction of oil by expeller pressing is simple and straightforward, but requires the algae to be fully dried, which is energy intensive. Furthermore, the extraction efficiency is not very high, and a substantial amount of unextracted oil is left behind. The benefit of solvent extraction is that the algae do not need to be fully dried, but common subcritical solvents, such as hexane, pose environmental, health, and safety concerns [19]. In addition, although most solvent is recovered and reused, its associated cost is also burdensome. Supercritical fluid extraction may be the most efficient method as it can extract almost all the oil and provide the highest purity inasmuch as supercritical fluids are selective [17]. Furthermore, extraction with supercritical CO_2 eliminates the use of harmful solvents. However, its high-pressure operation and required high-pressure equipment increase the overall process cost.

Algae oil has been proposed as both a sustainable and economically feasible solution to alternative liquid transportation fuels.

2.2.2 Microalgae and Growth

Microalgae or microphytes are a division of algal organisms encompassing diatoms, the green algae, and the golden algae. These microscopic organisms are incredibly efficient solar energy converters that perform photosynthesis and are capable of rapid growth in either freshwater or saline environments. Although this value varies from species to species and depends upon cultivation conditions, roughly 50% of the weight of algae is lipid oil. Algae are typically cultivated in either open or closed ponds, photobioreactors, or hybrid systems of both. Once the algae have matured they are harvested and processed to extract the algae's oil.

2.2.3 Algae Harvesting

Collecting, concentrating, and processing algae consists of separating algae from the growth medium, drying, and processing it to obtain the desired product. Separating algae from its growth medium is generally referred to as algae harvesting. The term *algae harvesting* technologically refers to the activity of concentration of a fairly diluted (ca. 0.02–0.06% total suspended solids, TSS) algae suspension until a slurry or paste containing 5–25% TSS or greater is obtained. Specific harvesting methods depend primarily on the type of algae and growth media. The high water content of algae must be removed to enable further processing. The most common harvesting processes include: (1) microscreening, (2) flocculation, and (3) centrifugation [20, 21]. The three methods represent different unit operations of filtration, flotation, and centrifugation, respectively. Therefore, these harvesting steps must be energy-efficient and relatively inexpensive; as such, selecting easy-to-harvest algae strains becomes quite important. Macroalgae harvesting requires substantial manpower, whereas microalgae can be harvested more easily using microscreens, centrifugation, flocculation, or by froth flotation.

2.2.3.1 Microscreening Harvesting of Algae

Membrane filtration is one of the algae harvesting methods and is usually aided by a vacuum pump. Membrane filtration provides well-defined pore openings to separate algal cells from the culture. An advantage of the membrane filtration harvesting method is that it is capable of collecting and concentrating microalgae or cells of very low initial density (concentration). However, concentration by membrane filtration is somewhat limited to small volumes and leads to the clogging and fouling of the filter (membranes) by the packed cells when vacuum is applied. Fouling and clogging of the membrane surface due to increased concentration of algal cells results in sharp declines in flux and requires maintenance.

A modified filtration method involves the use of a reverse-flow vacuum in which the pressure operates from above rather than below, making the process gentler and avoiding or alleviating the packing of cells on the membrane. This method itself has been modified to allow a relatively large volume of water to be concentrated in a short period of time (20 liters to 300 ml in three hours or a concentration of nearly 70 times in three hours) [21].

Cross-flow filtration is a purification separation technique, typically employed for submicron-sized materials, where the majority of the feed (algae–water suspension) flow travels tangentially across the surface of the filter rather than perpendicularly into the filter. It is advantageous over standard filtration, because the filter cake is being constantly washed away during the filtration process, thereby increasing the service time that the filtration device can be used without maintenance stoppage.

Cross-flow microfiltration (MF) was investigated by Hung and Liu [22] for separation of green algae, *Chlorella* sp., from freshwater under several different transmembrane pressures (TMP) and also with both laminar and turbulent flows by varying the cross-flow velocity. The study examined the hydrodynamic conditions and interfacial phenomena of microfiltration of green algae and revealed the interrelations among the cross-flow velocity, MF flux decline, and TMP [22].

Forward osmosis (FO) is an emerging membrane separation process, and it has recently been explored for microalgae separation. It is claimed that forward osmosis membranes use relatively small amounts of external energy compared to the conventional methods of algae harvesting. The driving force through a semi-permeable membrane for forward osmosis separation is an osmotic pressure gradient, such that a "draw" solution of high concentration (relative to that of the feed solution of dilute algae suspension) is used to induce a net flow of water through the membrane into the draw solution, thus effectively separating the feedwater from its solutes (microalgae). Zou et al. [23] studied the FO algae separation by comparing two different draw solutions of NaCl and $MgCl_2$ and also examining the efficacy as well as their membrane fouling characteristics.

Sometimes concentrated algae may be collected with a *microstrainer.* When a microstrainer is used to collect algae, the processed algae–water suspension may look faintly green, indicating that it could be further concentrated. However, due to its eventual clogging, a microscreen alone is usually insufficient for long-term continuous or large-scale operation; substantial energy and labor input are required to remove the clogging and reopen the flow channels.

A novel process for harvesting, dewatering, and drying (HDD) of algae from an algae–water suspension has been developed by Algaeventure Systems, LLC. In this process, a superabsorbent polymer (SAP) fabric belt is put in contact with the bottom of the screen (water meniscus), thereby enabling the movement of a vast amount of water without moving the algae and achieving dewatering. This is based on the fact that water–water hydrogen bonding is stronger than water-to-algae's weak intermolecular forces. As such, reduced surface tension, enhanced capillary effect, and modified adhesion effect can be built in and the system can be designed to be continuous. In a prototype testing, an exceptional rate of HDD was achieved with a very low power input. A schematic of Algaeventure Systems harvester is shown in Figure 2.2 [24].

Superabsorbant polymers (SAPs) can absorb and retain very large amounts of a liquid (such as an aqueous solution) in comparison to their own mass and have been widely used in baby diapers, personal hygiene and care products, and water soluble or hydrophilic polymer applications.

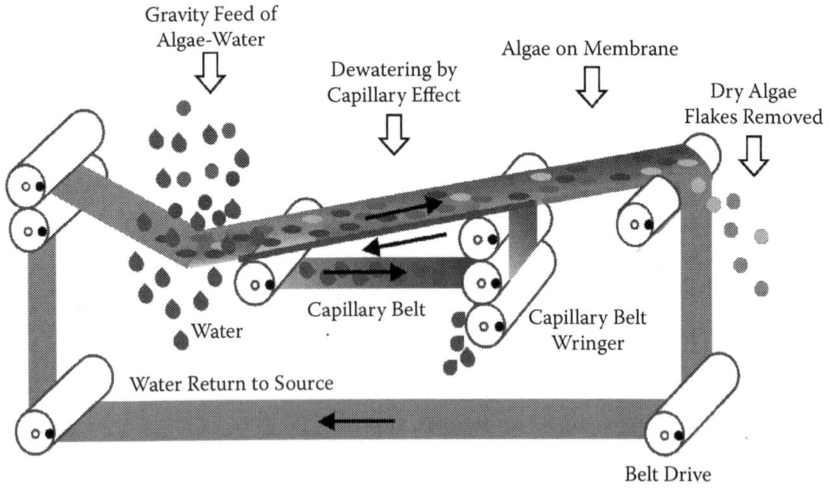

FIGURE 2.2
A schematic of the Algaeventure System harvester. (Modified from Algae Venture Systems. 2011. *Harvesting, Dewatering & Drying* (accessed June 2011). Available at: http://www.algaevs.com/harvesting-dewatering-and-drying.)

2.2.3.2 *Algae Harvesting by Flocculation*

Algae flocculation is a method of separating algae from its medium by using chemicals to force the algae to form lumps or aggregates. The main disadvantage of this separation method is that the additional chemicals are difficult to remove or recover from the separated algae, thus making it inefficient and uneconomical for commercial use. The cost to remove or recover these chemicals may be too expensive to be commercially viable, unless a technological breakthrough is achieved.

Flocculating agents, or *flocculants,* are chemicals that promote flocculation by causing colloids and other suspended particles in liquids to aggregate, forming a floc. In general, there are two types of flocculants commonly used: inorganic flocculants and organic polymer/polymer electrolyte flocculants. Alum (hydrated potassium aluminum sulfate ($KAl(SO_4)_2 \cdot 12H_2O$)) and ferric chloride ($FeCl_3$) are the chemical flocculants most frequently used to harvest algae [20]. A commercial product called "Chitosan," popularly used for water purification, can also be used as a flocculant but is far more expensive than other flocculants. The shells of crustaceans, such as shrimp, lobster, crabs, and crayfish, are ground into powder and processed to obtain chitin, a polysaccharide found in the shells, from which chitosan is derived via deacetylation. Water that is more brackish or saline requires additional chemical flocculant to induce flocculation [20]. High molecular weight organic polymers are considered good flocculants, because several segments of a polymer can attach themselves to the surface of a colloidal particle and the remainder of the segments are extended into the solution [25].

Harvesting via chemical flocculation alone, by current technoeconomic standards, is a method that may be too expensive for large operations. In addition to the chemical flocculation discussed above, there are different methods of flocculation of algae, including autoflocculation [26], bioflocculation [27], and electroflocculation. Bioflocculation has been studied and practiced in wastewater and sewage treatment. The efficacy of algae flocculation depends upon a large number of factors that are usually poorly understood or ignored, including cell size, cell shape, cell wall thickness, cell surface and interfacial properties, and so on. By considering these properties as well as appropriate combinations of flocculation with other concentrating techniques, there is room for substantial enhancement in algae flocculation practice.

2.2.3.3 *Algae Harvesting by Centrifugation*

The choice of a good harvesting method for algae is crucial to the efficiency of the entire process in terms of capital investments as well as operation costs. The key factors for comparison include high cake dryness in the separated algae and low specific energy demand during the process, that is, energy demand per unit mass of algae harvested.

In a *single-stage harvesting process* using disk stack centrifuges, the algae–water suspension is directly fed into the centrifuge. Inside the centrifuge the suspension is separated into a mostly clear water phase and an algae concentrate. The algae concentrate is drawn out periodically and has a fluid/creamy consistency. The whole suspension has to be put into rotation to create a centrifugal force up to 10,000 g's, thus the specific energy demand is relatively high. Therefore, this single-stage harvesting process is especially suitable for small and middle-sized facilities.

Disk stack centrifuges have been successfully operated for separation of two different liquid phases and solids from each other in a continuous process [28]. The operational principle of a disk stack centrifuge is described below. A good example of separation of two different liquid phases is separation of biodiesel methyl ester and by-product glycerin, whereas a good example of solid–liquid separation is dewatering algae from an algae–water suspension. Alfa Laval offers a variety of sizes and types of disk stack centrifuges for such industrial applications [29]. In a disk stack centrifuge used for liquid–solid separation, the denser solids are pushed outward by centrifugal forces against the rotating bowl wall, and the less dense liquid phases form inner concentric layers. By inserting specially designed disk stacks where the liquid phases meet, a very high separation efficiency is achieved. The solids such as algae cakes can be removed manually, intermittently, or fully continuously, depending upon the specific process design and application. The separated liquid phase overflows in the outlet area on top of the bowl into recovery vessels, which are sealed off from each other to prevent potential cross-contamination [29].

The *Flottweg enalgy process* [30] is a two-stage algae harvesting process consisting of (1) preconcentration via static settling, filtration, flocculation, or dissolved air flotation (DAF), and (2) bulk harvesting using the Flottweg Sedicanter® in order to dewater the algae suspension to concentrate. In contrast to the aforementioned single-stage process, only a small, predewatered, part of the algae suspension is separated by centrifugation in this two-stage process, thus reducing the energy demand drastically. Whereas the preconcentrator provides a clear water phase as an initial step, the Flottweg Sedicanter dewaters the algae concentrate to obtain a solids cake with 22–25% dry substance [30]. A schematic of the Flottweg two-stage harvesting process is shown in Figure 2.3.

2.2.4 Algae Oil Extraction

Three major methods of extracting oil from algae are: (1) solvent extraction with hexane, (2) expeller/press, and (3) supercritical fluid extraction (SFE). The extracted oil can then either be used as SVO or processed by a transesterification reaction to produce biodiesel. Algal oil is considered a "balanced" carbon neutral fuel, because any CO_2 taken out of the atmosphere by the algae is returned when the algae biofuels are burned. It has been estimated

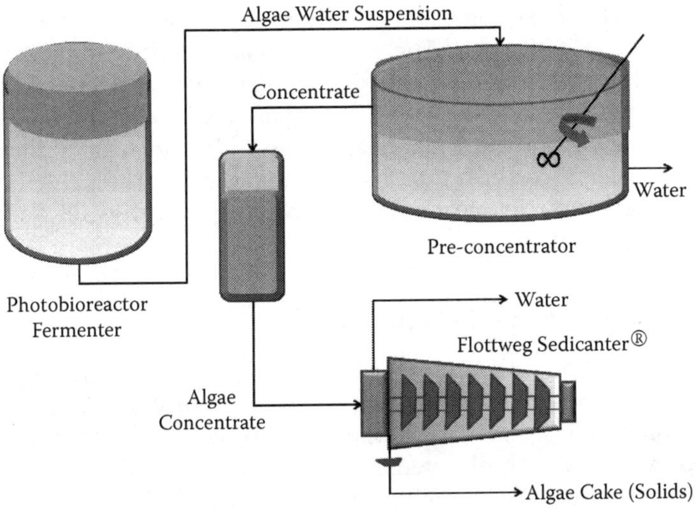

FIGURE 2.3
Flottweg two-stage enalgy process of algae harvesting.

by Weyer et al. [31] that the theoretical maximum limit of unrefined oil that could be produced from algae is 354,000 L · ha^{-1} · yr^{-1} or 38,000 gal · ac^{-1} · yr^{-1}, and limits for the practical cases examined in their study range from 46,300 to 60,500 L · ha^{-1} · yr^{-1}.

2.2.4.1 Expeller Pressing Extraction of Algae Oil

Expeller pressing methods have a long history of extracting vegetable oil from oil seeds; the methods involve some means of mechanically squeezing oil from crushed oil-containing seeds under an applied pressure. Most cooking oils are produced via expeller pressing of a variety of feedstock including maize, sunflower, soya, sesame, coconut, mustard seed, and groundnuts. Even though this mechanical extraction method of oil from oil seeds is technologically very simple, the process has been significantly enhanced with the development of energy-efficient and mechanically superior machinery. Furthermore, large processing plants have advantages over small plants in terms of reduced processing cost and extraction efficiency; most countries have such centralized large plants for cooking oil manufacture. These plants are also known as "oil refineries," that is, vegetable oil refineries. However, logistical burdens and transportation costs of raw materials and finished products also make small-scale highly efficient plants relevant and viable options. This is especially true for algae oil and biodiesel processing.

Oil presses are typically used for vegetable oils and biodiesel processing on both large and small scales. There are two different kinds of large-scale processing methods involving oil presses, one being hot processing and the other being cold processing. In hot processing, the system includes a steam cooker and oil press. The steam cooker is mainly for pretreatment of oil seeds. In cold processing, the machine operates at a low temperature (e.g., 80°C) when it presses the seeds. An advantage of cold processing is that the extraction environment is not destructive to the nutrients in the oil. Dry algae can also be processed, quite similarly to oil seeds, using an oil press for extraction of algae oil, by mechanically rupturing the cell walls and collecting the extracted oil.

2.2.4.2 Ultrasonically Assisted Extraction

Ultrasonic extraction or *ultrasonication* can enhance and accelerate the algae oil extraction processes. In an ultrasonic reactor, intense sonication of liquids generates ultrasonic waves that propagate into the liquid medium. During the low-pressure cycle, high-intensity small vacuum bubbles are created in the liquid due to the pressure imbalance. When these bubbles attain a certain critical dimension (cavity size), they collapse violently during a high-pressure cycle. As these bubbles collapse vigorously near the algae cell walls (i.e., implode) they create shock waves, and locally high-pressure and high-speed liquid jets. The resultant shear forces cause algae cell walls to mechanically rupture and release or help release their contents (algae lipids) into the solvent medium. This process of bubble formation and subsequent collapse is mechanistically called *cavitation* [32]. The advantages of the process include:

(a) Dry cake is not required for oil extraction.
(b) No caustic chemical is involved.
(c) The ultrasonication process can be used in conjunction with enzymatic extraction.
(d) Environmental impact is minimal.

However, the process is not yet proven on a large scale and the energy cost needs to be lower in order to be competitive. Ultrasonication can be employed in a solvent extraction process, however, the process is usually classified as mechanical extraction, largely based on its mechanically induced cell wall rupture mechanism.

2.2.4.3 Single-Step Extraction Process by OriginOil, Inc.

Although many new technologies have been developed to extract straight vegetable oil from algae, this section specifically describes the single-step extraction process developed by OriginOil, Inc.

The single-step extraction process begins with the mature algae entering the system as an algae and water suspension. Before entering the extraction tank, the stream is subjected to pulsed electromagnetic fields and pH modification in a process known as *quantum fracturing*. As the terminology implies, quantum fracturing creates a fluid-fracturing effect, thereby mechanically distressing algae cells [33]. The electromagnetic fields are generated using a low-voltage power input and the pH is modified using carbon dioxide, which helps optimize electromagnetic delivery and assists in cell degradation. The electromagnetic field created causes algae cells to release internal lipids. After quantum fracturing, the processed culture passes into a gravity clarifier and a return culture stream recycles into the inlet stream. The gravity clarifier separates the processed culture into layers of oil, water, and biomass. The lipid layer exit stream produces SVO and the water layer exits via a recycle stream to the bioreactor. The biomass can then be harvested for a number of purposes including livestock feed, ethanol processing, and biomass gasification. A schematic of the OriginOil single-step extraction process [34] is shown in Figure 2.4.

This innovative method can extract approximately 97% of the lipid oil contained in algae cells. Typical extraction values for expeller press extraction and supercritical fluid extraction are approximately 75% and 100%, respectively, meaning this method would be competitive with industrial significance. This process does not require heavy machinery, chemicals, or dewatering of the feedstock and is thus claimed to use much less energy than traditional extraction processes.

The OriginOil single-step extraction process is an enhancement to the traditional algae oil extraction methods. The unique quantum fracturing method of extraction utilized by the process reduces the cost of processing by eliminating the need for chemical input or energy-intensive machinery. As shown in the flowsheet, the OriginOil process uses a radically different approach in oil extraction which is based on a sequence of oil extraction→solids separation→dewatering basically in a single integrated

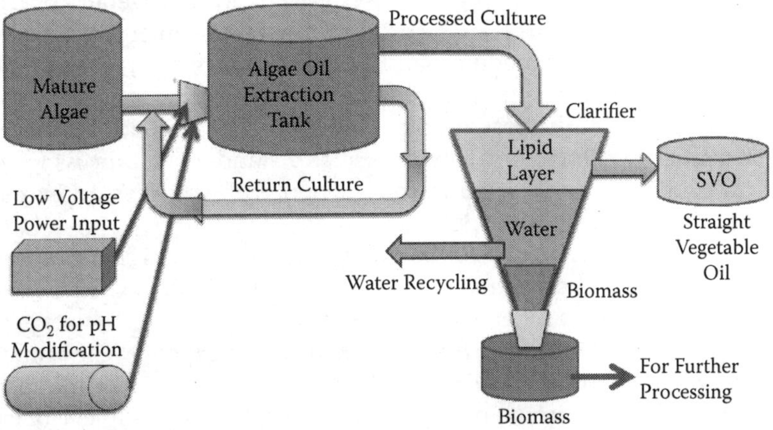

FIGURE 2.4
A schematic of the OriginOil single-stage algae oil extraction process.

step, whereas the conventional approach follows a sequential order of solids separation→dewatering→oil extraction [33]. As the result of this novel process alignment, no initial dewatering is required, substantial energy savings can be expected, and the capital expenditure becomes reduced. With the biodiesel and bio-oil industries' growth, single-step extraction technology will prove to be an invaluable tool for keeping algae-derived fuels competitive with petroleum.

2.2.4.4 Solvent Extraction of Algae Oil

Algal oil can be extracted from microalgae using an effective chemical solvent [35]. Hexane, cyclohexane, benzene, ether, acetone, and chloroform have proven to be effective in oil extraction of microalgae paste [36]. Among these, hexane has long been used as an oil extraction solvent in the food industry and is relatively inexpensive. A chemical method is usually faster in terms of extraction speed and requires a lower energy input for the extraction process itself. One of the drawbacks of using chemical solvents for algae oil extraction is the safety issues involved in working with the chemicals. Care must be taken to avoid exposure to chemical vapors and direct contact with the chemical, either of which can cause serious personal injury. Benzene is classified as a carcinogen, and most ethers are highly flammable. Another disadvantage of using a chemical is the additional cost of recovery of the chemical for reuse in the process.

In a broader classification of algae oil extraction technology, chemical extraction methods include: (1) hexane solvent method, (2) Soxhlet extraction, and (3) supercritical fluid extraction.

2.2.4.4.1 Hexane Solvent Method

Hexane solvent extraction has long been used effectively for vegetable oil extraction. A very good example is in the production of soybean oil, for which hexane solvent extraction is predominantly used in industrial production due to its lower energy consumption and higher extraction efficiency (oil yield) in comparison to hydraulic presses, that is, the expeller method. Furthermore, hexane extraction technology can be used as a stand-alone process for algae oil extraction or it can be used in conjunction with the physical extraction technology of the oil press/expeller method.

If a chemical extraction process based on cyclohexane as a chemical solvent is employed for algae oil extraction in conjunction with an expeller method, the envisioned process scheme is as follows.

The algae oil and lipids are first extracted using an expeller. The remaining pulp and biomass are then mixed with cyclohexane to extract the residual oil content remaining in the residue. The algae oil dissolves in cyclohexane, whereas the pulp and residues do not. The biomass is filtered out from the solution, and the biomass rejected here can be used for other energy generation processes such as gasification. As the last stage, the algae oil and cyclohexane are separated by distillation. This two-stage extraction process of combined cold expeller press and hexane solvent extraction is capable of achieving an extraction efficiency of higher than 95% of the total oil present in the algae [35].

2.2.4.4.2 Soxhlet Extraction

Soxhlet extraction is a distillative extraction method that uses a chemical solvent and the extraction unit is equipped with a heated solvent reflux mechanism. These days, Soxhlet extraction is very widely used in chemical laboratories for extraction of a variety of materials (chemical, biological, and polymeric), where the desired compound has a limited solubility in a chosen solvent. Interestingly, the original Soxhlet extraction process was invented in 1879 by Franz von Soxhlet for the extraction of a lipid from a solid material [37].

Oils are extracted from the algae via repeated washing, or percolation, with an organic solvent such as hexane or petroleum ether, under hot reflux in a specially designed glassware setup equipped with a condenser, a distillation path, a siphon arm, a thimble, and a distillation pot [35]. Soxhlet extraction is meant to be a laboratory process for a small-scale operation, but it is very useful for initial technology development. The extraction process is usually slow, taking an hour to several days, depending upon the specifics of the extraction process conditions as well as the specific solvent chosen for the extraction. The energy efficiency per unit mass of product yield is inevitably very low.

2.2.4.5 Supercritical Fluid Extraction of Algae Oil

Supercritical fluids have synergistic properties of both liquids and gases, such as low viscosity, high material diffusivity, and high solvent density, which make them good media for selective extraction [38]. The properties of low viscosity and high molecular diffusivity are gaslike properties, whereas high solvent density is more of a liquidlike property. A fluid is supercritical when both the temperature and pressure are above the critical point values. For example, the critical pressure and temperature of CO_2 are 72.8 atm and 31.16°C, respectively [39]. Supercritical carbon dioxide (sc-CO_2) has received a great deal of attention in many chemical process applications due to its low critical temperature and inexpensive abundance. As with most other supercritical solvents, the solvent power of sc-CO_2 increases as the solvent density of carbon dioxide is increased.

The process for supercritical fluid extraction of oil from algae is similar to that of any vegetable oil. The supercritical CO_2 acts as a selective solvent to extract the oil. The oil is soluble in supercritical CO_2, in particular very high pressure CO_2, but the proteins and other solids are not [39]. To further increase yield, a cosolvent, such as methanol or ethanol, can be used to increase the solubility of more polar components of the oil [38]. Generally speaking, supercritical fluid extraction has the capability of yielding high-quality oil and biomass and sc-CO_2 is considered an environmentally benign solvent that gives low environmental impact.

A semi-batch supercritical extraction process of vegetable oil is described in the book by McHugh and Krukonis [39]. In this process, the extraction vessel is filled with crushed algae. The algae must be crushed in order for the oil to be accessible to the sc-CO_2 because the extraction rate is limited by the mass transfer rate through the cell wall of a whole cell [38]. Supercritical CO_2 is passed through the algae, extracting the oil, and leaving the solid residues in the vessel. The pressure of the supercritical CO_2 and oil mixture is reduced so that the oil precipitates. The CO_2 is then repressurized and recycled back into the extractor vessel. Several vessels can be used to optimize the system efficiency so that while some are being charged, extraction is happening in others. A schematic of the process flow diagram is shown in Figure 2.5.

The solubility of all vegetable triglycerides is approximately the same and depends largely on the temperature and pressure conditions of a supercritical fluid. At 70°C and 800 atm, CO_2 and triglycerides become miscible, and by dropping the pressure by 200 atm, the oil will separate from the CO_2 [39]. This fairly low operating temperature allows for extraction of highly unsaturated triglycerides without degradation [38]. Although the process shown in Figure 2.5 was an early attempt at supercritical fluid extraction of algae oil, the pressures of 200 atm and 800 atm are excessively high for most chemical processing operations. In order to reduce the operating pressure, different solvent and cosolvent combinations can be developed for process synergism and implemented accordingly. Supercritical extraction of algae oil is still in the research stage. Research on making biodiesel from algae and

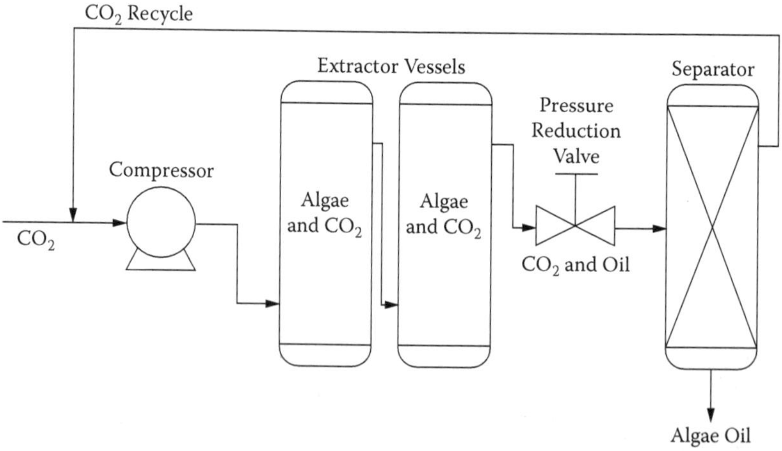

FIGURE 2.5
A schematic of the semi-batch supercritical algae oil extraction process. (Modified from McHugh and Krukonis, 1986. *Supercritical Fluid Extraction: Principles and Practice.* Boston: Butterworth.)

supercritical fluid extraction of algae oil is being done in many labs, including Sandia National Lab [18].

For the production of goods from algae oil to be feasible, the cost must be competitive with that of petroleum-based or other biomass-based products. The cost of the oil extraction process is a critical component of the overall cost of production [19]. The biggest obstacle facing the economical use of supercritical extraction technology is the possibility to feed and remove the solids continuously [39], as well as the ability to extract oil at a manageably low pressure by using an appropriate solvent combination. One possible option to overcome this obstacle is to have multiple vessels whose stages of operation can be alternated during the process cycles and stages. The other possibility is the use of a synergistically effective cosolvent supercritical fluid system, which has an enhanced solubility toward algae oil without requiring an excessively high pressure condition. Another economical obstacle currently preventing production of biofuels based on the supercritical fluid extraction from being competitive is the capital cost associated with use of expensive and energy-intensive equipment. Recent advances achieved in the supercritical fluid technology in other chemical, biological, and petrochemical areas, in particular, pressure vessel designs, advanced reactor materials, solids-handling capability, binary and ternary solvent systems for maximum process synergism, tunability of fluid properties, ingenious energy and process integration, and extraction and development of high-value by-products could offer substantially enhanced process options for algae oil extraction based on supercritical fluid technology.

2.2.4.6 Enzymatic Extraction

The enzymatic process uses select enzymes to degrade algae cell walls and in the process system water acts as a solvent medium for enzyme action. No additional solvent is involved, therefore this process facilitates an easier downstream fractionation. The advantages of the enzymatic extraction process include:

(a) The process does not require dry cakes for oil extraction.
(b) No caustic chemicals are required or involved.
(c) Mild process conditions are used.
(d) The process can be synergistically integrated with other processes such as ultrasonification.
(e) Environmental impact is minimal.

The process is also in a beginning stage and finding efficient and robust enzymes for the process is a challenge. Unless a cost-effective enzyme is developed and proven for the process, this process would be more expensive costwise than hexane solvent extraction. However, there are ample opportunities with this approach, inasmuch as the process can be utilized in conjunction with many other mechanical extraction technologies. In addition, a drastically different product portfolio may be developed for highly value-added coproducts and by-products.

2.2.5 By-Product Utilization

The technical report from the National Renewable Energy Laboratory discusses further uses for extraction process by-products [19]. The high-value products such as pigments and omega-3 fatty acids could be separated and sold. Other high-value by-products could be developed as coproducts and exploited for commercial potential. The remaining high-protein solids could be used for animal feed. The residual biomass could also be used for further energy generation such as biomass gasification followed by methanol and dimethylether synthesis or feedstock for a combined heat and power (CHP) process. Profitable utilization of by-products is very important for the overall process economics of the algae-based biofuel process. Development of specialty end-uses of algae oil can provide another valuable option for the algae industry. Such specialty end-uses include utilization of algae oil as raw material or feedstock for the manufacture of biobased solvents, biobased lubricants, bioplastics, biodegradable plastics, functional fillers, and more.

Although the cost to manufacture products from algae oil is not currently competitive with that of petroleum-based products, further research in the area could make the process and its products economically feasible, while contributing to the sustainability of the society.

2.3 Manufacture of Biodiesel

The name *biodiesel* comes from the fact that the fuel is derived from biological sources and the fuel is, at least originally, meant to be used in diesel engines. Biodiesel is considered a renewable fuel as it is produced from plant oils and animal fats [40]. However, it should be noted that "biodiesel" as a name potentially implies and encompasses more than what is technologically defined as biodiesel. In other words, biodiesel is far more specific than simply "a biologically derived fuel that can be used on diesel engines."

Biodiesel is a renewable alternative fuel for diesel engines comprised of a long-chain mono-alkyl ester and is the product of the transesterification reaction of triglycerides with low molecular weight alcohols such as methanol and ethanol. According to the National Biodiesel Board (NBB), which is the national trade association representing the biodiesel industry in the United States, biodiesel is defined as "a domestic, renewable fuel for diesel engines derived from natural oils like soybean oil, and which meets the specifications of ASTM D 6751" [41]. Although ASTM D 6751 provides the original specifications for 100% pure biodiesel (B100), there are other objective fuel standards and specifications for biodiesel fuel blends such as:

- Biodiesel blends up to 5% (B5) to be used for on- and off-road diesel applications (ASTM D975-08a)
- Biodiesel fuel blends from 6 to 20% (B6–B20; ASTM D7467-09)
- Residential heating and boiler applications (ASTM D396-08b)

There is a major global push enacted by the Kyoto Protocol (adopted on December 11, 1997; entered into force on February 15, 2005) to reduce emissions of greenhouse gases (GHGs), more particularly, carbon dioxide [42]. Currently, the world is seriously concerned with energy sustainability and affordability, as many industrialized and developing nations are economically hurting from escalating costs of energy and fuels, in particular, petroleum-based transportation fuels. Biodiesel is one of the alternative fuels that can help the world address these issues. Biodiesel is considered a mostly carbon-neutral fuel and is completely biodegradable.

2.3.1 Historical Background of Biodiesel Manufacture

Biodiesel has been around for quite some time, but it was not considered a viable fuel until recently. The transesterification of triglycerides was discovered by E. Duffy and J. Patrick as early as 1853; in fact, this happened many years before the first functional diesel engine was invented. It was not until 1893 that Rudolf Diesel invented the first diesel engine and designed it to run on peanut oil. Later in the 1920s, the diesel engine was redesigned to run on petrodiesel, a fossil fuel derived from petroleum crude [43]. Petrodiesel had been much cheaper to produce compared to any biofuel, thus there had not been many active developments in the biodiesel infrastructure.

It was not until 1977 that the first industrial biodiesel process using ethanol was patented. Later, in 1979, South Africa started research on the transesterification of sunflower oil. After four years, South African Agricultural Engineers published a process for fuel-quality, engine-tested biodiesel. An Australian company called Gaskoks used this process to build the first biodiesel pilot plant in 1987 and later built an industrial-scale plant in 1989. The industrial-scale plant was capable of processing 30,000 tons of rapeseed per year [43]. As a matter of fact, rapeseed oil has also become the primary feedstock for biodiesel in Europe (estimates for 2006: more than 4 million tons of rapeseed oil went into biodiesel) [44]. During the 1990s, many European countries such as Germany and Sweden started building their own biodiesel plants. By 1998 approximately 21 countries had some sort of commercial biodiesel production.

In September of 2005, the state of Minnesota became the first U.S. state to mandate that all diesel fuel sold in the state contain a certain part biodiesel, requiring a content of at least 2% biodiesel (B2 and up). This established that biodiesel blend fuel is no longer a choice, but a standard and mandate. On April 23, 2009, the European Union (E. U.) adopted the Renewable Energy Directive (RED) which included a 10% target for the use of renewable energy in road transport fuels by 2020. It also established the environmental sustainability criteria that biofuels consumed in the European Union have to comply with, covering a minimum rate of direct GHG emission savings as well as restrictions on the types of land that may be converted to production of biofuel feedstock crops [45, 46].

2.3.2 Transesterification Process for Biodiesel Manufacture

Biodiesel can be produced in a few different ways. The process can be operated either as a batch process or as a continuous process. It is usually performed catalytically, using a strong base or acid as the catalyst. Alternatively, it can be operated noncatalytically, using supercritical methanol. The most popular process in the industry currently uses methanol as the alcohol, sodium hydroxide (NaOH) as the base catalyst, and is a continuous process.

Biodiesel is most popularly produced by the transesterification reaction of triglycerides. Triglycerides are found in plant oils and animal fats and the molecular structure of a triglyceride is shown in Figure 2.5. Transesterification occurs when the triglycerides are mixed with an alcohol, typically either methanol or ethanol. As the hybridized terminology of "*trans*- + esterification" implies, transesterification is a chemical reaction in which the aliphatic organic group (R–) of an ester is exchanged with another aliphatic organic group (R′–) of an alcohol, thereby producing a different ester and a different alcohol. In other words, the starting ester (triglyceride) and monohydric alcohol (methanol) are converted by the transesterification reaction into a simpler form of esters (biodiesel) and a more complex form of alcohol, that is, trihydric alcohol (glycerin). In this reaction of transesterification of triglycerides, three alcohol molecules liberate the long-chain fatty acids from the glycerin backbone by bonding (i.e., esterification) with the carboxyl group carbons in the triglyceride molecule, as shown in Figure 2.6. The products of the transesterification reaction are a glycerin molecule and three long-chain mono-alkyl ester molecules, otherwise commonly known as biodiesel. More specifically, if a biodiesel is produced by transesterification reaction between soy triglycerides and methanol, the resultant biodiesel ester is often referred to as methyl soyate. The transesterification reaction is usually catalyzed using a strong base such as NaOH or KOH. The base helps to catalyze this reaction by removing the hydrogen in the hydroxyl group on the alcohol molecules.

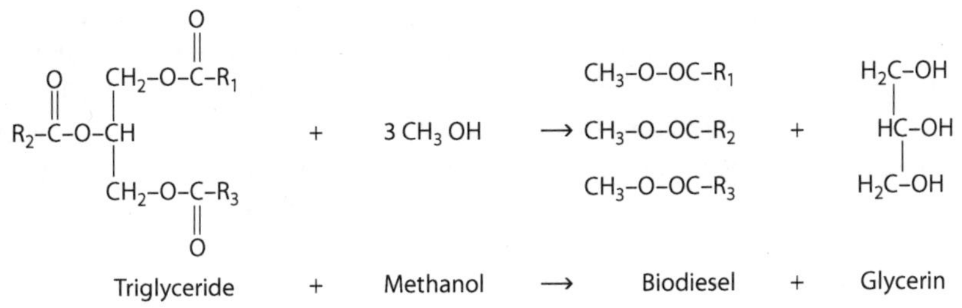

FIGURE 2.6
Stoichiometric reaction chemistry of transesterification.

Before oils and fats can react to form biodiesel they must go through a pretreatment process. The first stage of the pretreatment process involves filtering to remove dirt and other particulate matters from the oil. Next, water must be removed from the oil because it will hydrolyze the triglycerides to form fatty acid and glycerin instead of biodiesel and glycerin. Free fatty acids can directly react with base catalyst to form soap, which is certainly not desirable for biodiesel manufacture. If soap formation is active, the process would require an additional amount of base catalyst to compensate for the reactive depletion. Finally, the oil must be tested for free fatty acid (FFA) content. Typically, less than 1% FFA in oil is acceptable for processing without further provisional treatments. Free fatty acids are long-chain carboxylic acids that have broken free from the triglycerides, typically from thermal degradation of triglycerides as a result of prolonged exposure to heat. These acids can increase soap formation in the reactor, as mentioned earlier. Too much soap in the reactor causes substantial difficulties: (a) soap formation becomes a reason for an additional amount of base catalyst usage to overcome its reactive depletion in the soap formation; (b) additional problems arise in product separation; and (c) as an extreme case, the formed soaps mix with water from the fuel wash stage to create an emulsion that can seriously slow down or even prevent settling of the wash water layer from the product biodiesel layer. There are two ways to deal with the free fatty acids in the oil. An acid can be added to the oil to convert the free fatty acids into biodiesel; this is the case with an acid-catalyzed esterification reaction. Alternatively, they can be neutralized, turned into soap, and removed from the oil. After being pretreated the oil is then sent to the reactor. The methanol that reacts with the oil also has to go through some pretreatment. Before the methanol is sent to the reactor it goes through a mixer where it is combined with the sodium hydroxide catalyst. The oil and methanol/catalyst mixture are then fed into the reactor to undergo the transesterification reaction. Methanol is fed in excess of around 1.6 times the stoichiometric amount and the reactor is kept at around 60°C. With the aid of the base catalyst the reaction is able to proceed at up to 98% conversion. The exit stream from the reactor is fed into a separator. The glycerin by-product has a much greater density (glycerin specific gravity at

25°C = 1.263) than the biodiesel (specific gravity at 60°C = 0.880) and is therefore easily removed via gravity separation. After the biodiesel is separated from the glycerin by-product it goes through a purification process. The first step is to neutralize the remaining catalyst by adding an acid to the biodiesel. Then the biodiesel is sent through a stripper to remove any methanol left from the reactor. This methanol is then recycled back to the methanol/catalyst mixer.

After the methanol removal, the biodiesel goes through a water wash to remove all the soaps and salts (e.g., neutralized salt of NaOH catalyst) generated during transesterification and neutralization. The biodiesel is then dried and stored as the final product. The glycerin by-product, or crude glycerin, also goes through some purification. The crude glycerin contains a considerable amount of methanol, which comes out unreacted due to its excess amount of feed to the reactor. The glycerin goes through a distillation separation that recovers a great deal of the methanol for recycling. The recycled methanol collects most of the water that entered the process and therefore it must go through a separate distillation column to be purified. The glycerin that comes out of the distillation process is pure glycerin that can be marketed for other industries including the pharmaceutical and cosmetics industries. A schematic of the biodiesel manufacturing process via transesterification is shown in Figure 2.7.

As mentioned above, crude glycerin is a mixture of glycerin, methanol, and salts. Crude glycerin can be sold as is or further purified into pharmaceutical-grade glycerin. A marketable grade of crude glycerin is generally at least 80% glycerin with less than 1% methanol. Crude glycerin that has lower levels of glycerin or higher levels of methanol often has little or no value; this is especially true in the current era of an oversupply of glycerin on the market. Although glycerin is overly abundant in the world marketplace due to rapidly increased biodiesel production, the purification of crude glycerin into pure glycerin is quite energy-intensive and costly. Efficient chemical conversion of glycerin or crude glycerin into other value-added chemicals and petrochemicals, in addition to the conventional end-uses established in the food, pharmaceutical, and cosmetics sectors, would help stabilize the market price of glycerin and provide additional income to the biodiesel industry, thus improving the industry's gross margin and profitability.

2.3.3 Properties of Biodiesel

2.3.3.1 Cetane Rating (CR)

The two most beneficial properties of biodiesel are its higher cetane ratings (CR) and better lubricating properties than the ultra-low sulfur diesel (ULSD). The CR of biodiesel ranges between 50 and 60, whereas that of ULSD ranges between 45 and 50. On the other hand, the CR of vegetable oil ranges

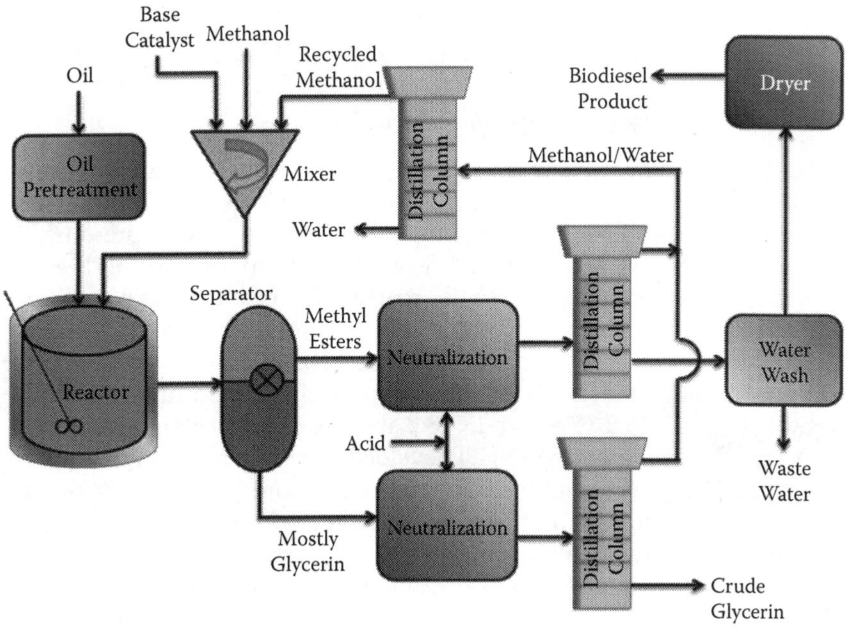

FIGURE 2.7
A schematic of biodiesel manufacturing process via transesterification.

between 35 and 45. Addition (or blending) of biodiesel in petrodiesel in a low concentration is reported to reduce fuel system wear, resulting in a beneficial effect to diesel engines.

2.3.3.2 Calorific Value (CV) or Heating Value (HV)

The lower calorific value (LCV) or lower heating value (LHV) of biodiesel is 37.37 MJ/kg, whereas that for low sulfur diesel is 42.612 MJ/kg. The higher calorific value (HCV) or higher heating value (HHV) of biodiesel is 40.168 MJ/kg, whereas that for low sulfur diesel is 45.575 MJ/kg. Both heating values of biodiesel are approximately 12% lower than those of low-sulfur diesel. Variations in the heating values for biodiesel are mainly from the variability of the biodiesel feedstock, that is, the source of triglycerides. A downside of biodiesel in comparison to petrodiesel is its lower calorific value, even though the difference is not very substantial. However, it has been claimed by fuel engineers that the ultimate fuel efficiency of biodiesel is comparable to that of petrodiesel, despite its lower energy density, thanks to several compensating factors including more complete combustion and better lubricity. The lower heating value of biodiesel is attributable to its oxygenated molecular structure in contrast to the nonoxygenated hydrocarbon structures of petrodiesel. Highly oxygenated structures of biodiesel adversely affect the cold flow properties of biodiesel, which are represented by the fuel's cloud point, pour point, and cold filter plugging point [47].

2.3.3.3 General Physical Properties of Biodiesel

Biodiesel has a color varying from golden or light brown to dark brown. The color depends upon the originating biodiesel feedstock. Biodiesel is immiscible with water, and has a high boiling point and a low vapor pressure. The boiling point data of biodiesel are very scarcely reported in the literature [48] because a smoke point is usually reached before a boiling point during its measurement, thus making the measurement itself quite difficult. At a smoke point, ingredients of biodiesel are degrading, that is, going through pyrolytic decomposition. Based on the published literature values, pure methyl esters of C18:0, C18:1, and C18:2 are 625, 622, and 639 K, respectively [49]. Yuan, Hansen, and Zhang [48] developed models, based on the Antoine equation and a group contribution method, for predicting vapor pressure and the normal boiling point of pure methyl esters and biodiesel fuels. The flash point of biodiesel (>130°C) is significantly higher than that of petroleum diesel (64°C) or gasoline (−45°C), which makes biodiesel substantially easier to handle and less flammable. Biodiesel has a density of approximately 0.880 g/cm^3, which is higher than that of petrodiesel (0.85 g/cm^3) or gasoline (0.71–0.77 g/cm^3).

2.3.3.4 Cold Flow Properties

Pure biodiesel (B100) has poor cold-temperature properties as straight diesel fuel. Biodiesel's *cloud point* (*CP*) and *cold filter plugging point* (*CFPP*) are both high, thus making pure biodiesel (B100) unsuitable as a cold-climate fuel without blending or additives. When biodiesel is cooled below a certain temperature, some ingredient molecules of biodiesel start to aggregate and form crystals. This temperature varies depending on the biodiesel feedstock, but is consistently quite high. As the biodiesel is further cooled and the crystals become larger than one quarter of the wavelength of visible light, the fuel system starts to look cloudy. This point is known scientifically as the cloud point. The cloud point measurement follows the ASTM 2500. The lowest temperature at which biodiesel can pass through a 45-micron filter is called the cold filter plugging point. A high cold filter plugging point tends to clog up vehicle engines more easily. As biodiesel is further cooled below the CFPP, it will gel and eventually solidify. This point is called the *gel point*. Another important cold flow property is the *pour point*, which is defined as the lowest temperature where the fuel is observed to flow. As all these temperatures are generally higher for biodiesel (B100) than petrodiesel, biodiesel freezes faster than most petrodiesels. Commercial additives developed for diesel to improve its cold flow properties are mostly applicable to biodiesel and biodiesel blends [47].

2.3.3.5 Material Compatibility with Biodiesel

As for the material compatibility, biodiesel is quite different from petrodiesel. Biodiesel is compatible with high-density polyethylene (HDPE), and it is incompatible with polyvinylchloride (PVC) and polystyrene (PS), as PS is readily soluble in biodiesel and PVC is slowly dissolved in biodiesel. Polypropylene (PP) is affected by biodiesel, showing a swell increase (by 8–15%) and reduced hardness (by about 10%). Polyurethane (PUR) is also affected showing some swell increase (by 6%). Biodiesel affects some natural rubber and all nitrile rubber products, and biodiesel is compatible with commonly used viton-type synthetic rubbers in modern vehicles. Studies indicate that viton types B and F (FKM-GBL-S, and FKM-GF-S) are more resistant to acidic biodiesel [50]. Biodiesel affects, and is affected by, many metals including copper, zinc, tin, lead, cast iron, and brass, whereas biodiesel does not affect stainless steel (304 and 316), carbon steel, and aluminum. Brass, bronze, copper, lead, tin, and zinc may accelerate, by catalytic activities, the oxidation process of biodiesel creating fuel insolubles or gels and salts. As such these metals should be avoided as materials of construction for piping, regulators, and fittings in applications where biodiesel is expected to be in contact with them.

Neat biodiesel may degrade some hoses, gaskets, seals, elastomers, glues, and plastics with prolonged exposure. Acceptable storage tank materials for biodiesel include aluminum, steel, fluorinated polyethylene, fluorinated polypropylene, Teflon, and most fiber glasses [51].

2.3.4 Prospects and Economics

Based on the current level of process economics associated with the transesterification of vegetable oil in the United States, biodiesel requires a subsidy from the government in order to compete with petrodiesel and other fossil fuels. The scheduled expiration and delayed but retroactive reinstatement of the biodiesel subsidy ($1 for each gallon of biodiesel blended in the United States as of 2010) in the United States has lately been intensely contested and debated by Congress. Biodiesel plant owners have to sell their crude glycerin by-product for a decent value to stay profitable. Or, they need to convert the glycerin by-product and coproduct into other value-added chemicals or products. Successful development of profitable markets for by-products and coproducts may be a key determinant of the overall success for biodiesel industries. However, as fossil fuels become more expensive, biodiesel becomes a more feasible fuel alternative. As of 2010, the United States produces around 315 million gallons of biodiesel per year. However, the current level of U.S. production is substantially below the industry capacity; that is, biodiesel manufacturing facilities are being underutilized.

The future of the biodiesel industry depends strongly on the cost of feedstock. The raw material cost is a substantial portion of the biodiesel manufacturing cost. The current biodiesel industry's gross margin is very poor without taking into account the governmental subsidy. Food costs increased substantially in recent years, which also escalated the raw material cost of vegetable oils for the conventional biodiesel industry, even though some blame the biofuels industry for the food price hike. In this regard, the algae biofuel option is very promising, inasmuch as it does not compete for food or require arable land for algae growth.

Biodiesel offers many benefits to the environment that are worthy of note. It is considered a mostly carbon neutral fuel because all the carbon dioxide emitted from burning biodiesel originally came from plants and animals that removed it from the air. However, it is not completely carbon neutral because to process biodiesel all the way from plant oil and animal fat some nonrenewable energy is inevitably required. It is, however, a huge improvement. It reduces carbon dioxide emissions by 78% compared to petrodiesel. Biodiesel even offers a reduction in carbon monoxide and sulfur emissions. It does, however, have slightly higher nitrous oxide (N_2O) emissions. Biodiesel is also biodegradable and nontoxic. Biodiesel is available today as a pure (or neat) fuel and also as a blended fuel with petrodiesel. Blends such as B20 (20% biodiesel and 80% petrodiesel) burn cleaner than petrodiesel alone, thus reducing emissions of harmful air pollutants such as carbon monoxide, volatile organic compounds (VOCs), soot, and particulate matter (PM). Over time, biodiesel has good potential to play a large role in the future fuel economy as well as in the transportation fuel sector.

References

1. Dijkstra, A.J. 2009. Edible oil modification processes. In S. Lee (Ed.), *Encyclopedia of Chemical Processing*. Boca Raton, FL: Taylor & Francis, pp. 1–9.
2. Dijkstra, A.J. 2007. Modification processes and food uses. In F.D. Gunstone, J.L. Harwood, and A.J. Dijkstra (Eds.),*The Lipid Handbook*, Boca Raton, FL: Taylor & Francis, pp. 263–353.
3. Food and Nutrition Board, Institute of Medicine of the National Academies, 2005. *Dietary Reference Intakes for Energy, Carbohydrate, Fiber, Fat, Fatty Acids, Cholesterol, Protein, and Amino Acids (Macronutrients)*. Washington, DC: National Academies Press.
4. Economic Research Service. 2009. *Oil Crops Yearbook*. U.S. Department of Agriculture. [http://usda.mannlib.cornell.edu/MannUsda/viewStaticPage.do?url=http://usda.mannlib.cornell.edu/usda/ers/89002/2009/index.html.].
5. USDA. 2009. *Oil seeds: World market and trade, FOP 1-09: Table 3*. Major vegetable oils: World supply and distribution of oil seeds; world markets and trade monthly circular. Available at: http://www.fas.usda.gov/oilseeds/circular/2009/January/Oilseedsfull0109.pdf.

6. Mathiyazhagan, M., Ganapathi, A., Jaganath, B., Renganayaki , N., and Sasireka, N. 2011. Production of biodiesel from non-edible plant oils having high FFA content, *Int. J. Chem. Env. Eng.*, 2: 119–122.
7. National Institute of Standards and Technology (NIST). NIST chemistry webbook. [http://webbook.nist.gov/chemistry/].
8. Speight, J. G. 1994. *The Chemistry and Technology of Coal*. New York: Marcel Dekker.
9. National Biodiesel Board (NBB). 2011. Biodiesel energy content. [http://www.biodiesel.org/pdf_files/fuelfactsheets/BTU_Content_Final_Oct2005.pdf] (accessed July, 2011).
10. Walker, P.L., Rusinko, F., and Austin, L.G. 1959. Gas reactions in carbon. In D.D. Eley, P.W. Selwood, and P.B. Weisz (Eds.), *Advances in Catalysis*. New York: Academic Press, pp. 133.
11. Lee, S., Speight, J.G., and Loyalka, S.K. 2007. *Handbook of Alternative Fuel Technology*. Boca Raton, FL: CRC Press.
12. Lee, S. 1990. *Methanol Synthesis Technology*. Boca Raton, FL: CRC Press.
13. Lee, S. 1997. *Methane and Its Derivatives*. New York: Marcel Dekker.
14. Ucuncu, A. and Vesilind, P.A. 1991. Energy recovery from mixed paper waste. [http://www.p2pays.org/ref/11/10059.pdf].
15. Bello, E., Magaji T.S., and Agge, M. 2011. The effects of transesterification on selected fuel properties of three vegetable oils, *J. Mech. Eng. Res.*, 3: 218–225.
16. U.S. Energy Information Administration (EIA). 2011. *International Energy Outlook 2011*, Tech. Rep. DOE/EIA-0484, September 19.
17. Demirbas, A. and Fatih Demirbas, M. 2011. Importance of algae oil as a source of biodiesel, *Energy Conversion Manage.*, 52: 163–170.
18. Kent M.S. and Andrews, K.M. *Biological Research Survey for the Efficient Conversion of Biomass to Biofuels*. Tech. Rep. SAND2006-7221, 2007.
19. Milbrandt A. and Jarvis, E. 2010. *Resource Evaluation and Site Selection for Microalgae Production in India*, Tech. Rep. NREL/TP-6A2-48380.
20. Oilgae. 2011. Algae harvesting - flocculation (accessed June 2011). Available at: http://www.oilgae.com/algae/har/flc/flc.html.
21. Oilgae. 2010. Algae harvesting - filtration (accessed June 2011). Available at: http://www.oilgae.com/algae/har/fil/fil.html.
22. Hung, M.T. and Liu, J.C. 2006. Microfiltration for separation of green algae from water, *Colloids Surfaces B: Biointerfaces*, 51: 157–164.
23. Zou, S., Gu, Y., Xiao, D., and Tang, C.Y. 2011. The role of physical and chemical parameters on forward osmosis membrane fouling during algae separation, *J. Membrane Sci.*, 368, 356–362.
24. Algae Venture Systems. 2011. Harvesting, dewatering & drying (accessed June 2011). Available at: http://www.algaevs.com/harvesting-dewatering-and-drying.
25. Shelef, G., Sukenik, A., and Green, M. 1984. *Microalgae Harvesting and Processing: A Literature Review*, Tech. Rep. SERI/STR-231-2396. Solar Energy Research Institute and U.S. Department of Energy, Golden, CO.
26. Sukenik, A. and Shelef, G. 1984. Algal autoflocculation—verification and proposed mechanism, *Biotechnol. Bioeng*. 26: 142–147.
27. Oilgae. 2011. Bio-flocculation of algae research project @ Newcastle University. [http://www.oilgae.com/blog/2008/12/bio-flocculation-of-algae-research-project-newcastle-university.html] (accessed June 2011).

28. Milledge, J.J. and Heaven, S. 2011. Disc stack centrifugation separation and cell disruption of microalgae: A technical note. *Environ. Nat. Resources Res.* 1: 17–24.
29. Alfa Laval. 2011. Alfa Laval - disc stack centrifuge technology. [http://local.alfalaval.com/en-us/key-technologies/separation/separators/dafrecovery/Documents/Alfa_Laval_disc_stack_centrifuge_techonology.pdf] (accessed June 25, 2011).
30. Flottweg Separation Technology. 2011. Flottweg centrifuges for efficient algae harvesting. [http://www.flottweg.de/cms/upload/downloads/old/algen_Internetversion_englisch.pdf] (accessed June 2011).
31. Weyer, K.M., Bush, D.R., Darzins A., and Wilson, B.D. 2009. Theoretical maximum algal oil production, *Bioenergy Res.*, 3: 204–213.
32. Hielscher Utrasonics. 2011. Biodiesel from algae using ultrasonication. [http://www.hielscher.com/ultrasonics/algae_extraction_01.htm] (accessed June 25, 2011).
33. OriginOil. 2010. OriginOil - Algae harvesting, dewatering, and extraction. In *World Biofuels Markets*, Amsterdam, Netherlands.
34. OriginOil. 2011. Single-step extraction process. [http://www.originoil.com/technology/single-step-extraction.html].
35. Oilgae.2011. Extraction of algal oil by chemical methods. Available at: http://www.oilgae.com/algae/oil/extract/che/che.html (accessed June 2011).
36. Mercer, P. and Armenta, R.E. 2011. Developments in oil extraction from microalgae, *Eur. J. Lipid Sci. Technol.*, 113: 539–547.
37. Soxhlet, F. 1879. Die gewichtsanalytische Bestimmung des Milchfettes, *Polytechnisches J. (Dingler's)*, 232: 461.
38. Cohen, Z. and Ratledge, C. 2005. Single cell oils. In *Supercritical Fluid Extraction of Lipids and Other Materials from Algae*. New York: Taylor & Francis, pp. 220.
39. McHugh, M. and Krukonis, V. 1986. *Supercritical Fluid Extraction: Principles and Practice*. Boston: Butterworth.
40. Gerpen, J. 2005. Biodiesel processing and production, *Fuel Process.Technol.* 86: 1097–1107.
41. National Biodiesel Board (NBB). 2011. Biodiesel: America's advanced biofuel. [http://www.biodiesel.org/] (accessed May 2011).
42. United Nations Framework Convention on Climate Change. 2011. *Kyoto Protocol.* Available at: http://unfccc.int/kyoto_protocol/items/2830.php.
43. Progressive Fuel Limited. 2011. Biodiesel history (June 2011). Available at: http://www.progressivefuelslimited.com/biodiesel.asp.
44. Soyatech. 2011. Rapeseed. [http://www.soyatech.com/rapeseed_facts.htm].
45. Al-Riffai, P.B., Betina Dimaranan, B., and Laborde, D. 2010. European Union and United States biofuel mandates: Impacts on world markets. Inter-American Development Bank. Available at: http://idbdocs.iadb.org/wsdocs/getdocument.aspx?docnum=35529623.
46. Al-Riffai, P., Dimaranan, B., and Laborde, D. 2010. Global trade and environmental impact study of the EU biofuels mandate; final report. ATLASS Consortium. Available at: http://trade.ec.europa.eu/doclib/docs/2010/march/tradoc_145954.pdf.
47. National Biodiesel Board (NBB). 2007. Biodiesel cold flow basics. (accessed December 2011).
48. Yuan, W., Hansen A.C., and Zhang, Q. 2005. Vapor pressure and normal boiling point predictions for pure methyl esters and biodiesel fuels, *FUEL* 84: 943–950.

49. Graboski , M.S. and McCormick, R.L. 1998. Combustion of fat and vegetable oil derived fuels in diesel engines, *Prog. Energy Combust. Sci.* 24: 125–164.
50. Thomas, E.W., Fuller R.E., and Terauchi, K. 2007. Fluoroelastomer compatibility with biodiesel fuels, DuPont Performance Elastomers, LLC, January.
51. National Biodiesel Board (NBB). 2011. Materials compatibility. [http://www.biodiesel.org/pdf_files/fuelfactsheets/Materials_Compatibility.pdf] (accessed December 2011).

3

Ethanol from Corn

3.1 Fuel Ethanol from Corn

Ethanol is one of the simplest alcohols that have long been consumed in human history. Ethanol can be readily produced by fermentation of simple sugars that are obtained from sugar crops or converted from starch crops. This has long been practiced throughout the world. Feedstock for such fermentation ethanol includes corn, barley, rice, and wheat. This type of ethanol may be called *grain ethanol*, whereas ethanol produced from cellulosic biomass such as trees and grasses is called *cellulosic ethanol* or *biomass ethanol*. Both grain ethanol and cellulosic ethanol are produced via biochemical processes, whereas *chemical ethanol* is synthesized by chemical synthesis routes that do not involve any fermentation step.

Ethanol, ethyl alcohol [C_2H_5OH], is a clear and colorless liquid. Ethanol has a substituted structure of ethane, with one hydrogen atom replaced by a hydroxyl group, –OH. Ethanol is a clean-burning fuel thanks to its oxygen content and has a high octane rating by itself (Research Octane Number (RON) of 108.6). Therefore, ethanol is most commonly used as an oxygenated blend fuel to increase the octane rating of blend gasoline as well as to improve the emission quality of gasoline engines. Due to the presence of the oxygen atom in its molecular structure, ethanol is classified as an oxygenated fuel. In many regions of the United States, ethanol is blended up to 10% with conventional gasoline. The blend between 10% ethanol and 90% conventional gasoline is called "E10 blend" or simply "E10". Ethanol is quite effective as an oxygenated blending fuel, because its Reid vapor pressure is marginally low; that is, it does not increase the volatility of the blend gasoline significantly, unlike methanol. Reid vapor pressure (RVP) is defined as the absolute vapor pressure exerted by a liquid at 100°F (37.8°C) as determined by the standard testing method following ASTM-D-323. The test method also applies to volatile crude oil and volatile nonviscous petroleum liquids, except liquefied petroleum gases (LPG). Even though the Reid vapor pressure of ethanol is lower than that of methanol, fuel experts still consider it higher than they desire.

Ethanol can be produced from any biological materials that contain appreciable amounts of sugars or feedstock that can be converted into sugars. The former include sugarbeets and sugarcane, whereas the latter include starch and cellulose. For example, corn contains starch that can be easily converted into sugar and is, therefore, an excellent feedstock for ethanol fermentation. Because corn can be grown and harvested repeatedly, this feedstock eminently qualifies as a *renewable feedstock,* that is, this feedstock is not going to be simply depleted by exhaustive consumption.

Fermentation of sugars produces ethanol and this process technology has been practiced for well over 2,000 years in practically all regions of the world. Sugars can also be derived from a variety of sources. In Brazil, as an example, sugar from sugarcane is the primary feedstock for the country's ethanol industry which has been very successful and active. In North America, the sugar for ethanol production is usually obtained via enzymatic hydrolysis of starch-containing crops such as corn or wheat. The enzymatic hydrolysis of starch is a simple, relatively inexpensive, and effective process, and is also a mature commercial technology. Therefore, this process is used as a baseline or a benchmark against which other hydrolysis processes can be compared. The principal merit of ethanol production by fermentation of sugar/starch is in its technological simplicity and efficiency, however, its demerit is that the feedstock tends to be expensive and also competitively used for other principal applications such as food. Therefore, "food versus fuel" or "food versus oil" is an unavoidable critical issue addressing the risk of diverting farmland or crops for production of biofuels including corn ethanol to the detriment of the food supply on a global or regional scale. It has also contributed to the increase in food price, which in turn raises the cost of feedstock and hurts the profitability of the ethanol industry.

Technoeconomically speaking, this high cost of feedstock can be favorably offset to a certain extent by the sale of by-products or coproducts such as dried distillers grains (DDGs), provided that the high oil price is sustained in the market. Many corn refineries produce both ethanol and other corn by-products such as cornstarches, sweeteners, and DDGs so that the capital and manufacturing costs can be kept as low as possible by maximizing the overall process revenue. While they are manufacturing ethanol, corn refiners also produce valuable co-products such as corn oil and corn gluten feed. The North American ethanol industry is, therefore, investing significant efforts in developing new value-added by-products (and coproducts) that are higher in value and minimizing the process wastes, thus constantly making the grain ethanol industry more cost-competitive.

Corn refining in the United States has a relatively long history going back to the time of the Civil War with the development of the cornstarch hydrolysis process. Before this event, the main sources for starch had been coming from wheat and potatoes. In 1844, the Wm. Colgate & Company's wheat starch plant in Jersey City, New Jersey, unofficially became the first dedicated cornstarch plant in the world. By 1857, the cornstarch industry accounted for a significant portion of the U.S. starch industry. However, for this early era of corn processing, cornstarch was the only principal product of the corn refining industry and its largest customer was the laundry business. Cornstarch has also been used as a thickening agent in liquid-based foods such as soups, sauces, and gravies.

The industrial production of dextrose from cornstarch started in 1866. This industrial application and subsequent scientific developments in the chemistry of sugars served as a major breakthrough in starch technology and its processing. Other product developments in corn sweeteners followed and took place with the first manufacture of refined corn sugar, or anhydrous sugar, in 1882. In the 1920s, corn syrup technology advanced significantly with the introduction of enzyme-hydrolyzed products. Corn syrups contain varying amounts of maltose (a disaccharide formed by a condensation reaction of two glucose molecules joined with an α $(1\rightarrow4)$ bond) and higher oligosaccharides. Even though the production of ethanol by corn refiners had begun as early as after World War II, major quantities of ethanol via this process route were not produced until the 1970s, when several corn refiners began fermenting dextrose to make beverage and industrial alcohol. As such, the corn refiners' entry into the fermentation business has become a significant milestone for major changes and transformation of the industry, especially in the fuel ethanol industry. The corn refining industry seriously began to develop an expertise in industrial microbiology, fermentation technology, separation process technology, energy integration and process design, and by-product and waste utilization.

As of today, starch and glucose (or dextrose) are still important products of the corn wet milling industry. However, the products of microbiology and biochemical engineering including ethanol, fructose, food additives, and target chemicals have gradually outpaced them. New research and developments have significantly expanded the industry's product/by-product/coproduct portfolio, thus making the industry more profitable, flexible to market demands, competitive, and technologically advanced.

Lignocellulosic materials such as agricultural, hardwood, and softwood residues are also potential sources of sugars for ethanol production. The cellulose and hemicellulose components of these materials are essentially long and high molecular weight chains of sugars. They are protected by lignin, which functions more like "glue" that holds all of these materials together in the structure. Therefore, the liberation of simple sugars from lignocellulosic materials is not as simple and straightforward as that from sugar crops or starch crops. However, the biggest undeniable advantage of cellulosic ethanol is its use of nonfood feedstock and no detrimental use of arable land for fuel production. Details of cellulosic ethanol technology are covered in Chapter 4 and, therefore, not repeated here.

Ethanol plays three principal roles in today's economy and environment and they are

1. Ethanol in the United States replaces a significant amount of imported oil with a renewable domestic fuel.
2. Ethanol is an important oxygenated component of gasoline reformulation to reduce air pollution in many U.S. metropolitan areas, which are not achieving air quality standards mandated by the Clean Air Act Amendments (CAAA) of 1990. Ethanol is a cleaner-burning fuel due to its oxygen-containing molecular structure and also a superior gasoline blend fuel due to its renewability as a fuel and relatively low Reid vapor pressure of the blended fuel.
3. Ethanol provides a major income boost to farmers and agricultural communities where most ethanol feedstock is produced. Global corn prices have escalated more sharply than other crops due to the increased demand and higher corn prices have in turn motivated farmers to increase corn acreage at the expense of other crops, such as soybeans and cotton, raising their prices as well.

Ethanol, blended with gasoline at a 10% level (E10) or in the form of ethyl tertiary-butyl ether (ETBE) synthesized from ethanol, is effective in reducing carbon monoxide (CO) emission levels, ozone pollution, and NO_x emissions from automobile exhaust. Two of the major barriers to the wide acceptance of ethanol as a gasoline blend fuel are: (1) its Reid vapor pressure being not low enough, and (2) its high moisture-absorbing (hygroscopic) characteristics. As mentioned earlier, the Reid vapor pressure of ethanol is lower than that for methanol; however, it is still marginally high.

In its early years the U.S. fuel ethanol industry was expanding to meet the increased demand for oxygenated fuel that resulted from a withdrawal of methyl-tertiary-butyl ether (MTBE) from the domestic gasoline marketplace. In response to sharply rising national concern about the presence of MTBE in groundwater as well as potential risk to public health and the environment, the U.S. Environmental Protection Agency (EPA) convened a Blue Ribbon Panel to assess policy options regarding MTBE. The Blue Ribbon Panel recommended that the use of MTBE be dramatically reduced or eliminated. The EPA has subsequently stated that MTBE should be removed from all gasoline. Many U.S. states including California and New York mandated their own schedules of MTBE phase-outs and bans. As of September 2005, 25 states had signed legislation banning MTBE. According to a survey conducted in 2003, 42 states reported that they had action levels, cleanup levels, or drinking water standards for MTBE [1]. It is a remarkable turnaround in the chemical and petrochemical marketplace considering that MTBE used to be the fastest growing chemical in the United States in the 1990s. Recovering or retrofitting the MTBE plant investments would become an issue for this industry for years to come. Even with a rapid decline and disappearance of MTBE in the U.S. market, global production of MTBE has remained relatively constant at about 18 million tons/year, as of 2005, mainly due to the growth in Asian markets, where the use of ethanol or other oxygenated replacements is not established and ethanol subsidies are not provided.

United States fuel ethanol production has been increasing very rapidly for the first decade of the twenty-first century. According to the Renewable Fuels Association (RFA) [2], the U.S. ethanol production in 2002, 2003, and 2004 was 2.13, 2.80, and 3.40 billion U.S. gallons, respectively. Considering the production level of 2000 being 1.63 billion gallons, this is more than a twofold increase over five years. The U.S. production of ethanol in 2006, 2007, 2008, 2009, and 2010 was 4.9, 6.5, 8.9, 10.75, and 13.2 billion U.S. gallons, respectively. Comparing between the 2007 and 2010 statistics, it took only four years to double U.S. production. The trend in ethanol production in the United States is presented in Figure 3.1, which shows an exponential growth in ethanol production in the United States for the first decade of the twenty-first century. Due to the high cost of petroleum crude in recent years, the role of ethanol has realistically expanded far beyond the oxygenated fuel additive into that of a true alternative renewable transportation fuel. The increased use of ethanol in the United States has significantly contributed to the alleviation of dependence on imported petroleum.

Corn refining has also become America's premier by-products industry, and its success has set a desirable business model for future biofuel industries. Increased production of amino acids, proteins, antibiotics, and biodegradable plastics has added further value to the U.S. corn crop. In addition to cornstarches, sweeteners, and grain ethanol, corn refiners also produce corn

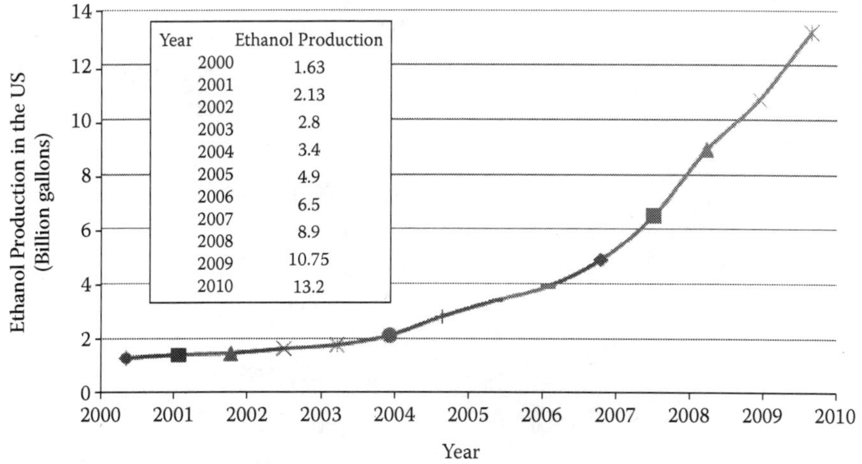

Year	Ethanol Production
2000	1.63
2001	2.13
2002	
2003	2.8
2004	3.4
2005	4.9
2006	6.5
2007	
2008	8.9
2009	10.75
2010	13.2

FIGURE 3.1
Ethanol production in the United States.

oils as well as a variety of important feed products. Corn products in the modern world are found in a large variety of areas and applications: (a) livestock feed grains; (b) food ingredients including sweeteners, starches, and polyols; (c) oil products including corn oil, acid oil, middlings, and corn wax oil; (d) cornstarches for papermaking and corrugated products; (e) personal care products utilizing natural polymers; (f) health and nutrition including sugar-free and low-sugar foods; (g) animal feeds including corn gluten meal, corn germ meal, and steepwater grain solubles; (h) pharmaceutical products including anhydrous dextrose; (i) manufacture of biodegradable polymer, poly(lactic acid; PLA), using cornstarch; and more.

Corn is the most traded crop product in the world with the United States being the leading exporter and Japan being the largest importer. The U.S. annual export of corn was about 50 million metric tons in the fiscal year 2010. Since 1980, the annual amount of U.S. export of corn has been fluctuating between 35 and 60 million metric tons. Even though the United States dominates the global trading market of corn, it accounts for about 15.2% of the total U.S. corn production. As such, corn prices are largely determined by supply-and-demand relationships in the U.S. market. The U.S. corn crop was valued at $66.7 billion in the fiscal year 2010 and production for the year was 331 million metric tons, which is equivalent to 12.1 billion bushels. The U.S. corn growth/production accounted for about 39% of the world production [3]. About 80 million acres were planted to grow corn and most corn production was in the heartland of the United States. Of the total corn produced in the United States in 2010, about 34.9% or 116 million metric tons were used for corn ethanol production. Figure 3.2 shows the breakdown of 2010 end-uses of corn by end-use categories and sectors in the United States.

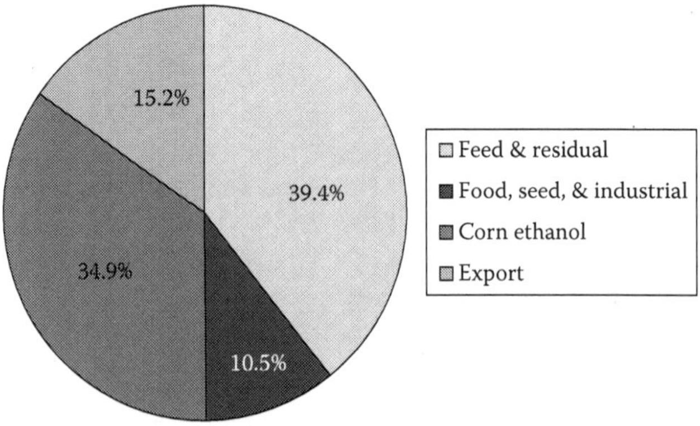

FIGURE 3.2
Breakdown by categories of 2010 end-uses of corn in the United States. (Source: U.S. Grains Council, 2011.)

The production of ethanol from starch and sugar-based resources in the United States reached 13.2 billion U.S. gallons in 2010. The amount of gasoline used by the United States for transportation was approximately 140 billion U.S. gallons per year in 2011, with ethanol used as a blend stock of up to 10% in marketed gasoline (E10) and also with a smaller E85 (85% ethanol) market. As ethanol production increases, the demand for ethanol in the fuel supply chain has nearly reached the 10% "blend wall" of 14 billion gallons. The Energy Independence and Security Act (EISA) of 2007 requires a mandatory Renewable Fuel Standard (RFS) requiring transportation fuels sold in the United States to contain a minimum of 36 billion gallons of renewable biofuels by 2022, including advanced and cellulosic biofuels and biomass-based diesel. The EISA further specifies that 21 billion U.S. gallons of the 2022 total biofuel blends in gasoline must be derived from noncornstarch products. Certainly, biomass-derived methanol, biomass-derived diesel, cellulosic ethanol, algae biodiesel, vegetable oil biodiesel, and others qualify as noncornstarch based biofuels.

3.2 Corn Ethanol as Oxygenated Fuel

3.2.1 Industrial Significance of Grain Ethanol

Ethanol production and utilization as automotive fuel received a major boost with the enforcement of the Clean Air Act Amendments (CAAA) of 1990. Blending gasoline with ethanol has become a popular method for gasoline producers to meet the new oxygenate requirements mandated by the CAAA. Provisions of the CAAA established the Oxygenated Fuels Program (OFP) and the Reformulated Gasoline (RFG) Program in an attempt to control carbon monoxide (CO) emissions and ground-level ozone problems. Both programs require certain oxygen levels in gasoline, viz., 2.7% by weight for oxygenated fuel and 2.0% by weight for reformulated gasoline. Public policies aimed at encouraging ethanol development/production were largely motivated by the nation's desire to improve air quality as well as to enhance future energy supply security. In addition, agricultural policymakers keenly see the expansion of the ethanol industry as a means of stabilizing farm income and reducing farm subsidies. Increasing ethanol production induces a higher demand for corn crops and raises the average corn price. Higher corn prices and stronger demand for corn reduce farm commodity program payments and the participation rate in the Acreage Reduction Program. From technical and scientific viewpoints as well as energy and environmental viewpoints, the use of ethanol as a motor fuel or blend fuel makes sense, inasmuch as ethanol can be produced in a renewable manner, that is, as a nondepletable energy source.

The rapidly growing corn ethanol industry in the first decade of the twenty-first century has increased the demand for corn very highly, which resulted in a significant price increase for corn. Although the market's corn prices are determined largely by the supply-and-demand relationship, several important observations about the market responses involving corn prices can be made, from both technoeconomic and socioeconomic viewpoints.

1. The tight supply of corn pushed the corn price on the market to escalate faster than other crops, thus hampering the affordability of corn for food purposes and creating a fuel versus food dilemma.
2. The corn price's escalation on the market has affected other food prices to varying extents. When the corn price went up, the poultry price also went up, due to the increase in the price of poultry feed which was also corn-based. Prices of other grain products were also affected, because higher corn prices have motivated farmers to increase corn acreage at the expense of other food crops.
3. When the crude petroleum price on the market went up, so did the corn ethanol price, although this trend is not unique for corn ethanol only. Furthermore, higher corn demand by the ethanol industry pushed the corn price higher, which in turn made ethanol production costlier due to higher feedstock prices [4].
4. Due to the large global market share of the U.S. corn trade, worldwide corn prices were largely affected by the U.S. domestic corn prices.

3.2.2 Clean Air Act Amendments of 1990

The Clean Air Act Amendments of 1990 target automobile emissions as a major source of air pollution. The Act mandates the use of cleaner burning fuels in U.S. cities with smog and air pollution problems. The oxygen requirements of CAAA spurred a market for oxygenates and created new market opportunities for ethanol. The Oxygenated Fuels Program targets 39 cities that do not meet National Ambient Air Quality Standards (NAAQS) for carbon monoxide (CO). CAAA mandates the addition of oxygen to gasoline to reduce CO emissions. It requires an oxygen level in gasoline of 2.7% by weight. Control periods vary by cities because most CO violations occur during the winter season. The average control period is about 4 months. The most widely used oxygenate in the market had been a methanol-derived ether, MTBE, which was made mostly from natural gas feedstock and was rapidly being phased out from the U.S. market in the twenty-first century, as explained in the previous section.

Most major gasoline refiners are using ethanol more and more to meet gasoline oxygenate content requirements. In 1993, about 300–350 million gallons of ethanol were blended with gasoline and sold in markets covered by the OFP. In 2004, fuel ethanol consumption reached 3.4 billion gallons in the United States. In about 10 years, the U.S. production of grain ethanol has seen a tenfold increase. In 2010, corn ethanol production in the United States reached 13.2 billion gallons, which is nearly a fourfold increase in the six-year period. The CAAA also requires the use of oxygenated fuels as part of the Reformulated Gasoline Program for controlling ground-level ozone formation. This program requires an oxygen level in gasoline of 2.0% by weight. Beginning in January 1995, reformulated gasoline was required to be sold in nine ozone nonattainment areas year-round. Other provisions in the act allow as many as 90 other cities with less severe ozone pollution to "opt-in" to the RFG program. Under a total opt-in scenario, as much as 70% of the nation's gasoline could be reformulated.

An oxygen level of 2.0% by weight in gasoline means that at least 5.75% by weight of ethanol needs to be blended in gasoline, based on the stoichiometric calculation of $2.0 \times (46/16) = 5.75$. Therefore, 2.7% oxygen requirement pushes the required level of ethanol in gasoline to 7.76 wt% as a minimum. Thus, 10% ethanol blended gasoline, E10, sold in gas stations is consistent with the result of such calculation. Even though ethanol is clean burning and has a relatively low Reid vapor pressure of blending, it has a substantially lower heating value than conventional gasoline. The higher heating value (HHV) and lower heating value (LHV) of gasoline are 47.3 and 44.4 kJ/g, respectively, whereas those for ethanol are 29.7 and 28.9 kJ/g, respectively. However, at a level of 10% ethanol blending, the reduced energy output is much less appreciable and could be compensated for by better engine performance.

3.2.3 Energy Independence and Security Act (EISA) of 2007

In 2007, the U.S. Congress enacted the Energy Independence and Security Act as a result of President Bush's Advanced Energy Initiative (AEI) challenging the United States to change the way its citizens fuel their vehicles to improve the nation's energy security. The important message delivered by the AEI was "Keeping America competitive requires affordable energy." The EISA of 2007 is an energy policy act designed to improve energy efficiency and to increase the supply of clean renewable fuels, thereby reducing U.S. energy consumption by 7% and greenhouse gas (GHG) emissions by 9% by 2030. The initiative requires a mandatory Renewable Fuel Standard requiring transportation fuels sold in the United States to contain a minimum of 36 billion gallons of renewable biofuels by 2022, including advanced and cellulosic biofuels and biomass-based diesel. The EISA further specifies that 21 billion gallons of the 2022 total biofuel blends in gasoline must be derived from noncornstarch products. Additionally, the EISA requires that the Corporate Average Fuel Economy (CAFE) standard increase to 35 miles per gallon by the year 2020. Although the EISA somewhat appears to discourage further growth of conventional corn-based ethanol as a blend fuel for the future and instead promotes rapid growth and market expansion of cellulosic ethanol, it still provides ample room for advancement of the U.S. ethanol industry.

3.2.4 Net Energy Balance of Corn Ethanol Production

In analyzing and discussing the energy balance of alternative fuel production, a term called *net energy value* (*NEV*) is often used. The net energy value is the difference between the energy content of product ethanol and the total energy used/consumed in producing and distributing ethanol. Higher corn yields of modern agricultural industry, lower energy consumption per unit of output in the fertilizer industry, and recent advances in fuel conversion technologies have significantly enhanced the economic and technical feasibility of producing ethanol from corn, when compared with that of just a decade ago. Therefore, studies based on the older data may tend to overestimate energy use (input), because the efficiency of growing corn as well as converting it to fuel ethanol has improved significantly over the past decade [5].

A large number of studies have been conducted to estimate the NEV of ethanol production. However, variations in data and model assumptions resulted in a widely differing range of estimated values (conclusions), ranging from a very positive to a negative value. A negative net energy value would mean that it takes more energy to produce the energy content of product ethanol. According to the study by Shapouri, Duffield, and Graboski [5] of the USDA, the net energy value of corn ethanol was calculated as +16,193 BTU/gal, assuming that fertilizers were produced by modern (1995 or so) processing plants, corn was converted in modern (also about 1995) ethanol facilities, farmers achieved normal corn yields, and energy credits were allocated to coproducts. Updated values for the NEV of corn ethanol by Shapouri et al. show +21,205 BTU/gal (July 2002) [6] and +30,528 BTU/gal (October 2004), respectively [7]. The first value of 21,205 BTU/gal was based on the higher heating value of ethanol, whereas the second value of 30,528 BTU/gal was based on the lower heating value of ethanol. However, another study conducted by Pimentel and Patzek (2005) showed that the NEV of ethanol was –1,467 kcal/liter (equivalent to –16,152 BTU/gal), which was based on the LHV [8]. A recent study thoroughly conducted by Argonne National Laboratory (2005) shows that ethanol generates 35% more energy than it takes to generate [9].

As shown, sharp differences in the calculated NEV of ethanol production among the studies still existed and they stemmed from several factors, which were comprehensively identified and directly compared in a report by MathPro Inc. [10]. According to MathPro's analysis, the differences in the NEV reflected sharp differences in four energy usage categories and they are [10]

1. Energy used in corn production: The USDA estimates (20.2 K BTU/gal in 2002 and 18.7 BTU/gal in 2004) are about half of Pimentel–Patzek's (37.9 K BTU/gal).
2. Energy used in corn transport: The USDA estimates (2.1 K BTU/gal in 2002 and 2004) are less than half of Pimentel–Patzek's (4.8 K BTU/gal).
3. Energy used in ethanol production: The USDA estimates (46.7 K BTU/gal in 2002 and 49.7K BTU/gal in 2004) are substantially lower than that of Pimentel–Patzek (56.4 K BTU/gal).
4. Coproduct energy credit: The 2002 USDA estimate (–13.5 K BTU/gal) is twice that of Pimentel–Patzek (–6.7 K BTU/gal). The 2004 USDA estimate (–26.3 K BTU/gal) is twice the 2002 USDA estimate and four times Pimentel–Patzek's estimate [10].

The "true" and "actual" value of ethanol's NEV would depend on various factors that involve the geographical region, agricultural productivity, efficiency of the ethanol production process, energy efficiency of fertilizer manufacture, and much more. It has been observed that the ethanol proponents have claimed positive NEVs, whereas the ethanol critics have referred to negative NEVs. As such, this subject has been controversial, from analytical and technoeconomical standpoints. However, it is certain that modern corn ethanol plants use substantially less energy and produce more ethanol per bushel of corn than older plants, and it also appears certain that the claims of negative NEVs have been based on obsolete material and energy balance data of the corn ethanol industry [11]. In 2008, Mueller conducted a very extensive milestone survey of the nation's ethanol plants in terms of new energy use and coproduct data as well as land use and his conclusions published in 2010 clearly showed significant improvements over the 2001 data [12].

The 2001 survey by BBI International [42] found that dry mill plants use, on average, 36,000 BTU of thermal energy and 1.09 kWh of electrical energy per gallon of ethanol produced, while producing an average of 2.64 gallons of ethanol per bushel of corn. However, ethanol plants in 2008 used an average of 25,859 BTU of thermal energy and 0.74 kWh of electricity per gallon of ethanol produced, which is 28.2 and 32.1% lower than the values of 2001, respectively. Ethanol produced per bushel of corn, meanwhile, increased by 5.3% to 2.78 gallons per bushel in 2008 [12]. This survey clearly supports that the NEV of ethanol production based on modern technology data is on the positive side.

3.2.5 Food versus Fuel

Although corn is an excellent source of starch and is heavily grown in the United States, its traditional use and value as a major food resource inevitably triggers a controversial debate of "food versus fuel." The supply-and-demand dynamics of corn in the U.S. marketplace for both fuel and food end-uses has significantly contributed to the recent escalation of corn prices, which in turn increased the production cost of ethanol as well as the price of corn-derived foods. This has been one of the principal reasons that drive the commercialization efforts of cellulosic ethanol production which is based on nonedible renewable feedstock.

3.2.6 Corn Ethanol Production Technologies

3.2.6.1 Dry Mill Process versus Wet Mill Process

Ethanol production facilities can be classified into two broad types: wet milling and dry milling operations. As the term "dry" implies, the dry milling process first grinds the entire corn kernel into flour which is referred to as "meal" or "corn meal." Dry mills are usually smaller in size (capacity) and are built primarily to manufacture ethanol only. The remaining stillage from ethanol purification undergoes a different process treatment to produce a highly nutritious livestock feed. Wet mill facilities are called "corn refineries," and also produce a list of high-valued coproducts such as high-fructose corn syrup (HFCS), dextrose, and cornstarch. Both wet and dry milling operations are currently used to convert corn to ethanol. Wet milling is usually a larger and more versatile process, and could be valuable for coping with volatile energy markets. Wet milling can be used to produce a greater variety of products such as cornstarch, corn syrup, ethanol, dry distillers grains, artificial sweeteners like Splenda®, and more. Although wet milling is a more versatile process and offers a more diverse product portfolio than dry milling, when producing fuel ethanol, dry milling has higher efficiency and lower capital and operating costs than wet milling. Most of the recent ethanol plants built in the United States are based on dry milling operations [13]. As of the end of 2008, a total of 86% of corn ethanol in the United States was commercially produced using the dry mill process using a total of 150 dry milling plants [14].

3.2.6.2 Ethanol Plant Energy Generation and Supply

Thermal energy and electricity are the main types of energy used in both types of milling plants. Dry milling corn ethanol plants have traditionally used natural gas as their process fuel for production. The choice of natural gas as a process fuel may turn out to be advantageous costwise due to the sharp decrease in natural gas price in recent years as well as the shale gas boom in the United States. Natural gas is used to generate steam for mash cooking, distillation, and evaporation as well as also being used directly in DGS dryers and thermal oxidizers that destroy the volatile organic compounds (VOCs) present in the dryer exhaust [15]. DGS stands for "distillers grain with solubles." Due to increased production efficiencies and expanded fuel capabilities, combined heat and power (CHP) has become increasingly popular as an efficient energy option for many new ethanol plants. CHP is an efficient, clean, and reliable energy services alternative, based on cogeneration of electricity and thermal energy on site. Therefore, CHP achieves avoiding line losses, increases reliability, and captures much of the thermal energy otherwise normally wasted in power generation to supply steam and other thermal energy needs at the plant site.

A CHP system typically achieves a total system efficiency of 60–80% compared to only about 50% for conventional separate generation of electricity and thermal energy [15]. By efficiently providing electricity and thermal energy from the same fuel source at the point of use, CHP significantly reduces the total fuel usage for a commercial ethanol plant, along with reductions in corresponding emissions of carbon dioxide (CO_2) and other pollutants. Generally speaking, electrical energy is used mostly for grinding and drying corn, whereas thermal energy is used for fermentation, ethanol recovery, and dehydration. On the other hand, flue gas is used for drying and stillage processing as part of waste heat recovery and energy integration efforts. The carbon dioxide generated from the fermentation process is also recovered and utilized to make carbonated beverages as well as to aid in the manufacture of dry ice as a by-product of the ethanol process. As mentioned earlier, based on the 2008 survey of 150 dry milling corn ethanol plants in the United States [14], ethanol plants in 2008 used an average of 25,859 BTU of thermal energy and 0.74 kWh of electricity per gallon of ethanol produced, which was 28.2 and 32.1% lower than the 2001 values of 36,000 BTU and 1.09 kWh, respectively. Ethanol productivity per bushel of corn also increased by 5.3% from 2.64 gallons in 2001 to 2.78 gallons per bushel in 2008 [12, 14]. It was also found that on average 5.3 pounds of dried distillers grains and 2.15 pounds of wet distillers grains (WDGs) as well as 0.06 gallons of corn oil per every gallon of ethanol are also produced as process coproducts. One U.S. bushel as a volume unit is equivalent to 35.23907 liters. Even though the U.S. corn ethanol industry has been considered a mature industry, the recent enhancements made on their process and energy efficiencies as well as the overall profitability are quite remarkable.

3.2.6.3 Ethanol Fermentation and Feedstock

The ethanol fuel manufacturing process is a combination of biochemical and physical processes based on traditional unit operations. Ethanol is produced by fermentation of sugars with yeast. The fermentation crude product is concentrated to fuel-grade ethanol via distillation. The organisms of primary interest to industrial fermentation of ethanol include *Saccharomyces cerevisiae, S. uvarum, Schizosaccharomyces pombe,* and *Kluyveromyces* spp., among which *Saccharomyces cerevisiae* is most commonly utilized.

Feedstock for ethanol fermentation are either sugar or starch-containing crops. These "biomass fuel crops" (tubers and grains) typically include sugarbeets, sugarcane, potatoes, corn, wheat, barley, Jerusalem artichokes, and sweet sorghum. Sugar crops such as sugarcane, sugarbeets, or sweet sorghum are extracted to produce a sugar-containing solution or syrup that can be directly fermented by yeast. Starch feedstock, however, must go through an additional step that involves starch-to-sugar conversion, as is the case for grain ethanol. Needless to say, sugar crops are simpler to convert to ethanol than starch crops. Therefore, the ethanol production cost, excluding the feedstock cost, is substantially lower for sugar crops than for starch crops.

3.2.6.4 Starch Hydrolysis

Starch may be regarded as a long-chain polymer of glucose (i.e., many glucose molecular units are bonded in a polymeric chain similar to a condensation polymerization product [16]). As such, macromolecular starches cannot be directly fermented to ethanol via conventional fermentation technology. They must first be broken down into simpler and smaller glucose units through a chemical process called *hydrolysis*. In the hydrolysis step, starch feedstock is ground and mixed with water to produce a mash typically containing 15 to 20% starch. The mash is then cooked at or above its boiling point and treated subsequently with two enzyme preparations. The first enzyme hydrolyzes starch molecules to short-chain molecules, and the second enzyme hydrolyzes the short chains to glucose. The first enzyme is amylase. Amylase liberates "maltodextrin" by the liquefaction process. Such maltodextrins are not very sweet inasmuch as they contain dextrins (a group of low molecular weight carbohydrates) and oligosaccharides (a saccharide polymer containing a small number of simple sugars, monosaccharides). The dextrins and oligosaccharides are further hydrolyzed by enzymes such as pullulanase and glucoamylase in a process known as *saccharification*. Complete saccharification converts all the limit dextrans (complex branched polysaccharides of many glucose molecules) to glucose, maltose, and isomaltose. The mash is then cooled to 30°C, and at this point yeast is added for fermentation.

3.2.6.5 Yeast Fermentation

Yeasts are capable of converting sugar into alcohol by a biochemical process called fermentation. The yeasts of primary interest to industrial fermentation of ethanol include *Saccharomyces cerevisiae, Saccharomyces uvarum, Schizosaccharomyces pombe,* and *Kluyveromyces* spp. Under anaerobic conditions, yeasts metabolize glucose to ethanol primarily via the Embden–Meyerhof pathway. The Embden–Meyerhof pathway of glucose metabolism is the series of enzymatic reactions in the anaerobic conversion of glucose to lactic acid (or ethanol in this case), resulting in energy in the form of adenosine triphosphate (ATP) [17]. The overall net reaction represented by a stoichiometric equation involves the production of two moles of ethanol from each mole of glucose as shown below. However, the yield attained in practical fermentation attempts does not usually exceed 90–95% of the theoretical value. In this case, the theoretical value (i.e., 100% of yield) means that exactly two moles of ethanol are produced from each mole of glucose input to the fermenter. Therefore, this 100% yield is equivalent to the mass conversion efficiency of 51%, which is defined later in this section. The following stoichiometric equation shows the basic biochemical reaction in the conversion by fermentation of glucose to ethanol, carbon dioxide, and endothermic heat.

$$C_6H_{12}O_6 = 2\ C_2H_5OH + 2\ CO_2$$

$$\Delta H^{\circ}_{298} = 92.3\ \text{kJ/mol}$$

Theoretically, the maximum conversion efficiency of glucose to ethanol is 51% on a weight basis, which comes from a stoichiometric calculation of:

$$2 * (\text{Molecular wt of Ethanol})/(\text{Molecular wt of Glucose}) = (2 * 46)/(180) = 0.51$$

However, some glucose is inevitably used by the yeast for production of cell mass and for metabolic products other than ethanol, thus reducing the conversion efficiency from its theoretical maximum of 51%. In practice, 40 to 48% of glucose, on a weight basis, is actually converted to ethanol. With a 46% fermentation efficiency, 1,000 kilograms of fermentable sugar would produce about 583 liters of pure ethanol, after taking into account the density of ethanol (specific gravity at 20°C = 0.789), or

$$(1{,}000\ \text{kg sugar}) * (0.46\ \text{kg ethanol/kg sugar})/(0.789\ \text{kg ethanol/L}) = 583\ \text{L ethanol}$$

Conversely, about 1,716 kilograms of fermentable sugar are required to produce 1,000 liters of ethanol, when a 46% mass conversion efficiency is assumed. Mash typically contains between 50 and 100 grams of ethanol per liter (about 5 to 10% by weight) when the fermentation step is complete. This is called *distilled mash* or *stillage*, which still contains a large amount of non-fermentable portions of fibers or proteins.

3.2.6.6 Ethanol Purification and Product Separation

Ethanol is subsequently separated from the mash by distillation, in which the components of a solution (in this case, water and ethanol) are separated by differences in boiling point (or individual vapor pressure). Separation is technically limited by the fact that ethanol and water form an *azeotrope*, or a constant boiling solution, of 95.63 wt% alcohol and 4.37 wt% water. This azeotrope is a minimum boiling mixture (or a positive azeotrope), for which the boiling temperature of the azeotrope is lower than that of the individual pure components, that is, water and ethanol. The minimum boiling temperature at the azeotropic concentration is 78.2°C, whereas the normal boiling points of ethanol and water are 78.4°C and 100°C, respectively.

The 5%, more precisely 4.37 wt%, water cannot be separated by conventional distillation, because the minimum boiling temperature is attainable at the azeotropic concentration, not at the pure ethanol concentration. Therefore, production of pure, water-free (anhydrous) ethanol requires an additional unit operation step following distillation. Dehydration, a relatively complex step in ethanol fuel production, is accomplished in one of two methods. The first method uses a third liquid, most commonly benzene, which is added to the ethanol/water mixture. This third component changes the boiling characteristics of the solution (now a ternary system instead of a binary system), allowing separation of anhydrous ethanol. In other words, this third component is used to break the azeotrope, thereby enabling conventional distillation to achieve the desired goal of separation. This type of distillation is also called *azeotropic distillation*, because the operation separates mixtures that form azeotropes. The second method employs *molecular sieves* that selectively absorb water based on the molecular size difference between water and ethanol. Molecular sieves are crystalline metal aluminosilicates having a three-dimensional interconnecting network of silica and alumina tetrahedra. Molecular sieves have long been known for their drying capacity (even to 90°C). There are different forms of molecular sieves that are based on the dimension of effective pore opening, and they include 3A, 4A, 5A, and 13X. Commercial molecular sieves are typically available in powder, bead, granule, or extrudate forms.

3.2.6.7 By-Products and Coproducts

As mentioned earlier, the nonfermentable solids in distilled mash (stillage) contain variable amounts of fiber and protein, depending on the feedstock. The liquid also contains soluble protein and other nutrients and as such remains valuable. The recovery of the protein and other nutrients in stillage for use as livestock feed can be essential to the overall profitability of ethanol fuel production. Protein content in stillage varies with feedstock. Some grains such as corn and barley yield solid by-products that are called dried distillers grains (DDGs). Protein content in DDG typically ranges from 25 to 30% by mass and makes an excellent feed for livestock.

3.2.6.8 Potential Environmental Issues of Liquid Effluents

The production of ethanol also generates liquid effluent, which may render a potential pollution problem or a concern of environmental stress on water systems. About nine liters of liquid effluent are generated for each liter of ethanol produced, which varies depending on the specific process adopted. Some of the liquid effluent may be recycled. Effluent can have a high level of biochemical oxygen demand (BOD), which is a measure of organic water pollution potential, and it is also acidic. Therefore, the liquid effluent must be treated before being discharged into the water stream. Specific treatment requirements depend on both feedstock quality (and type) as well as local pollution control regulations. Due to the acidity of the effluent, precautions and care must also be taken if the effluent is directly spread over fields [18].

3.3 Chemistry of Ethanol Fermentation

3.3.1 Sugar Content of Biological Materials

Figure 3.3 shows a much generalized view of plant cell wall composition. The base molecules that give plants their structure can be processed to produce sugars, which can be subsequently fermented to ethanol. As such, feedstock that can generate sugars more readily and cost-effectively automatically become prime candidates for ethanol fermentation.

The principal components of most plant materials are commonly described as *lignocellulosic biomass*. This type of biomass is mainly composed of the compounds called cellulose, hemicellulose, and lignin. *Cellulose* is a primary component of most plant cell walls and is made up of long chains of

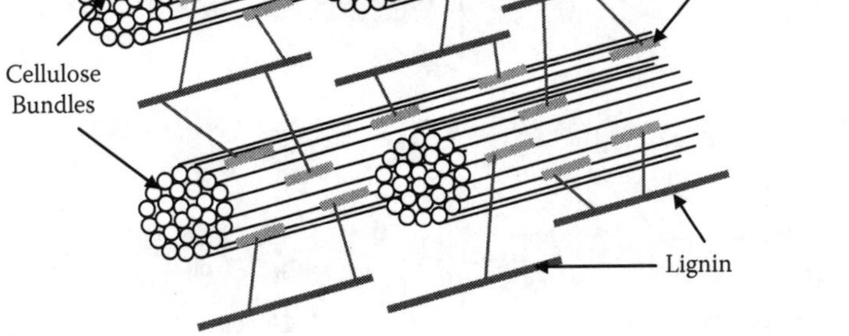

FIGURE 3.3
A generalized description of a plant cell wall.

the 6-carbon sugar, glucose, that are arranged in bundles (often described as crystalline bundles). The cellulose molecules in the plant cell wall are interconnected by another molecule called *hemicellulose*. The hemicellulose is primarily composed of the 5-carbon sugar, xylose. In addition to cellulose and hemicellulose, another macromolecule called *lignin* is also present in significant amounts and provides the structural strength for the plant. As also explained in Chapter 4, lignin is not easily converted into sugars or starches by current technology and therefore, has not been a target for alcohol fermentation. However, economically beneficial utilization of lignin is quite important in enhancing the overall process economics of ethanol production as well as minimizing process wastes. Technological developments have recently introduced a variety of processes of extracting and dissolving the cellulose and hemicellulose to produce sugars in such a form that can be readily fermented to ethanol. Generally speaking, efficient pretreatment can liberate the cellulose and hemicellulose from the plant material. Further treatment using chemicals, enzymes, or micro-organisms can also be applied to liberate simple sugars from the cellulose and hemicellulose, thus making them available to micro-organisms for fermentation to ethanol.

3.3.2 Conversion of Sugars to Ethanol

Figure 3.4 illustrates the stoichiometric conversion of cellulose to ethanol [19]. The first step involves cellulose hydrolysis that is essentially cleaving the chemical bonds in the cellulose to produce glucose.

Once the large molecules are extracted from plant cells, they can be broken down into their component sugars using enzymes or acids. The

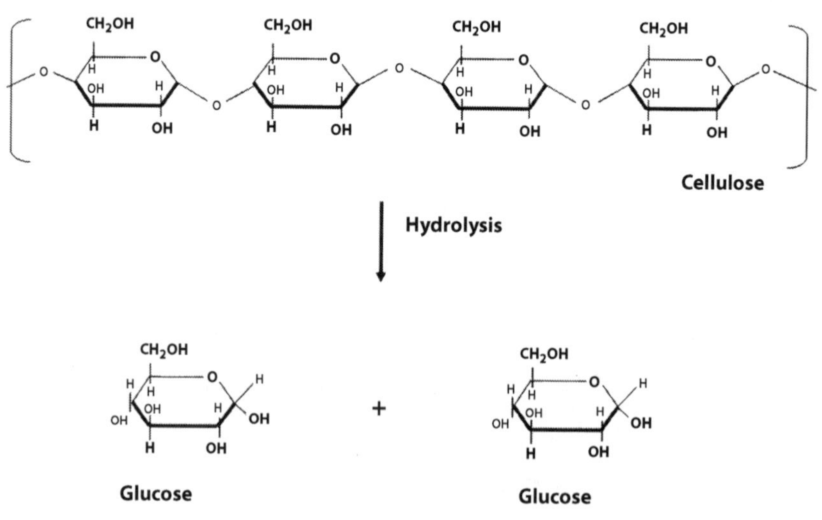

FIGURE 3.4
Hydrolysis of cellulose.

sugars can be subsequently converted to ethanol using appropriately selected micro-organisms via a process called fermentation. The fermentation of ethanol from 6-carbon sugars (such as D-glucose) follows the stoichiometric equation.

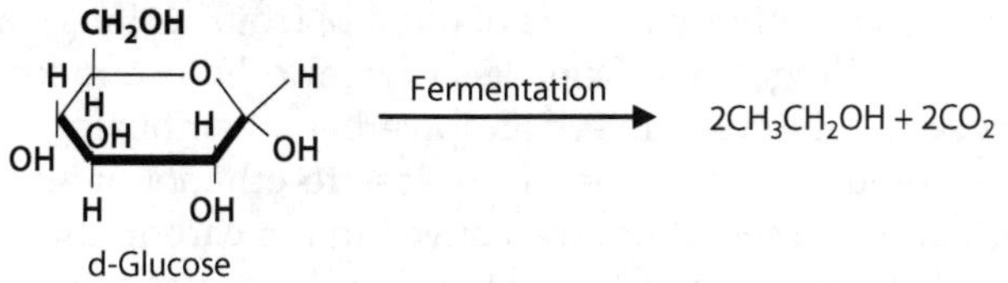

d-Glucose

According to the stoichiometric equation, one mole of D-glucose produces 2 moles of ethanol and 2 moles of carbon dioxide. Considering the molecular weights of glucose, ethanol, and carbon dioxide being 180, 46, and 44, respectively, the maximum theoretical yield of ethanol by weight % from the process would be 92/180 = 51%. Nearly half the weight of the glucose 88/180 (49%) is converted to carbon dioxide at its theoretical maximum. As such, a significant amount of carbon dioxide is generated by the fermentation step, which needs to be captured or utilized for economically beneficial purposes.

Hemicellulose is made up of the 5-carbon sugar, xylose, arranged in chains with other minor 5-carbon sugars interspersed as side chains. Similarly to the cellulose case, the hemicellulose can also be extracted from the plant material and treated to liberate xylose that, in turn, can be fermented to produce ethanol. However, xylose fermentation is not as straightforward or efficient as glucose fermentation based on currently available technology. Depending on the micro-organism and conditions employed, a number of different fermentation paths are possible or conceivable. The array of products can include ethanol, carbon dioxide, and water as

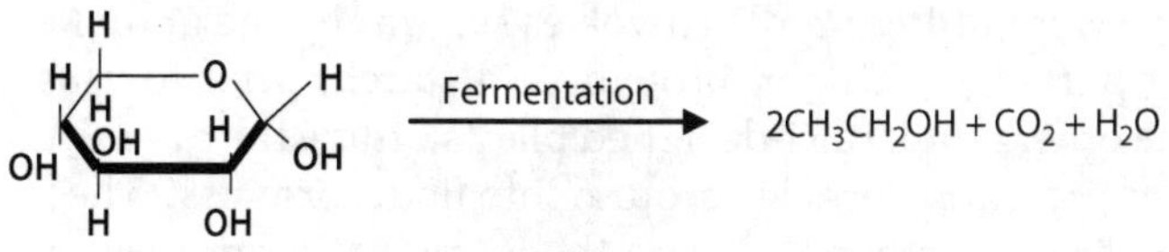

Xylose

Actually, three different reactions have been documented with yields of ethanol ranging from 30 to 50% of the weight of xylose as the starting material (i.e., weight ethanol produced/weight xylose). They are:

$$3 \text{ Xylose} \rightarrow 5 \text{ Ethanol} + 5 \text{ Carbon Dioxide}$$

$$\text{Xylose} \rightarrow 4 \text{ Ethanol} + 7 \text{ Carbon Dioxide}$$

$$\text{Xylose} \rightarrow 2 \text{ Ethanol} + \text{Carbon Dioxide} + \text{Water}$$

The first reaction yields a maximum of 51% (= 5 * 46/(3 * 150)), the second 41% (= 4 * 46/(3 * 150)), and the third 61% (= 2 * 46/150), respectively. Although the maximum theoretical ethanol yields from these fermentation reactions range between 41 and 61%, the practical yields of ethanol from xylose as starting material are in the range of 30 to 50%.

In the discussion of potential yields of ethanol from various starting materials, two different ranges of efficiencies of hemicellulose-to-xylose conversion and xylose-to-ethanol conversion have been combined to provide an overall conversion efficiency of hemicellulose to ethanol of about 50%. Just as with the glucose fermentation, the conversion of carbon dioxide to value-added products would vastly improve the overall process economics of ethanol production, because the yield of carbon dioxide is not only significant in amounts but also inevitable. It must be noted that even though xylose fermentation to ethanol is also mentioned in this chapter, the main focus of this chapter is on glucose fermentation, more particularly corn sugars into ethanol. Ethanol-from-corn technology involves glucose fermentation, not xylose fermentation, as required in cellulosic ethanol technology. Xylose fermentation or hemicellulose fermentation is treated in depth in Chapter 4.

3.4 Corn-to-Ethanol Process Technology

Fermentation of sugars to ethanol, using commercially available fermentation technology, provides a fairly simple, straightforward means of producing ethanol with little technological risk. The system modeled assumes that the molasses is clarified, and then fermented via cascade fermentation with a yeast recycle. The stillage is concentrated by multiple-effect evaporation and a molecular sieve is used to dehydrate the ethanol. Corn ethanol is commercially produced in one of two ways, using either the wet mill or dry mill process. The wet milling process involves separating the grain kernel into its component parts (germ, fiber, protein, and starch) prior to yeast fermentation. On the other hand, ICM-designed plants utilize the dry milling process, where the entire grain kernel is ground into flour form first. The starch in the flour is converted to ethanol during the fermentation process, also creating carbon dioxide and distillers grain as principal by-products.

3.4.1 Wet Milling Corn Ethanol Technology

For the past two centuries in the United States, corn refiners have been developing, improving, and perfecting the process of separating corn into its component parts to create a variety of value-added corn products and by-products. The *corn wet milling process* separates corn into its four basic components, viz., starch, germ, fiber, and protein. There are eight basic steps involved to accomplish this corn refining and alcohol fermentation process [20].

1. First, the incoming corn is *visually inspected and cleaned*. Corn refiners use #2 yellow dent corn, which is removed from the cob during harvesting. One bushel of yellow dent corn weighs about 56 pounds on average. Refinery people inspect arriving corn shipments and clean them two or three times to remove cob, dust, chaff, and any other foreign unwanted materials before the next processing stage of steeping. An effective screening process can save a great deal of trouble in the subsequent stages. The inspected and screened corn is then conveyed to storage silos holding up to 350,000 bushels.
2. Second, it is steeped to initiate polymeric bond cleavage of starch and protein into simpler molecules. *Steeping* is typically carried out in a series of stainless steel tanks. Each steep (or steeping) tank may hold about 2,000–13,000 bushels of corn soaked in water at 50–52°C for 28–48 hours. During steeping, the kernels (as shown in Figure 3.5) absorb water, thereby increasing their moisture levels from 15% to 45% by weight and also more than doubling in size by swelling [20]. The addition of 0.1% sulfur dioxide (SO_2) to the water suppresses excessive bacterial growth in the warm water environment. As the corn swells and softens, the mild acidity of the steeping water begins to loosen the gluten bonds within the corn and eventually release the starch [20]. A *bushel* is a unit of volume measure used as a dry measure of grains and produce. A bushel of corn or milo weighs about 56 pounds, a bushel of wheat or soybeans weighs about 60

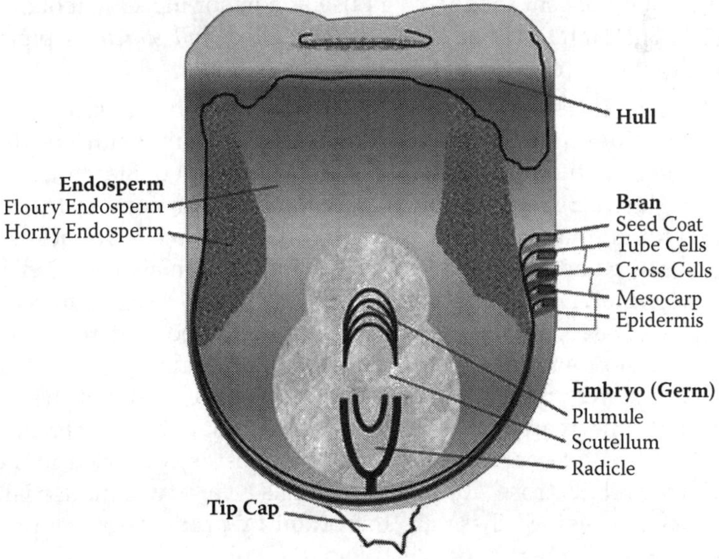

FIGURE 3.5
Corn kernel.

pounds, and a bushel of sunflowers weighs about 25 pounds. Or, a U.S. bushel is equivalent to 35.23907 liters as a volume unit.

3. The third step is the germ separation. It starts with coarse grinding of the corn in the slurry to separate/break the germ from the rest of the kernel. The germ is the embryo of a kernel of grain, as shown in Figure 3.5. This germ separation is accomplished in cyclone separators, which spin the low-density corn germ out of the slurry. Therefore, this cyclone separator is called a *germ separator.* It is also called a *degerminating mill.* The germs, which contain about 85% of corn's oil, are pumped onto screens and washed repeatedly to remove any starch left in the mixture [20]. A combination of mechanical and solvent processes extracts the oil from the germ. The oil is then refined and filtered into finished corn oil. The germ residue is saved as another useful component of animal feed. Both corn oil and germ residues are important by-products of this process.
4. As the fourth step, the remaining slurry, consisting of fiber, starch, and protein, is finely ground and screened to separate the fiber from the starch and protein. After the germ separation step described in Step 3, corn and water slurry goes through a more thorough grinding in an impact or attrition-impact mill to release the starch and gluten from the fiber in the kernel. The suspension of starch, gluten, and fiber flows over fixed concave screens, which catch fiber but allow starch and gluten to pass through. The fiber is collected, slurried, and screened again to reclaim any residual starch or protein, then piped or sent to the feed house as a major ingredient of animal feed. The starch–gluten suspension, called *mill starch,* is piped or sent to the starch separators [20].
5. Fifth, starch is separated from the remaining slurry in *hydrocyclones.* By centrifuging mill starch, the gluten is readily spun out due to the density difference between starch and gluten. Starch is denser than gluten. Separated gluten, a type of protein composite, can be used for animal feed. *Corn gluten meal* (*CGM*) is a by-product of corn processing and is used as animal feed. CGM can also be used as an organic herbicide. The starch, now with just 1–2% protein remaining, is diluted, washed 8 to 14 times, rediluted and rewashed in hydrocyclones to remove the last trace of protein and produce high-quality starch, typically more than 99.5% pure. Some of the starch is dried and marketed as unmodified cornstarch, another portion is modified into specialty starches, but most is converted into corn syrups and dextrose [20]. Cornstarch has a variety of industrial and domestic uses. All these are important by-products of the process that contribute to the corn distillers' profitability.
6. Sixth, the cornstarch then is converted to syrup (corn syrup) and this stage is called the starch conversion or starch-to-sugar conversion

step. The starch–water suspension is liquefied in the presence of acid or enzymes. Enzymes help convert the starch to dextrose that is soluble in water as an aqueous solution. Treatment with another enzyme is usually carried out, depending upon the desired process outcome. The process of acid and enzyme reactions can be stopped or terminated at key points throughout the process to produce a proper mixture of sugars such as dextrose (a monosaccharide, $C_6H_{12}O_6$) and maltose (a disaccharide, $C_{12}H_{22}O_{11}$) for syrups to meet desired specifications [20]. For example, in some cases, the conversion of starch to sugars can be halted at an early stage to produce low-to-medium sweetness syrups. In other situations, however, the starch conversion process is allowed to proceed until the syrup becomes nearly all dextrose. After this conversion process, the syrup is then refined in filters, centrifuges, or ion-exchange columns, and excess water is evaporated to result in concentrated syrup. Syrup can be sold directly as is, crystallized into pure dextrose, or processed further to produce high-fructose corn syrup (HFCS). Across the corn wet milling industry, about 80% of starch slurry goes to corn syrup, sugar, and alcohol fermentation.

7. Seventh, the concentrated syrups can be made into several other products through a fermentation process. Dextrose is one of the most fermentable forms of all of the sugars. Dextrose is also called *corn sugar* and *grape sugar*, and dextrose is a naturally occurring form of glucose, that is, D-glucose. Dextrose is better known today as glucose. Following the conversion of starch to dextrose, dextrose is piped and sent to fermentation reactors/units/facilities where dextrose is converted to ethanol by traditional yeast fermentation. Using a *continuous* process, the fermenting mash is allowed to flow, or cascade, through several fermenters in series until the mash is fully fermented and then leaves the final tank. In a *batch* fermentation process, the mash stays in one fermenter for about 48 hours before the distillation process for alcohol purification is initiated. Generally speaking, a continuous mode is more effective with a higher fermenter throughput, whereas higher-quality product may be obtained from a batch mode.
8. As the eighth step, ethanol separation or purification follows the fermentation step. The resulting broth is distilled to recover ethanol or concentrated through membrane separation to produce other by-products. Carbon dioxide generated from fermentation is recaptured for sale as dry ice and nutrients still remaining in the broth after fermentation are used as components of animal feed ingredients. These by-products also contribute significantly to the overall economics of the corn refineries.

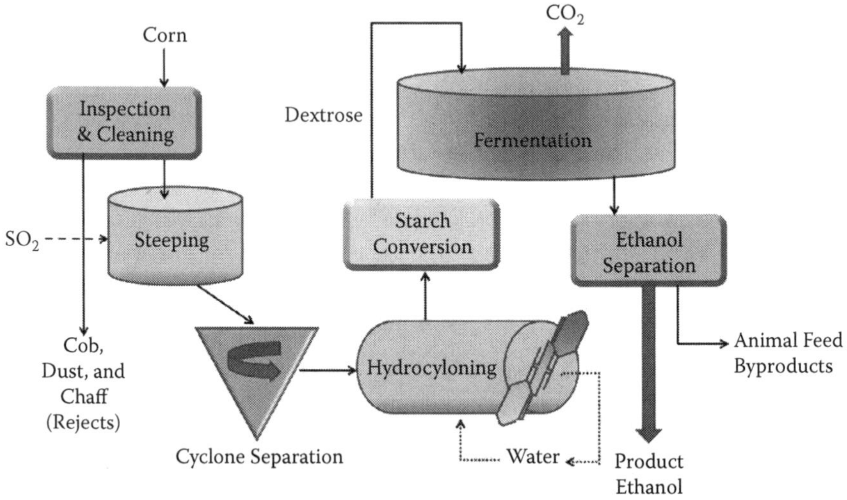

FIGURE 3.6
A schematic of a typical wet-milling corn-to-ethanol process.

Even though the term "by-product" was used throughout the process description, "coproduct" may be a better term in corn refining technology, inasmuch as these products are not only valuable but also targeted in the master plan of corn distillers. The corn-to-alcohol process detailed above can be summarized in a schematic process diagram, as shown in Figure 3.6.

3.4.2 Dry Milling Corn Ethanol Process

In comparison to the wet milling ethanol process where the corn kernel has to be separated into its components of germ, fiber, gluten, and starch prior to the fermentation step, the dry milling ethanol process first grinds the entire corn kernel into coarse flour form and then ferments the starch in the flour directly into ethanol. The dry milling corn ethanol process by ICM, Inc. is outlined below [21].

1. In the first step, corn receiving and storage, the corn grain is delivered by truck or rail to the corn ethanol fermentation plant. Grains are loaded in storage bins (silos) designed to hold sufficient amounts of grain to supply the plant operation continuously for 7–12 days.
2. The second step, milling, is where the grain is inspected and screened to remove debris (including corn cobs, stalks, finer materials, stones, and foreign objects) and ground into coarse flour. The screening is usually done using a blower and screen. Coarse grinding is typically

performed using a hammer mill. The feed rate from the milling step to the next stage of hot slurrying is typically controlled by the use of weighing tanks.

3. During the so-called cooking process, also called hot slurry, primary liquefaction, and secondary liquefaction, the starch in the flour is physically prepared and chemically modified for fermentation.
   a. In the hot slurry method the coarsely ground grain is soaked in hot process water, the pH of the solution is adjusted to about 5.8, and an alpha-amylase enzyme is added. The slurry is heated to 180–190°F (82–88°C) for 30–60 minutes to reduce its viscosity. Agitation needs to be provided.
   b. Primary liquefaction follows, where the slurry is then pumped through a pressurized jet cooker at 221°F (105°C) and held for 5 minutes. The jet cooker is also known as a steam injection heater. The mixture is then cooled by an atmospheric or vacuum flash condenser. The jet cooker is a critical component as steam helps to evenly hydrolyze and rapidly heat the slurry. The fluid dynamic relationship between the jet cooker's steam injector and condensing tube produces a pressure drop to help maximize shear action to improve starch conversion [22].
   c. The third process, secondary liquefaction, occurs after the flash condensation cooling. The mixture is held for 1–2 hours at 180–190°F (82–88°C) to give the alpha-amylase enzyme sufficient time to break down the starch into short-chain, low-molecular-weight dextrins. This chemical conversion is called *gelatinization*. Generally speaking, during the gelatinization step, there is a sharp increase in the slurry viscosity that is rapidly decreased as the α-amylase hydrolyzes the starch into lower molecular weight dextrins. Dextrins are a group of low-molecular-weight carbohydrates produced by the hydrolysis of starch and are mixtures of polymers of D-glucose units linked by α-(1→4) or α-(1→6) glycosidic bonds. After pH and temperature adjustment, a second enzyme, glucoamylase, is added as the mixture is pumped into the fermentation tanks. Glucoamylase is an amylase enzyme that cleaves the last alpha-1,4-glycosidic linkages at the nonreducing end of amylase and amylopectin to yield glucose. In other words, glucoamylase is an enzyme that cleaves the chemical bonds near the ends of long-chain starches (carbohydrates) and releases maltose and free glucose. Maltose, or malt sugar, is a disaccharide that is formed from two units of glucose joined with an α(1→4)bond.

4. The fourth step is called simultaneous saccharification fermentation. Once inside the fermentation tanks, the mixture is now referred to as mash, because it is an end product of mashing (which involves mixing of the milled kernel and water followed by mixture heating). The glucoamylase enzyme breaks down the dextrins, oligosaccharides, to form simple sugars, that is, monosaccharides. Yeast is added at this stage to convert the sugar to ethanol and carbon dioxide via an alcohol fermentation reaction. The mash is then allowed to ferment for 50–60 hours, resulting in a mixture that contains about 15% ethanol as well as the solids from the grain and added yeast [21, 23].
5. In the distillation step the fermented mash is pumped into a multicolumn distillation system. The distillation columns utilize the boiling point difference between ethanol and water to distill and separate the ethanol from the solution. By the time the product stream is ready to leave the distillation columns, it contains about 95% ethanol by volume (which is 190-proof). This point is just immediately below the azeotropic concentration of the ethanol–water binary system, as explained in Section 3.2.6.6. To overcome this azeotropic limitation of maximum achievable ethanol concentration via straight distillation, several optional methods are being used, including jumping over the azeotropic point or bypassing the distillation. The residue from this process, called *stillage*, contains nonfermentable solids and water and is pumped out from the bottom of the distillation columns into the centrifuges.
6. The sixth step is that of dehydration. The 190-proof ethanol still contains about 5 vol.% water. This near-azeotropic binary mixture is passed through a molecular sieve to physically separate the remaining water from the ethanol based on the size difference between the two molecules [21]. This dehydration step produces 200-proof anhydrous (waterless) ethanol, that is, near 100% ethanol.
7. Product ethanol storage is the seventh step. Before the purified ethanol is sent to storage tanks, a small amount of denaturant chemical is added, making it unsuitable for human consumption. There are so many different kinds of denaturants available on the market for diverse purposes other than fuel ethanol. However, only certain gasoline-compatible blendstocks are suitable as denaturants for fuel ethanol. Some ethanol refineries also sell their denaturants for other ethanol industries. The ASTM D4806-11a specification covers nominally anhydrous denatured fuel ethanol intended for blending with unleaded or leaded gasolines for use as a spark-ignition automotive engine fuel. According to this specification, the only denaturants used for fuel ethanol shall be natural gasoline, (also known as natural gas liquid [NGL]), gasoline components,

or unleaded gasoline at the minimum concentration prescribed. Methanol, pyrroles, turpentine, ketones, and tars are explicitly listed as prohibited denaturants for fuel ethanol meant to be used as gasoline blendstock [24]. Most ethanol plants' storage tanks are sized to allow storage of 7–12 days' production capacity.

8. During the ethanol production process, two valuable coproducts are created: carbon dioxide and distillers grains. Their recoverable values are very important to the overall process economics and this is why they are called coproducts rather than simply by-products.

During yeast fermentation, a large amount of carbon dioxide gas is generated. Because CO_2 is a major greenhouse chemical, its release into the atmosphere is not desirable. The carbon dioxide generated by fermentation is of high concentration and its purification is relatively straightforward. Therefore, carbon dioxide from ethanol fermentation is commonly captured and purified with a scrubber so it can be marketed to the food processing industry for use in carbonated beverages and flash-freezing applications. Dry ice is a common coproduct of the ethanol refineries.

The stillage from the bottom of the distillation columns contains solids derived from the grain and added yeast as well as liquid from the water added during the process. The stillage is sent to centrifuges for separation into thin stillage (a liquid with 5–10% solids) and wet distillers grains [21].

Some of the thin stillage is recycled back to the cook/slurry tanks as makeup water, reducing the amount of fresh water required by the cooking (hot slurry) process. The rest is sent through a multiple-effect evaporation system where it is concentrated into a syrup containing 25–50% solids. This syrup, which is high in protein and fat content, is then mixed back in with the wet distillers grains [21]. This is a step intended to recover most of the nutritive components from the stillage. With the added syrup, the WDG still contains most of the nutritive value of the original feedstock plus the added yeast and as such it makes excellent cattle feed. After the addition of the syrup, it is conveyed to a wet cake pad, where it is loaded for transport.

Many ethanol refinery facilities do not have enough nearby cattle farms or established markets to utilize all of their WDG products. However, WDG must be used soon after it is produced, because it gets spoiled rather easily. Therefore, WDG is often sent through an energy-efficient drying system to remove moisture and extend its shelf life. Dried distillers grains are commonly used as a high-protein ingredient in cattle, swine, poultry, and fish diets. Modified forms of DDGs are also being researched for human consumption due to the outstanding nutritive values. In more practical senses, DDG is better known as a corn ethanol coproduct than WDG.

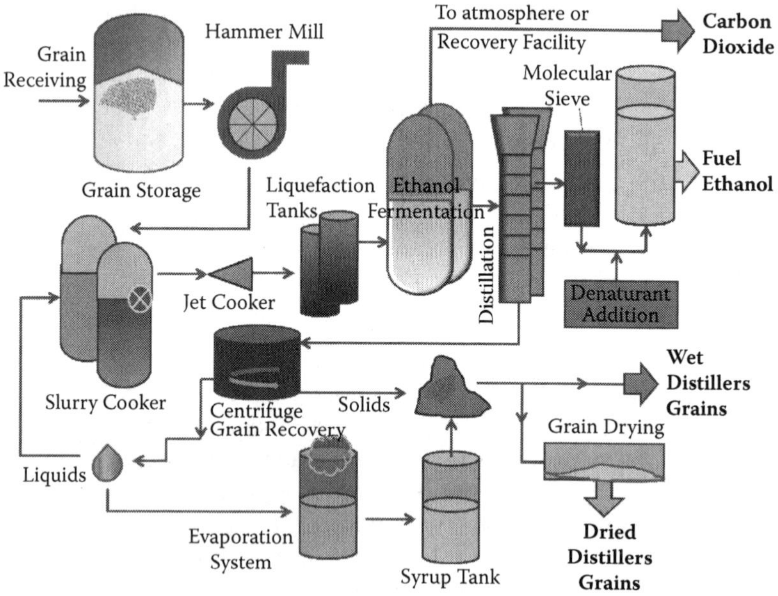

FIGURE 3.7
A schematic of ICM's dry milling ethanol production system. (Modified from ICM. ICM's dry milling ethanol production.)

A schematic of ICM's dry milling corn ethanol process [21] is shown in Figure 3.7.

3.4.3 Industrial Cleaning of Ethanol Plant

Periodic shutdowns due to maintenance and repairs as well as emergency stoppages are inevitable with all fuel and chemical production plants. In this regard, an ethanol plant is no exception. Certain pieces of machinery involving ducts, pipes, headers, stacks, and valves tend to foul more quickly and frequently than other pieces, thereby plugging up the associated equipment, developing leaks, and potentially disabling the entire process. Therefore, highly efficient industrial cleaning of ethanol plants during periods of shutdown becomes an important issue. Several cleaning options are available and also conceivable, including chemical-based cleaning, hydro-cleaning, compressed air cleaning, and dry ice blasting. The dry ice blasting process is used to clean the ethanol plant and its vital components during the shutdown periods [25]. One of its advantages over hydrocleaning is that there is no water left behind where the system has to be dry; other advantages include the nontoxicity of carbon dioxide as well as the availability of carbon dioxide on the plant site if captured during the fermentation stage. The need for ethanol plant cleaning is universally applicable to both wet milling and dry milling plants.

3.5 By-Products/Coproducts of Corn Ethanol

Ethanol by-products (coproducts) include distillers dried grains or dried distillers grains, distillers dried grains with solubles, wet distillers grains, corn bran, corn gluten feed, corn gluten meal (CGM), corn germ meal, and condensed fermented corn extractives. A bushel of corn produces about 2.78 gallons of ethanol. About 5.3 pounds of DDGs, 2.15 pounds of WDGs, and 0.06 gallons of corn oil are also produced per gallon of ethanol as coproducts [14]. Per bushel of corn, the amounts of DDGs and WDGs coproduced are on average about 14.7 and 6.0 pounds, respectively. As of 2010, nearly 3.8 million tons of DDGs (including both DDG and DDGS [dried distillers grains with solubles]) are produced in domestic dry grind ethanol production, that is, dry milling ethanol production. This accounts for more than 98% of the total U.S. DDG and DDGS production and the remaining 1–2% comes from the alcohol beverage industries. DDG is nearly identical to DDGS except that the former does not contain the distillers solubles, which is a "sticky" syrup.

Carbon dioxide is also becoming an important coproduct, as mentioned earlier. Instead of becoming a greenhouse gas emitting industry, carbon dioxide generated in the alcohol fermentation reaction can be relatively easily captured due to its high concentration and purified for manufacture of dry ice or compressed carbon dioxide gas for the food and beverage industries.

Corn oil is very valuable in both food and fuel applications, however, a corn kernel contains only about 3.6–4.0 wt% of corn oil/fat, Due to its low level, any process targeting direct extraction of corn oil alone from the corn kernel would not be a cost-effective solution. In this regard, recent dry milling ethanol plants potentially offer a good opportunity as potentially large sources of corn oil, as long as an economical separation process can be developed and implemented. Coproduction of corn oil is one of the promising options for the corn ethanol refinery to improve the gross margins in the industry.

Commercial processes for separation of corn oil are currently being developed [26, 27]. POET, the largest ethanol producer in the world has been producing Voila™ corn oil since the beginning of 2011 for the biodiesel and feed markets. POET in Iowa had been predicted to produce corn oil as feedstock for 12 million gallons of biodiesel per year by the end of 2011 [28]. SunSource technology produces corn oil as an additional coproduct available to ethanol producers, most likely to be used in biodiesel production. The process uses centrifuge technology to extract the oil from the distillers grains in the evaporation step [27]. They also claim that removing the corn oil from the distillers grains does not lower the value of the grain feed coproduct and instead makes it easier to handle. They claim that the process also reduces volatile organic compounds emission from the dryers [27].

R&D efforts in cost-effective corn oil extraction and purification are under way. Corn oil can be extracted from dry-milled or wet-milled germ by crushing for $35–45/tonne or by hexane extraction for $20–40/tonne, which are significant after considering all other associated costs as well as the market price for unrefined corn oil [29]. Quick germs [30] and enzymatically milled germs [31] have been successfully produced in laboratory quantities with 30% and 39% oil, respectively. Oil yields of 65 wt% can be recovered from wet-milled or dry-milled germ by expeller pressing [32]. Oil separation from corn germ using aqueous extraction (AE) and aqueous enzymatic extraction (AEE) was studied and the efficiency of the process evaluated by Dickey, Kurantz, and Parris [33]. A recent AEE study [34] reports 90 wt% oil recovery from wet-milled corn germ, at a 24 g scale.

3.6 Ethanol as Oxygenated and Renewable Fuel

Oxygenated fuel is conventional gasoline that has been blended with an oxygenated hydrocarbon to achieve a certain desired concentration level of oxygen in the blended fuel. Oxygenated fuel is required by the Clean Air Act Amendments of 1990 for areas that do not meet federal air quality standards, especially those for carbon monoxide. The oxygen present in the blended fuel helps the engine to burn the fuel more completely, thus emitting less carbon monoxide. Extra oxygen already present in situ in the oxygenated fuel formulation helps efficient conversion into carbon dioxide rather than carbon monoxide. Gasoline blends of at least 85% ethanol are considered alternative fuels under the Energy Policy Act of 1992 (EPAct). E85 is used in flexible fuel vehicles (FFVs) that are currently offered by most major automobile manufacturers. FFVs can run on 100% gasoline, E85, or any combination of the two and qualify as alternative fuel vehicles under EPAct regulations.

Reformulated gasoline is a formulation of gasoline that has lower controlled amounts of certain chemical compounds that are known to contribute to the formation of ozone (O_3) and toxic air pollutants. It is less evaporative than conventional gasoline during the summer months, thus reducing evaporative fuel emission and leading to reduced volatile organic compound emission. It also contains oxygenates, which increase the combustion efficiency of the fuel and reduce carbon monoxide emission. The Clean Air Act Amendments of 1990 require RFG to contain oxygenates and have a minimum oxygen content of 2.0% oxygen by weight. RFG is required in the most severe ozone nonattainment areas of the United States. Other areas with ozone problems have voluntarily opted into the program. The U.S. EPA has implemented the RFG program in two phases: Phase I for 1995 to 1999 and Phase II having begun in 2000.

To be more specific, the Clean Air Act Amendments mandated the sale of reformulated gasoline in the nine worst ozone nonattainment areas beginning January 1, 1995. Initially, the U.S. EPA determined the nine regulated areas to be the metropolitan areas of Baltimore, Chicago, Hartford, Houston, Los Angeles, Milwaukee, New York City, Philadelphia, and San Diego. The important parameters for RFG by the Clean Air Act Amendments of 1990 were

1. At least 2% oxygen by weight
2. A maximum benzene content of 1% by volume
3. A maximum of 25% by volume of aromatic hydrocarbons

As of 2011 in the United States, RFG is required in cities with high smog levels and is optional elsewhere. RFG is currently used in 17 states and the District of Columbia. About 30% of gasoline sold in the United States as of 2011 is reformulated.

Methyl tertiary-butyl ether was one of the most commonly used oxygenated blend fuels until recent claims arose of health and environmental problems associated with MTBE used as a blending fuel. Tertiary-amyl methyl ether (TAME), ethyl tertiary-butyl ether, and ethanol have also been used in oxygenated and reformulated fuels. Responding to the rapid phase-out of MTBE in the United States, ethanol has gained the most popularity as a blending fuel, based on its clean burning, relatively low Reid vapor pressure, the renewable nature of the fuel, minimal or no health concerns, and relatively low cost.

The RFG should have no adverse effects on vehicle performance or the durability of engine and fuel system components. However, there may be a slight decrease in fuel mileage (1 to 3% or 0.2–0.5 mile/gallon) in the case with well-tuned automobiles due to the higher concentrations of oxygenates that inherently have lower heating values. However, RFG burns more completely, thereby reducing formation of engine deposits and often boosting the actual gas mileage, particularly for older engines.

The Reid vapor pressure is crucially important information for blended gasoline from practical and regulatory standpoints. Evaporated gasoline compounds combine with other pollutants on hot summer days to form ground-level ozone, commonly referred to as *smog*. Ozone pollution is of particular concern because of its harmful effects on lung tissue and breathing passages. Therefore, the government, both federal and state, imposes an upper limit as a requirement, which limits the maximum level reformulated gasoline can have as its Reid vapor pressure. By such regulations, the government not only controls the carbon monoxide emission level, but also limits the evaporative emission of the fuel. Due to this limit, certain oxygenates may not qualify as a gasoline blending fuel even if they may possess excellent combustion efficiency and high octane rating. One such example is methanol. Furthermore, the legal limits for the Reid vapor pressure depend upon many factors including current environmental conditions, geographical regions, climates, time of the year (such as summer months vs. winter months), and the like. It should also be noted that ground-level ozone is harmful to humans, whereas stratospheric ozone is essential and beneficial for global environmental safety.

The oxygenated fuel program (OFP) is a winter-time program for areas with problems of carbon monoxide air pollution. The oxygenated winter fuel program uses normal gasoline with oxygenates added. On the other hand, the reformulated gasoline program is for year-round use to help reduce ozone, CO, and air toxins. Although both programs use oxygenates to reduce CO, RFG builds on the benefit of oxygenated fuel and uses improvements in the actual formulation of gasoline to reduce pollutants including volatile organic compounds [35].

Although methyl-tertiary-butyl ether was once credited with significantly improving the nation's air quality, it has been found to be a major contributor to groundwater pollution. Publicity about the leaking of MTBE from gasoline storage tanks into aquifers as well as its adverse health effects has prompted legislators from the midwestern United States to push for a federal endorsement of corn-derived ethanol as a substitute oxygenate. Many U.S. states including California and New York mandated their own schedules of MTBE phase-outs and bans. This MTBE phase-out has served as an incentive for corn ethanol industries for marketing their products as being environmentally more acceptable than other alternatives and at the same time renewable.

The Energy Policy Act of 2005 (EPAct 2005, P.L. 110-58), established the first-ever renewable fuels standard (RFS) in federal law, requiring increasing volumes of ethanol and biodiesel to be blended with the U.S. fuel supply between 2006 and 2012. The Energy Independence and Security Act of 2007 (P.L. 110-140, H.R. 6) amended and increased the RFS, requiring 9 billion

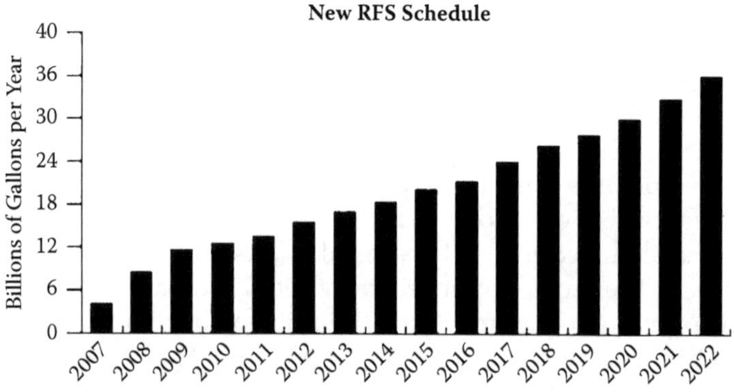

FIGURE 3.8
New renewable fuels standard (RFS) indicating the total amount of renewable fuel use for 2008 through 2022. (Courtesy of the American Coalition for Ethanol. 2010. *All About Ethanol* (October). Available at: http://www.ethanol.org/.)

gallons of renewable fuel use in 2008, stepping up to 36 billion gallons by 2022, as shown in Figure 3.8. A major portion of the increase is expected to come from cellulosic ethanol.

Considering the annual gasoline consumption in the United States to be approximately 140 billion gallons and also assuming that all gasoline sold in the United States is blended with ethanol up to 10% (i.e., E10), the total annual demand for ethanol by E10 in the United States would be about 15.5 million gallons. One can readily notice that this estimated saturation point for ethanol demand in the United States for E10 blend is not far from the 2010 total U.S. ethanol production from corn, which was 13.2 million gallons. Thus, it is evident that the RFS numbers for future years are based on (a) expanded use of nonethanol renewable fuels such as biodiesel, (b) increased availability of cellulosic ethanol, (c) expanded adoption of alternative fuel vehicles (AFVs) and flexible fuel vehicles (FFVs), and more.

3.7 Ethanol Vehicles

Fuel ethanol is most commonly used as a fuel for internal combustion, four-cycle, spark-ignition engines in transportation and agriculture. It can be used as a direct replacement fuel for gasoline, or can be blended with gasoline as an extender and octane enhancer. The research octane number (RON) of ethanol is about 113 and as such ethanol blending enhances the octane rating of the conventional fuel [37]. The octane number is a quantitative measure of the maximum compression ratio at which a particular fuel can be utilized in an engine without some of the fuel/air mixture "knocking." By defining an octane number of 100 for iso-octane and 0 for *n*-heptane, linear combinations of these two components are used to measure the octane number of a particular fuel. Therefore, a fuel with an octane number of 90 would have the same ignition characteristics at the same compression ratio as a 90/10 mixture of iso-octane and *n*-heptane. It should be noted that there are several different rating schemes for octane numbers of fuels: research octane number, motor octane number (MON), and the average of the two ((R + M)/2) that is often called the anti-knock index (AKI) or pump octane number (PON). The research octane number (RON or F1) simulates fuel performance under low severity engine operation, whereas the motor octane number (MON, or F2) simulates more severe operation that might be incurred at high speed or high load. Therefore, RON is nearly always higher in value than MON for the very same fuel. In the United States, the octane of a gasoline is usually reported as the average of RON and MON, that is, (R + M)/2.

The use of ethanol to replace gasoline requires modifications to the carburetor, fuel injection system components, and often the compression ratio. Therefore, efficient and safe conversion of existing gasoline engines is a complex matter. Engines specifically designed and manufactured to operate on ethanol fuel, or predominantly ethanol fuel, will generally be more efficient than modified gasoline engines. Ethanol concentrations of between 80 and 95% can be used as fuel, which eliminates the need for cumbersome dehydration processing steps thus simplifying the distillation step. This complication comes from the fact that the ethanol–water solution makes an azeotropic mixture at 95.4% of ethanol (by mass), a minimum boiling mixture. In many cases, the conversion of engines to operate on azeotropic ethanol may be simpler and more cost-effective than ethanol dehydration as an effort to produce 99+% purity of ethanol.

In the United States, E85 is a federally designated alternative fuel that contains 85% ethanol and 15% gasoline. As of 2003, there were hundreds of thousands of E85 vehicles on the roads in the United States. As of 2010, almost 8 million vehicles on U.S. highways were flexible fuel vehicles [38]. E85 vehicles are flexible fuel vehicles that can run on a very wide range of fuels, ranging from 100% gasoline (with 0% ethanol) to 85% ethanol (with 15% gasoline), however, they run best on E85 [36]. Nearly all the major automobile makers offer many models of passenger cars and sports utility vehicles (SUVs) with E85 engines.

In the United States, the National Ethanol Vehicle Coalition (NEVC) is actively promoting expanded use of 85% ethanol motor fuel based on its clean burning as well as renewability of the fuel. E85 fuel can achieve a very high octane rating of 105. As an extra incentive plan for E85 users, the U.S. federal government provides federal income tax credits for the use of E85 as a form of alternative transportation fuel. The E85 vehicles undoubtedly help alleviate the petroleum dependence of the world by using renewable alternative fuel source.

In unmodified engines, ethanol can replace up to 20% of the gasoline, that is, E20. In the United States, up to 10% blend of ethanol, E10, is quite popularly used. Blending ethanol with gasoline extends the gasoline supply, and improves the quality of gasoline by increasing its octane value as well as adding clean burning properties of oxygenates. There are advantages to using gasoline/ethanol blends rather than pure (or very high concentration) ethanol. Blends do not require engine modification. Therefore, ethanol can be integrated rapidly with the existing infrastructure including gasoline supply and distribution systems.

Even though the use of ethanol in specially designed two-cycle engines has been demonstrated on a number of occasions, it is not yet commercialized. One of the major issues has been the fact that ethanol does not mix well with lubricating oil typically used for such engines. Therefore, development of lubricating oils that are not affected by ethanol is an important step for this application.

Similarly, the use of ethanol in diesel-fueled engines is quite feasible, but is not practiced much, due to a number of technical difficulties. These limitations are based on ethanol's inability to ignite in compression ignition engines as well as poor miscibility with diesel. However, ethanol can be used in supercharged diesel engines for up to about 25% of the total fuel, preferably the rest being diesel. This can be achieved by delivering ethanol from a separate fuel tank and injecting it into the diesel engine through a supercharger air stream. This mode of fuel delivery system may be called a "dual fuel system" in comparison to blended fuel that is delivered as a preblended fuel from a single fuel tank. Ethanol can also replace aviation fuel in aircraft engines, even though this potential is not commercially exploited.

As a recent effort, a dual-fuel internal combustion engine (ICE) technology has been developed and demonstrated, in which ethanol is used as a cofuel with acetylene (C_2H_2) that is the principal fuel in this specific application. The dual-fuel system has been favorably demonstrated on modified gasoline and diesel engines originally designed for cars, trucks, forklifts, tractors, and power generators. Up to 25% of ethanol in acetylene-based dual-fuel systems has been successfully tested. The role of ethanol was found very effective in eliminating knocking/pinging and lowering the combustion temperatures thus reducing NO_x emissions from combustion [39, 40].

3.8 Other Uses of Ethanol

In the presence of an acid catalyst (typically, sulfuric acid) ethanol reacts with carboxylic acids to produce ethyl esters. The two largest-volume ethyl esters are ethyl acrylate (from ethanol and acrylic acid) and ethyl acetate (from ethanol and acetic acid).

Ethyl acetate is a common solvent used in paints, coatings, and in the pharmaceutical industry. The most familiar application of ethyl acetate in the household is as a solvent for nail polish. A typical reaction that synthesizes ethyl acetate is based on esterification:

$$C_2H_5OH + CH_3COOH = C_2H_5OOCCH_3 + H_2O$$

This chemical reaction follows very closely a second-order reaction kinetics, an often-used example problem for second-order elementary reactions in chemical reaction engineering textbooks.

Recently, Kvaerner Process Technology developed a process that produces ethyl acetate directly from ethanol without acetic acid or other cofeeds. Considering that both acetic acid and formaldehyde can also be produced from ethanol, this ethanol-to-ethyl acetate process idea is quite innovative and significant. The Kvaerner process allows the use of fermentation ethanol, produced from biorenewable feedstock, as a sustainable single-source feed, which is remarkable. Furthermore, the process elegantly combines both dehydrogenation and selective hydrogenation in its process scheme, thus producing hydrogen as a process by-product which makes the process economics even better.

Ethyl acrylate, which is synthesized by reacting ethanol and acrylic acid, is a monomer used to prepare acrylate polymers for use in coatings and adhesives. Ethanol is a reactant for ethyl-t-butyl ether, as is the case for methanol to methyl-t-butyl ether. ETBE is produced by reaction between isobutylene and ethanol as

$$C_2H_5OH + CH_3C(CH_3){=}CH_2 \rightleftharpoons C(CH_3)_3\,OC_2H_5$$

Vinegar is a dilute aqueous solution of acetic acid prepared by the action of *Acetobacter* bacteria on ethanol solutions. Ethanol is used to manufacture ethylamines by reacting ethanol and ammonia over a silica- or alumina-supported nickel catalyst at 150–220°C. First, ethylamine with a single amino group in the molecule is formed and further reactions create diethylamine and triethylamine. The ethylamines are used in the synthesis of pharmaceuticals, agricultural chemicals, and surfactants.

Ethanol can also be used, instead of methanol, for transesterification of triglycerides in biodiesel production using vegetable oils or algae oils, as discussed in Chapter 2. In the United States, methanol is currently more popularly used for this purpose, mainly due to its more favorable process economics.

In addition, ethanol can be used as feedstock to synthesize petrochemicals that are also derived from petroleum sources. Such chemicals include ethylene and butadiene, but are not limited to these. This option may become viable for regions and countries where the petrochemical infrastructure is weak but agricultural produce is vastly abundant. This is particularly true for the times when petroleum prices are very high. Ethanol can also be converted into hydrogen via reforming reaction, that is, chemical reaction with water at an elevated temperature typically with the aid of a catalyst. Even though this method of hydrogen generation may be economically less favorable than either steam reforming of methane or electrolysis, the process can be used for special applications, where specialty demands exist or other infrastructure is lacking.

More recently, supercritical water reformation of crude ethanol beer was developed for hydrogen production [41]. The process utilizes supercritical water ($T > 374$ C and $P > 218$ atm) functioning both as a highly energetic reforming agent and as a supercritical solvent medium, thus effectively eliminating the service of any noble metal catalyst or the need of pure ethanol. Furthermore, its direct noncatalytic reformation of unpurified crude ethanol beer alleviates the need for any energy-intensive predistillation or distillation of a water–ethanol solution, thereby achieving overall energy savings.

Poly(lactic acid) or polylactide (PLA) is a thermoplastic aliphatic polyester derived from cornstarch. Poly(lactic acid) is one of the leading biodegradable polymers, which is derived from renewable biosources, more specifically corn in the United States. A variety of applications utilizing poly(lactic acid) are being developed, wherever biodegradability of plastic materials is desired. PLA can be used by itself, blended with other polymeric materials, or as composites. As biodegradable polymer technology further develops, the PLA market is also expected to grow and so is the cornstarch market.

References

1. U.S. Environmental Protection Agency (EPA). 2008. *Regulatory Determinations Support Document for CCL 2: Chapter 13. MTBE*, Tech. Rep. EPA – OGWDW, June.
2. Lichts, F.O. 2011. Industry statistics: 2010 world fuel ethanol production. Renewable Fuels Association. Available at: http://www.ethanolrfa.org/pages/statistics#E.
3. U.S. Grains Council. 2011. World corn production and trade. http://www.grains.org/corn (October).
4. Energy Policy Research Inc. (EPRINC). Implementation issues for the renewable fuel standard - part I: Rising corn costs limit ethanol's growth in the gasoline pool. http://eprinc.org/pdf/EPRINC-CornLimitsEthanol.pdf.
5. Shapouri, H., Duffield J.A., and Graboski, M.S. 1995. *Estimating the Net Energy Balance of Corn Ethanol,* United States Department of Agriculture, Tech. Rep. Agricultural Economic Report Number 721, July.
6. Shapouri, H., Duffield, J.A., McAloon, A., and Wang, M. *The 2001 Net Energy Balance of Corn Ethanol.* U.S. Department of Agriculture, Tech. Rep. AER-814, July 2002.
7. Shapouri, H., Duffield J.A., and Wang, M. 2004. The energy balance of corn ethanol: An update, U.S. Department of Agriculture, October.
8. Pimentel, D. and Patzek, T.W. 2005. Ethanol production using corn, switchgrass, and wood, *Nat. Resources Res.,* 14: 65–75.
9. Office of Energy Efficiency and Renewable Energy. U.S. Department of Energy, 2005. *The Net Energy Value of Corn Ethanol,* vol. 2010.
10. I. MathPro. 2005. The net energy value of corn ethanol: Is it positive or negative? November.
11. Nebraska Corn Board. 2010. Corn ethanol plants using less energy but producing more ethanol per bushel. http://www.nebraskacorn.org/news-releases/corn-ethanol-plants-using-less-energy-but-producing-more-ethanol-per-bushel/. (Accessed December 2011).
12. Mueller, S. and Copenhaver, K. 2010. News from corn ethanol: Energy use, co-products, and land use. In *Near-Term Opportunities for Biorefineries Symposium,* October 11–12, Champaign, IL.
13. Dale, R.T. and Tyner, W.E. 2006. *Economic and Technical Analysis of Ethanol Dry Milling: Model Description.* Purdue University, College of Agriculture, Department of Agricultural Economics. West Lafayette, IN.

14. Mueller, S. 2008. *Detailed Report: 2008 National Dry Mill Corn Ethanol Industry*. University of Illinois-Chicago;. http://ethanolrfa.3cdn.net/2e04acb7ed88d08d21_99m6idfc1.pdf.
15. Energy and Environmental Analysis, Inc. 2007. *Impact of Combined Heat and Power on the Energy Use and Carbon Emissions in the Dry Mill Ethanol Process*, Report to U.S. Environmental Protection Agency, Combined Heat and Power Partnership, November.
16. Odian, G. 2004. *Principles of Polymerization*. Hoboken, NJ: Wiley-Interscience.
17. Dorland, W.A.N. 2003. *Dorland's Illustrated Medical Dictionary*, 30th edition. Philadelphia: W.B. Saunders, Elsevier Health Sciences Division.
18. Bradley, C. and Runnion, K. 1984. Understanding ethanol fuel production and use. In *Volunteers in Technical Assistance*. Anonymous.
19. State of Hawaii. 1994. *Ethanol Production in Hawaii Report*.
20. Corn Refiners Association. 2005. The corn refining process, vol. 2010.
21. ICM. ICM's dry milling ethanol production.
22. Prosonix. 2011. AP-40 *Starch Processing for Wet Milling*. http://www.pro-sonix.com/files/AP-40_Starch_-_Wet_Milling_20101210.pdf (November).
23. Knauf, M. and Krau, K. 2006. Specific yeasts developed for modern ethanol production, *Sugar Industry*, 131: 753–775.
24. ASTM International. 2011. ASTM standards D4806-11a. In *ASTM Standards: Petroleum Standards*, ASTM Technical Committees, Ed. West Conshohocken, PA: ASTM International.
25. Midwest Dry Ice Blasting. 2011. Dry ice industrial cleaning for ethanol plants. http://www.midwestdryiceblasting.com/dry-ice-industrial-cleaning-for-ethanol-plants. December.
26. Schill. S.R. Plymouth oil to extract corn oil, germ at iowa plant. *Ethanol Producer Magazine* http://www.ethanolproducer.com/articles/4365/plymouth-oil-to-extract-corn-oil-germ-at-iowa-plant.
27. Rendleman, C.M. and Shapouri, H. New technologies in ethanol production. http://www.usda.gov/oce/reports/energy/aer842_ethanol.pdf.
28. POET. By year end, POET to produce enough corn oil for 12 million gallons per year of biodiesel. *Renewable Energy World.Com* http://www.renewableenergyworld.com/rea/partner/poet-7042/news/article/2011/12/by-year-end-poet-to-produce-enough-corn-oil-for-12-million-gallons-per-year-of-biodiesel.
29. Foster, G. 2005. Corn fractionation for the ethanol industry, *Ethanol Producer Mag.*, 11: 76–78.
30. Singh, V.J. and Eckhoff, S. 1996. Effect of soak time, soak temperature, and lactic acids on germ recovery parameters. *Cereal Chem.*, 73: 716–720.
31. Johnston, D., McAloon, A.J., Moreau, R.A., Hicks, K.B., and Singh V. 2005. Composition and economic comparison of germ fractions from modified corn processing technologies, *J. Am. Oil Chem. Soc.*, 82: 603–608.
32. Dickey, L.C., Cooke, P.H., Kurantz, M.J., McAloon, A.J., Parris N., and Moreau, R.A. 2007. Using microwave heating and microscopy to study optimal corn germ yield with a bench-scale press. *J. Am. Oil Chem. Soc.*, 84: 489–495.
33. Dickey, L.C., Kurantz, M.J., and Parris, N. 2008. Oil separation from wet-milled corn germ dispersions by aqueous oil extraction and aqueous enzymatic oil extraction, *Ind. Crops Products*, 27: 303–307.

34. Moreau, R.A., Johnston, D.B., Powell M.J., and Hicks, K.B. 2004. A comparison of commercial enzymes for the aqueous enzymatic extraction of corn oil from corn germ, *J. Am. Oil Chem. Soc.*, 81: 77–84.
35. Speight, J.G. and Lee, S. 2000. *Handbook of Environmental Technologies*. New York: Taylor & Francis.
36. American Coalition for Ethanol. 2010. *All About Ethanol* (October). Available at: http://www.ethanol.org/.
37. Lee, S., Speight J.G., and Loyalka, S.K. 2007. *Handbook of Alternative Fuel Technology*. Boca Raton, FL: CRC Press.
38. U.S. Department of Energy, Energy Efficiency & Renewable Energy, Vehicle Technologies Program. 2011. Flexible fuel vehicles: Providing a renewable fuel choice.
39. Wulff, J.W., Hulett, M., and Lee, S. 2000. Internal combustion system using acetylene fuel, U.S. Patent 6,076,487, June 20.
40. Wulff, J.W., Hulett, M., and Lee, S. 2001. A dual fuel composition including acetylene for use with diesel and other internal combustion engines, U.S. Patent 6,287,351, September 11.
41. Wenzel, J.E., Picou, J., Factor, M., and Lee, S. 2007. Kinetics of supercritical water tion of ethanol to hydrogen. In *Energy Materials, 2007 Proceedings of the Materials Science and Technology Conference*. Detroit, MI, Minerals, Metals, and Materials Society, pp. 1121–1132.
42. BBI International. 2003. *The Ethanol Plant Development Handbook, (Fourth Edition)*. Salida, CO: BBI International.

4

Ethanol from Lignocellulose

4.1 Lignocellulose and Its Utilization

4.1.1 Lignocellulose

The structural materials that are produced by plants to form cell walls, leaves, stems, and stalks are composed primarily of three different types of biobased macromolecular chemicals, which are typically classified as cellulose, hemicellulose, and lignin. These biobased chemicals are collectively called lignocellulose, lignocellulosic biomass, or lignocellulosic materials. As shown in Figure 4.1 here and also in Chapter 3, a generalized plant cell wall structure is like a composite material in which rigid cellulose fibers are embedded in a cross-linked matrix of lignin and hemicellulose that binds the cellulose fibers.

Generally speaking, the dry weight of a typical cell wall consists of approximately 35–50% cellulose, 20–35% hemicellulose, and 10–25% lignin [1]. Others claim that cellulose typically accounts for 40–50% of woody biomass, whereas lignin and hemicellulose each account for about 20–30%. Although lignin comprises only 20–30% of typical lignocellulosic biomass, it provides 40–50% of the overall heating value or total available energy of the biomass, due to its higher calorific value (CV) than cellulose and hemicellulose. This explains why chemical conversion or beneficial use of lignin is very important in fuel/energy utilization of lignocellulosic resources.

Lignocellulosic biomass structures also contain a variety of plant-specific chemicals in the matrix; these include extractives (such as resins, phenolics, and other chemicals) and minerals (calcium, magnesium, potassium, and others) that will leave behind ash when the biomass is combusted. The trace minerals and major elements in lignocellulosic materials display a high degree of variability for most of the elements between different species, between different organs within a given plant, and also depending on the growing conditions including the soil characteristics [2]. In addition to their potential health and environmental effects, trace minerals can exhibit nontrivial effects on the next stage chemical treatment, including catalytic conversion of thermochemical intermediates of lignocellulose.

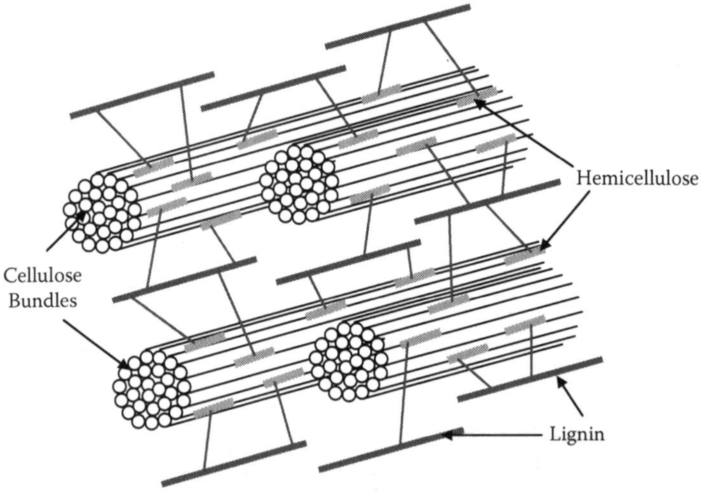

FIGURE 4.1
A generalized representation of plant cell wall structure.

Cellulose is a large polymeric molecule composed of many hundreds or thousands of monomeric sugar (glucose) molecules, and in this regard it may be considered a polysaccharide. The molecular linkages in cellulose form linear chains that are rigid, highly stable, and resistant to chemical attack. Due to its linear polymeric structure, cellulose exhibits crystalline properties [3]; for example, cellulose may be somewhat soluble in a suitable solvent. However, cellulose molecules in their crystalline form are packed so tightly that even small molecules such as water cannot easily permeate the structure. Logically, it would be even more difficult for larger enzymes to permeate or diffuse into the cellulose structure. Cellulose exists within a matrix of other polymers, mainly hemicellulose and lignin, as illustrated in Figure 4.1.

On the other hand, hemicellulose consists of short and highly branched chains of sugar molecules. It contains both five-carbon sugars (usually D-xylose and L-arabinose) and six-carbon sugars (such as D-galactose, D-glucose, and D-mannose) as well as uronic acid. For example, galactan, found in hemicellulose, is a polymer of the sugar galactose, whose solubility in water is 68.3 g per 100 grams of water at room temperature. Uronic acid is a sugar acid with both a carbonyl and a carboxylic function. Hemicellulose is amorphous due to its highly branched macromolecular structure [3] and is relatively easy to hydrolyze to its constituent simple sugars, both five-carbon and six-carbon sugars. When hydrolyzed, the hemicellulose from hardwoods releases sugary products high in xylose (a five-carbon sugar), whereas the hemicellulose contained in softwoods typically yields more six-carbon sugars. Even though both five-carbon and six-carbon sugars, illustrated in Figure 4.2, are simple fermentable sugars, there is a discerning difference

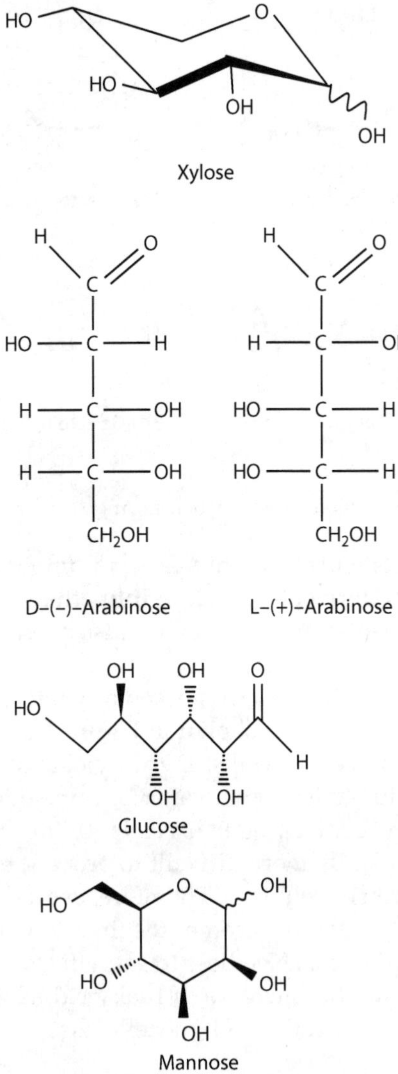

FIGURE 4.2
Molecular structures of five-carbon and six-carbon sugars. (Continued)

in their fermentation chemistry and process characteristics with regard to specific yeasts and enzymes involved.

Humans have had far more extensive and successful experience in the fermentation of six-carbon sugars (hexoses) than five-carbon sugars (pentoses or xyloses), as well evidenced by a long history of manufacturing alcoholic beverages throughout the world. This statement is still valid for fuel ethanol fermentation as well. Many years ago, it was believed that xylose could not

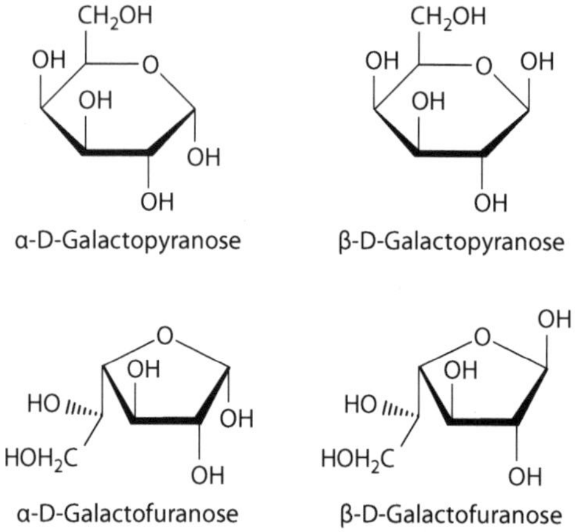

FIGURE 4.2 (CONTINUED)
Molecular structures of five-carbon and six-carbon sugars.

be fermented by yeasts, but in recent years, a number of yeasts have been found to be capable of fermenting xylose into ethanol [4, 5]. Genetic engineering of xylose fermentation in yeasts has also been carried out with successful outcomes [6].

Lignin is a complex and highly crosslinked aromatic polymer that is covalently linked to hemicellulose, as shown in Figure 4.1. Lignin contributes to the stabilization of mature cell walls. Lignin yields more energy than cellulose when burned due to its higher calorific value. Lignin is a macromolecule whose typical molecular weight exceeds 10,000. Due to its crosslinked structure, lignin is generally more difficult to process, extract, hydrolyze, or react than cellulose or hemicellulose. Therefore, degradation or biodegradation of the crosslinked structure becomes the first step for biofuel production from the cellulosic feedstocks. Needless to say, efficient conversion of lignin would result in a substantial increase in fuel yield as well as an enhanced economic outlook with utilization of lignocelluloses.

4.1.2 Cellulose Degradation, Conversion, and Utilization

Developing countries with characteristically weak economies and insufficient industrial infrastructures have been more seriously hurt by the energy crises. To fully exploit the potential of available fossil-based fuels and other alternative renewable resources, significantly large capital funds, which developing countries may be lacking in, are required. The trend has thus been shifted toward the small-scale or localized utilization of energy resources. One potentially promising area where developing countries can achieve relatively quick success is the supplementing of their fossil fuel supplies with alternative renewable fuels derived from food and agricultural crops such as sugarcane, cassava, maize, and sorghum. This option is also viable for developed economies.

Focused primarily upon petroleum as a primary source of transportation fuels, ethanol has garnered a great deal of attention as a liquid fuel source alternative to gasoline or as a gasoline blend to reduce the consumption of conventional gasoline. The ethanol alternative fuel program has been most seriously pursued by Brazil and the United States. In Brazil, all cars are run on either gasohol (a 22 vol% mixture of ethanol with gasoline, or E22, mandated in 1993; a 25% blend, or E25, mandated since 2007) or pure ethanol (E100). In Brazil, the National Program of Alcohol, PROALCOOL, started in November 1975, was created in response to the first oil crisis of 1973. This program effectively changed the consumption profile of transportation fuels in the country. In 1998, these ethanol-powered cars consumed about 2 billion gallons of ethanol per year and about 1.4 billion gallons of ethanol were additionally used for producing gasohol (E22, i.e., 22 vol% ethanol and 78 vol% gasoline) for other cars [7]. In March 2010, a milestone of 10 million flex-fuel ethanol-powered vehicles produced in Brazil was achieved. The Brazilian program has successfully demonstrated large-scale production of ethanol from sugarcane and the use of ethanol as a sustainable motor fuel. In 2010, Brazil produced about 6.92 billion gallons [8] of ethanol, whereas the United States produced just over 13.2 billion gallons; 2010 U.S. production was more than two times greater than that of 2007 (6.5 billion gallons) and about eight times greater than that of 2000 (1.62 billion gallons). As shown, ethanol has been the fastest growing chemical in the United States for the past decade. These two countries alone were responsible for about 90% of the world's industrial ethanol production in 2010.

Regarding the atmospheric concentrations of greenhouse gases (GHGs), the National Research Council (NRC), responding to a request from Congress and with funding from the U.S. Department of Energy, emphasizes the need for substantially more research and development on renewable energy sources, improved methods of utilizing fossil fuels, energy conservation, and energy-efficient technologies. The Energy Policy Act (EPAct) of 1992 was passed by the U.S. Congress to reduce the nation's dependence on imported petroleum by requiring certain fleets to acquire alternative fuel vehicles, which are capable of operating on nonpetroleum fuels. Alternative fuels for vehicular purposes, as defined by the Energy Policy Act, include ethanol, natural gas, propane, hydrogen, biodiesel, electricity, methanol, and p-series fuels. *P-Series fuels* are a family of renewable fuels that can substitute for gasoline. The Energy Policy Act of 2005 changed U.S. energy policy by providing tax incentives and loan guarantees for energy production of various types, which included tax reductions for alternative motor vehicles and fuels including bioethanol [9].

The United States does not suffer from a lack of energy resources (it has plenty of coal and oil shale reserves), but it is in need of conventional liquid transportation fuels. The market for transportation fuels has been dominated by petroleum-based fuels and that trend is expected to continue for a while. A great many researchers of the world have worked on the biological production of liquid fuels from biomass and coal [10]. They have found micro-organisms that can produce ethanol from biomass, convert natural gas into ethanol, and convert syngas derived from coal gasification into liquid fuels. These micro-organisms are found to be energy efficient and promising for industrial production. The microbial process works at ordinary temperature and pressure and offers significant advantages over chemical processes, such as direct coal liquefaction and Fischer–Tropsch synthesis, which operate under severe conditions to produce liquid fuels from coal.

Researchers have focused on using lignin as a renewable source to derive traditional liquid fuel. Lignins are produced in large quantities in the United States as by-products of the paper and pulp industry. As a consequence, the prices of some lignin products, such as lignosulfonates or sulfonated lignins, are relatively low. Lignosulfonates are used mainly as plasticizers in making concrete and also used in the production of plasterboard. Global production of lignin for various industrial applications is estimated to be quite high, even though reliable statistical data are unavailable. In China, the national lignin production has grown from 32 million metric tons in 2006 to 45 million metric tons in 2010, at a relatively fast growth rate.

4.2 Lignocellulose Conversion

4.2.1 Ethanol

4.2.1.1 Ethanol as Chemical and Fuel

Ethanol, C_2H_5OH, is one of the most significant oxygenated organic chemicals because its unique combination of physical and chemical properties make it suitable as a solvent, a fuel, a germicide, a beverage, and an antifreeze; its versatility as an intermediate to other chemicals and petrochemicals also contributes to its significance. Ethanol is one of the largest bulk-volume chemicals used in industrial and consumer products. The main uses for ethanol are as an intermediate in the production of other chemicals and as a solvent. As a solvent, ethanol is second only to water. Ethanol is a key raw material in the manufacture of plastics, lacquers, polishes, plasticizers, perfume, and cosmetics. The physical and chemical properties of ethanol are primarily dependent upon the hydroxyl group which imparts the polarity to the molecule and also gives rise to intermolecular hydrogen bonding. In the liquid state, hydrogen bonds are formed by the attraction of the hydroxyl hydrogen of one molecule and the hydroxyl oxygen of another molecule. This makes liquid alcohol behave as though it were largely dimerized. Its association is confined to the liquid state, whereas it is monomeric in the vapor state.

Another important property of ethanol in its fuel application is that the ethanol–water binary system forms an azeotrope at a binary concentration of 95.63 wt% ethanol and 4.37 wt% water and this azeotropic mixture boils at 78.15°C, which is lower than ethanol's normal boiling point (NBP) of 78.4°C. Therefore, straight distillation cannot boil off ethanol at a concentration higher than this azeotropic concentration. Most industrial grade ethanol has 95 wt% ethanol and 5 wt% water (190 proof). Therefore, fuel-grade ethanol is commonly produced by a combinatory process between distillation and zeolite-based absorption/adsorption in order to overcome the azeotropic concentration barrier encountered in the distillation separation process.

4.2.1.2 Manufacture of Industrial Alcohol

Industrial alcohol can be produced (1) synthetically from ethylene, (2) as a by-product of certain industrial operations, or (3) by the fermentation of sugars, starch, or cellulose. There are two principal processes for the synthesis of alcohol from ethylene. The original method (first carried out in the 1930s by Union Carbide) was the indirect hydration process, alternately referred to as the strong sulfuric acid–ethylene process, the ethyl sulfate process, the esterification hydrolysis process, or the sulfation hydrolysis process. The other synthetic process, designed to eliminate the use of sulfuric acid, is the direct hydration process, where ethanol is manufactured by directly reacting ethylene with steam. The hydration reaction is exothermic and reversible; that is, the maximum conversion is limited by chemical equilibrium.

$$CH_2 = CH_{2(g)} + H_2O_{(g)} \rightleftharpoons CH_3CH_2OH_{(g)} \qquad (-\ H^0_{298}) = 45\ kJ\ /\ mol$$

Only about 5% of the reactant ethylene is converted into ethanol per each pass through the reactor. By selectively removing the ethanol from the equilibrium product mixture and recycling the unreacted ethylene, it is possible to achieve an overall 95% conversion. Typical reaction conditions are: 300°C, 6–7 MPa, and employing phosphoric (V) acid catalyst adsorbed onto a porous support of silica gel or diatomaceous earth material. This catalytic process was first utilized on a large scale by Shell Oil Company in 1947.

In addition to the direct hydration process, the sulfuric acid process, and fermentation routes to manufacture ethanol, several other processes have also been suggested [11–14]. However, none of these has been successfully implemented on a commercial scale.

4.2.1.3 Fermentation Ethanol

Fermentation, one of the oldest chemical processes known to humans and most widely practiced by them, is used to produce a variety of useful products and chemicals. In recent years, however, many of the products that can be made by fermentation are also synthesized from petroleum feedstock, often at lower cost or more selectively. It is also true that modern efforts of exploiting renewable biological resources rather than nonrenewable petroleum resources as well as focusing on green technologies, thereby alleviating the process involvement of harmful chemicals, are strong drivers for biological treatment processes such as fermentation. The future of the fermentation industry, therefore, depends on its ability to utilize the high efficiency and specificity of enzymatic catalysis to synthesize complex products and also on its ability to overcome variations in the quality and availability of the raw materials.

Ethanol can be quite easily derived by fermentation processes from any material that contains sugar(s) or sugar precursors. The raw materials used in the manufacture of ethanol via fermentation are classified as sugars, starches, and cellulosic materials [15, 16]. Sugars can be directly converted to ethanol by simple chemistry, as fully discussed in Chapter 3. Starches must first be hydrolyzed to fermentable sugars by the action of enzymes. Likewise, cellulose must first be converted to sugars, generally by the action of mineral acids (i.e., inorganic acids such as the common acids sulfuric acid, hydrochloric acid, and nitric acid). Once the simple sugars are formed, enzymes from yeasts can readily ferment them to ethanol.

4.2.1.4 Fermentation of Sugars

A widely used form of sugar for ethanol fermentation is blackstrap molasses, which contains about 30–40 wt% sucrose, 15–20 wt% invert sugars such as glucose and fructose, and 28–35 wt% of nonsugar solids. The direct fermentation of sugarcane juice, sugarbeet juice, beet molasses, fresh and dried fruits, sorghum, whey, and skimmed milk have been considered, but none of these could compete economically with molasses. From the viewpoint of industrial ethanol production, sucrose-based substances such as sugarcane and sugarbeet juices present many advantages, including their relative abundance and renewable nature. Molasses, the noncrystallizable residue that remains after sucrose purification, has additional advantages: it is a relatively inexpensive raw material, readily available, and already used for industrial ethanol production. Molasses is used in dark brewed alcoholic beverages such as dark ales and also for rum. Bioethanol production in Brazil uses sugarcane as feedstock and employs first-generation technologies based on the use of the sucrose content of sugarcane. The enhancement potential for sugarcane ethanol production in Brazil was discussed by Goldemberg and Guardabassi [17] in the two principal areas of productivity increase and area expansion.

Park and Baratti [18] studied the batch fermentation kinetics of sugarbeet molasses by *zymomonos mobilis*, a rod-shaped gram-negative bacterium that can be found in sugar-rich plant saps. *Z. mobilis* degrades sugars to pyruvate using the Entner–Doudoroff pathway [19]. The pyruvate is then fermented to produce ethanol and carbon dioxide as the only products. This bacterium has several interesting and advantageous properties that make it competitive with the yeasts and, in some aspects, superior to yeasts; important examples include higher ethanol yields, higher sugar uptake, higher ethanol tolerance and specific productivity, and lower biomass production.

When cultivated on molasses, however, *Z. mobilis* generally shows poor growth and low ethanol production as compared to cultivation in glucose media [18]. The low ethanol yield is explained by the formation of by-products such as levan and sorbitol. Other components of molasses such as organic salts, nitrates, or the phenolic compounds could also be inhibitory for growth [20]. As such, its acceptable and utilizable substrate range is restricted to simple sugars such as glucose, fructose, and sucrose. Park and Baratti [18] found that in spite of good growth and prevention of levan formation, the ethanol yield and concentration were not sufficient for the development of an industrial process.

In a study by Kalnenieks et al., potassium cyanide (KCN) at submillimolar concentrations (20–500 μM) inhibited the high respiration rates of aerobic cultures of *Z. mobilis* but, remarkably, stimulated culture growth [21]. Effects of temperature and sugar concentration on ethanol production by *Z. mobilis* have been studied by scientists. Cazetta et al. [22] investigated the effects of temperature and molasses concentration on ethanol production. They used factorial design of experiments (DOE) in order to study varied conditions concurrently; the different conditions investigated included varying combinations of temperature, molasses concentration, and culture times. They concluded that the optimal conditions found for ethanol production were 200 g/L of molasses at 30°C for 48 hours and this produced 55.8 g ethanol/L.

Yeasts of the "saccharomyces genus" are mainly used in industrial processes for ethanol fermentation. One well-known example is *Saccharomyces cerevisiae*, which is most widely used in brewing beer and wine. However, *S. cerevisiae* cannot ferment D-xylose, the second most abundant sugar form of the sugars obtained from cellulosic materials. One micro-organism that is naturally capable of fermenting D-xylose to ethanol is the yeast *Pichia stipitis*, however, this yeast is not as ethanol- and inhibitor-tolerant as traditional ethanol-producing yeast, that is, *Saccharomyces cerevisiae*. Therefore, its industrial application is impractical, unless significant advances are made. There have been efforts that attempt to generate *S. cerevisiae* strains that are able to ferment D-xylose by means of genetic engineering [23]. Scientists have been working actively to ferment xylose with high productivity and yield by developing variants of *Z. mobilis* that are capable of using C_5-sugars (pentoses or xyloses) as a carbon source [24]. Advances with promising results are being reported in the literature.

As a significant advance in metabolistic changes brought about by genetic engineering, Tao [25] altered an *Escherichia coli* B strain, which is an organic acid producer, to *E. coli* strain KO11, which is an ethanol producer. The altered KO11 strain yielded 0.50 g ethanol/g xylose using 10% xylose solution at 35°C and pH of 6.5. This result provides an example of how the output of a microbe can be altered.

Utilizing a combination of metabolic engineering and systems biology techniques, two broad methods for developing more capable and more tolerant microbes and microbial communities are the recombinant industrial and native approaches [26]. The two methods differ as follows:

1. Recombinant industrial host approach: Insert key novel genes into known robust industrial hosts with established recombinant tools.
2. Native host approach: Manipulate new microbes with some complex desirable capabilities to develop traits needed for a robust industrial organism and to eliminate unneeded pathways.

The research on yeast fermentation of xylose to ethanol has been very actively studied; particular emphasis has been placed on genetically engineered *Saccharomyces cerevisiae. S. cerevisiae* is a safe micro-organism that plays a traditional and major role in modern industrial bioethanol production [27]. *Saccharomyces cerevisiae* has several advantages including its high ethanol productivity as well as its high ethanol and inhibitor tolerance. Unfortunately, this yeast does not have the capability of fermenting xylose. A number of different strategies based on genetic engineering and advanced microbiology have been applied to engineer yeasts to become capable of efficiently producing ethanol from xylose. These novel strategies included: (a) the introduction of initial xylose metabolism and xylose transport, (b) changing the intracellular redox balance, and (c) overexpression of xylulokinase and pentose phosphate pathways [27]. One of the pioneering studies involves the development of genetically engineered Saccharomyces yeasts that can co-ferment both glucose and xylose to ethanol by Sedlak et al. [28]. Even though their recombinant yeast *Saccharomyces cerevisiae* with xylose metabolism added was found to be the most effective yeast, they still utilized glucose more efficiently than xylose.

According to their experimental results, following rapid consumption of glucose in less than 10 hours, xylose was metabolized more slowly and less completely. In fact, xylose was not totally consumed even after 30 hours. Ideally, xylose should be consumed simultaneously [26] with glucose at a similar efficiency and speed; however, the newly added capability of co-fermentation of both glucose and xylose was a ground-breaking discovery. Furthermore, they found that although ethanol was the most abundant product from glucose and xylose metabolism, small amounts of the metabolic by-products of glycerol and xylitol also were obtained [28]. The above two issues, viz. higher efficiency for xylose fermentation and optimization and by-product control, are the subjects of intense research investigation.

4.2.2 Sources for Fermentable Sugars

4.2.2.1 Starches

The grains generally provide cheaper ethanol feedstock in most regions of the world and industrial conversion may be kept relatively inexpensive because they can be stored more easily than most sugar crops, which often must be reduced to a form of syrup prior to storage. Furthermore, the grain milling ethanol process produces a by-product that can be used for protein meal in animal feed [29]. Fermentation of starch from grains is inherently more complex, involving more steps than sugars because the starch must first be converted to sugar and then to ethanol. A simplified equation for the conversion of starch to ethanol can be written as

$$C_6H_{10}O_5 + H_2O \xrightarrow[\text{Fungal amylase}]{\text{Enzyme}} C_6H_{12}O_6 \xrightarrow{\text{Yeast}} 2C_2H_5OH + 2CO_2$$

As shown in Figure 4.3, in making grain alcohol, the distiller produces a sugar solution from feedstock, ferments the sugar to ethanol, and then separates the ethanol from water through distillation.

Among the disadvantages in the use of grain are its fluctuations in price. Critics of corn ethanol have made remarks in relation to "fuel versus food" and stated that the recent food price increase has something to do with corn ethanol manufacture, whereas others strongly oppose this view with statistical data and logical reasons [31]. Ethanol in the gasoline boosts the fuel's octane rating and also helps cleaner burning. In the United States, ethanol is currently the most popular oxygenated fuel additive as discussed in Chapter 3.

4.2.2.2 Cellulosic Materials

Cellulose from wood, agricultural residue, and wastes from pulp and paper mills must first be converted to sugar before it can be fermented. Enormous amounts of carbohydrate-containing cellulosic waste are generated every year throughout the world. Cellulosic ethanol is claimed to reduce greenhouse gas emissions by more than 90% over conventional petroleum-based fuels [32]. In addition, cellulosic ethanol is free from the criticism of "food versus fuel" because it is not derived from food crops. Based on these reasons, lignocellulosic ethanol is classified as a second-generation biofuel. New ways of reducing the cost of cellulosic ethanol production include the development of effective pretreatment methods, replacement of acidic hydrolysis with efficient enzymatic hydrolysis, commercialization of robust enzymes,

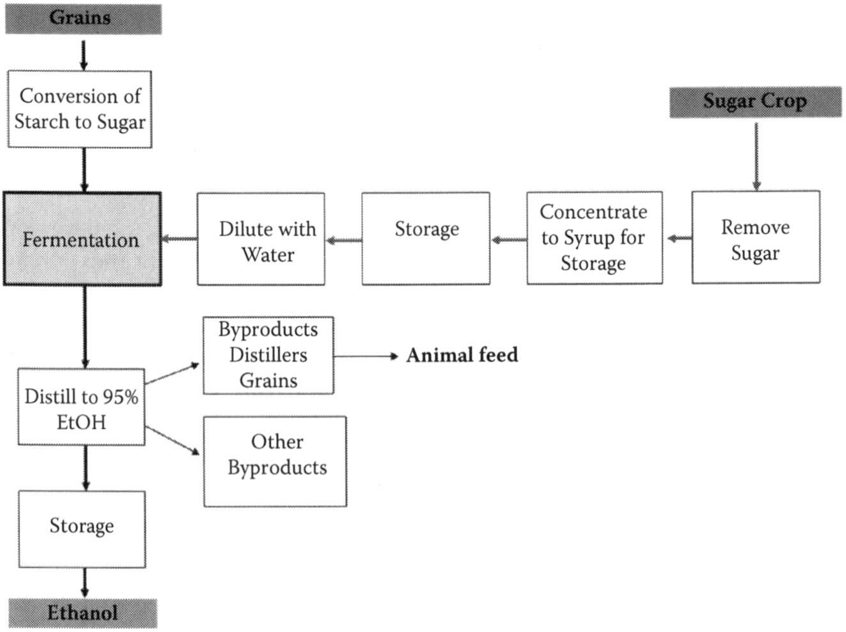

FIGURE 4.3
Synthesis of ethanol from grains and sugar crops. (Courtesy of the U.S. Congress Office of Technology Assessment, 1980. *Energy from Biological Process*, U.S. Government Printing Office, Washington, D.C., Tech. Rep. 2, pp. 142–177.)

and fine-tuning of enzymatic hydrolysis and fermentation times, in addition to the fermentation selectivity and effectiveness for both C_6 and C_5 sugars.

4.3 Historical Perspective of Alcohol Fermentation Technology

Although less heralded, cellulosic ethanol has a fairly rich history. One of the first recorded attempts at commercializing a cellulosic ethanol process was made in Germany as early as in 1898. The process was based on the use of dilute acid to hydrolyze the cellulose to glucose and the subsequent fermentation of glucose to ethanol. The reported productivity was 7.6 liters of ethanol per 100 kg of wood waste, equivalent to 18 gallons per short ton. As an early process, the conversion of wood waste into ethanol was quite remarkable; the process was further enhanced in Germany to yield about 50 gallons of ethanol per short ton of biomass. In the United States this process was further enhanced during World War I by adopting a single-stage dilute sulfuric acid hydrolysis process, by which the overall ethanol yield per input biomass was about 50% lower than the original German version, but the throughput of the process was much higher. This American process was short-lived, due to a significant decrease in lumber production in the post-war era. However, this process was brought to commercial operation again during World War II for production of butadiene by ethanol conversion to ultimately produce synthetic rubber. Even though the process achieved an ethanol yield of 50 gallons/dry ton of wood cellulose, this level of productivity was far from profitable and the process was halted after the war. Even though commercial production had been stopped, active research on cellulosic ethanol continued throughout the world, intensifying even more as a result of several rounds of petroleum crises, booming bioethanol consumption, and rapid advances in biotechnology.

In 1978, Gulf Oil researchers [33] designed a commercial-scale plant producing 95×10^6 liters per year of ethanol by simultaneous enzymatic hydrolysis of cellulose and fermentation of resulting glucose as it is formed, thereby overcoming the problem of product inhibition. The process consisted of a unique pretreatment which involved the grinding and heating of the feedstock followed by hydrolysis with a mutant bacterium, also specially developed for this purpose. Mutated strains of the common soil mold *trichoderma viride* were able to process 15 times more glucose than natural strains. Simultaneous hydrolysis and fermentation reduced the time requirement for the separate hydrolysis step, thus reducing the cost and increasing the yield. Also, the process did not use acids which would increase the equipment costs. The sugar yields from the cellulose were about 80% of what was theoretically achievable, but the small amount of hemicellulose in the sawdust was not converted. This fact demonstrated a need for an effective pretreatment to cause hemicellulose separation.

As advances in enzyme technology have been realized, the acid hydrolysis process has been gradually replaced by a more efficient enzymatic hydrolysis process. In order to achieve efficient enzymatic hydrolysis, chemical or biological pretreatment of the cellulosic feedstock has become necessary to prehydrolyze hemicelluloses in order to separate them from the lignin. The researchers of the Forest Products Laboratory of the U.S. Forest Service (USFS) and the University of Wisconsin–Madison developed the sulfite pretreatment to overcome recalcitrance of lignocellulose (SPORL) for robust enzymatic saccharification of lignocellulose [34].

Cellulase is a class of enzyme that catalyze cellulolysis, which breaks cellulose chains into glucose molecules. In recent years, various enzyme companies and biotechnology industries have contributed significant technological breakthroughs in cellulosic ethanol technology through the development of highly potent cellulase enzymes as well as the mass production of these enzymes for enzymatic hydrolysis with economic advantages. Research, development, and demonstration (RD&D) efforts in cellulase enzyme by many international companies such as Novozymes, Genencor, Iogen, SunOpta, Verenium, Dyadic International, and national laboratories such as National Renewable Energy Laboratory (NREL), are quite significant.

Cellulosic ethanol garnered strong endorsements and received significant support from the U.S. President George W. Bush in his State of the Union address, delivered on January 31, 2006, that proposed to expand the use of cellulosic ethanol. In this address, President Bush outlined the Advanced Energy Initiative (AEI 2006) to help overcome America's dependence on foreign energy sources and the American Competitiveness Initiative (ACI 2006) to increase R&D investment and strengthen education. The Renewable Fuel Standard (RFS) program was originally enacted under the Energy Policy Act of 2005 (EPAct 2005) and established the first renewable fuel volume mandate in the United States. The original RFS is referred to as RFS1. RFS1 required 7.5 billion gallons of renewable fuel to be blended into gasoline by 2012. The original timeline and renewable fuel volume mandate were revised and expanded.

The Energy Independence and Security Act (EISA) of 2007 established long-term renewable-fuels production targets through the second Renewable Fuel Standard (RFS2). The RFS2 expanded upon the initial corn-ethanol production volumes and timeline of the original RFS, under which the U.S. EPA is responsible for implementing regulations to ensure that increasing volumes of biofuels for the transportation sector are produced. The U.S. EPA released its final rule for the expanded RFS2 in February 2010, through which its statutory requirements established specific annual volumes, for the total renewable fuel volume, from all renewable fuel sources [35]. As a mandate potentially affecting the long-term future of corn ethanol, the RFS2 mandates that the country as a whole is required to blend 36 billion gallons of renewable fuels into the transportation fuel sector by 2022, of which 16 billion gallons is expected to come from non-corn based ethanol. The U.S. EPA implementation of the RFS2 would position the United States for making significant improvements in the greenhouse gas footprint due to the transportation sector. In February 2010, the White House under the leadership of President Barack Obama released "Growing America's Fuel," which is a comprehensive roadmap to advanced fuels deployment [36].

In 2004, the researchers at the National Renewable Energy Laboratory, in collaboration with two major industrial enzyme manufacturers (Genencor International and Novozymes Biotech), achieved a dramatic reduction in cellulase enzyme costs, which was one of the major stumbling blocks in the commercialization of cellulosic ethanol. Cellulases belong to a group of enzymes known as glycosyl hydrolases, which cleave (hydrolyze) chemical bonds linking a carbohydrate to another molecule. The novel technology involves a cocktail of three types of cellulases: endoglucanases, exoglucanases, and beta-glucosidases. These enzymes synergistically work together to attack cellulose chains, pulling them away from the crystalline structure, breaking off cellobiose molecules (two linked glucose residues), splitting them into individual glucose molecules, and making them available for further enzymatic processing. This breakthrough work is claimed to have resulted in twenty- to thirtyfold cost reduction and earned NREL and collaborators an R&D 100 Award [37].

This is certainly a milestone accomplishment in cellulosic ethanol technology development; however, further cost reductions are required in cellulase enzyme manufacture, new routes need to be developed to enhance enzymatic efficiencies, the development of enzymes with higher heat tolerance and improved specific activities is highly desired, better matching of enzymes with plant cell-wall polymers needs to be achieved, a high-solid enzymatic hydrolysis process with enhanced efficiency needs to be developed, and more. The U.S. Department of Energy Workshop Report summarizes, in scientific detail, the identified research needs in the area (Biofuels Joint Roadmap [26]).

In recent years, major advances have also been made utilizing genetic engineering and advanced microbiology in the development of robust microbe systems that are capable of efficiently co-fermenting both C_5 and C_6 sugars and that are resistant to inhibitors and tolerant against process variability.

Another effort for production of cellulosic ethanol is via catalytic conversion of gaseous intermediates produced by thermochemical conversion of cellulosic materials without the use of enzymes. Certainly, there is a trade-off between the purely chemical route and the enzymatic route in various aspects, including conversion efficiency, product selectivity, reaction speed, capital cost, overall energy efficiency, raw material flexibility, and more. A large commercialization effort launched by Range Fuels in 2007, based on catalytic conversion of thermochemical intermediates derived from biomass, was shut down in 2011 without meeting its original goals.

4.4 Agricultural Lignocellulosic Feedstock

One reason that, until now, the world has depended so heavily on natural gas and petroleum for energy and the manufacture of most organic materials is that gases and liquids are relatively easy to handle. Solid materials such as wood, on the other hand, are difficult to collect, transport, and process into components that can make desired products for energy. As such, solid materials seriously lack in continuous processability and render logistical problems in their utilization.

Simply speaking, agricultural lignocellulose is inexpensive and renewable because it is made via photosynthesis with the aid of solar energy. In addition, the quantity of biological materials available for conversion to fuel, chemicals, and other materials is virtually unlimited. Greater biomass utilization can also help ameliorate solid waste disposal problems. In 2009, 243 million tons of municipal solid waste (MSWs) were generated in the United States, which is equivalent to about 4.3 pounds of waste per person per day. Of this waste, 28.2% was paper and paperboard, 13.7% yard clippings, 6.5% wood, and 14.1% food scraps [38]. Considering that some food scraps contain cellulosic materials, about 50% of the total municipal solid wastes is cellulosic and could be converted to useful chemicals and fuels [39].

Although lignocellulose is inexpensive, it involves transformational efforts to convert to fermentable sugars. Furthermore, as shown in Figure 4.4, lignocellulose has a complex chemical structure with three major components, each of which must be processed separately to make the best use of high efficiencies inherent in the biological process. The three major components of lignocellulose are crystalline cellulose, hemicellulose, and lignin.

A general scheme for the conversion of lignocellulose to ethanol is shown in Figure 4.5. The lignocellulose is pretreated to separate the xylose and, sometimes, the lignin from the crystalline cellulose. This step is very important, because the efficiency of pretreatment affects the efficiency of the ensuing steps. The xylose can then be fermented to ethanol, whereas the lignin can be further processed to produce other liquid fuels and valuable chemicals. Crystalline cellulose, the largest (around 50%) and most useful fraction, remains behind as a solid after the pretreatment and is sent to an enzymatic

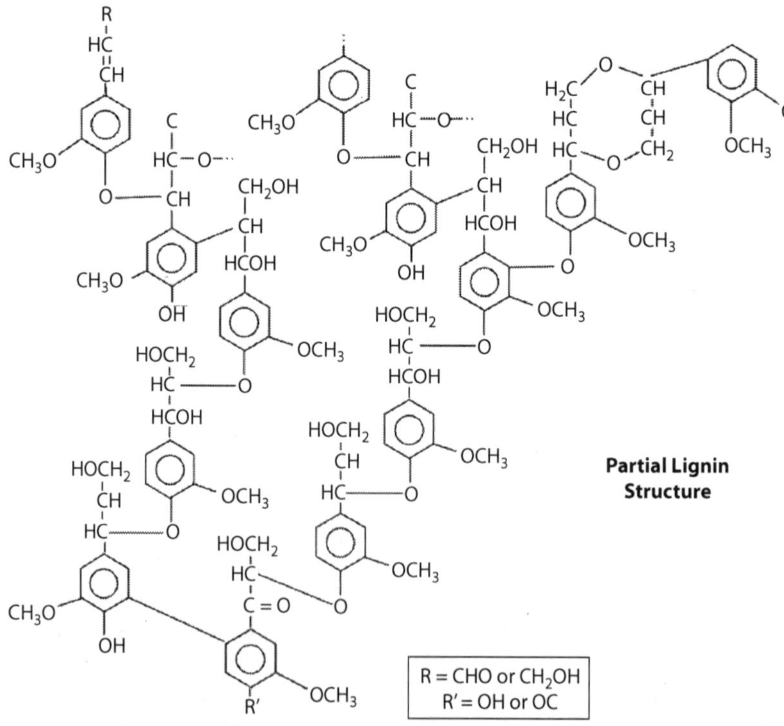

FIGURE 4.4
Major polymeric components of plant materials. (Adapted from C&EN, 1990. Major polymeric components of plant materials, *Chem. Eng. News*, September 10.)

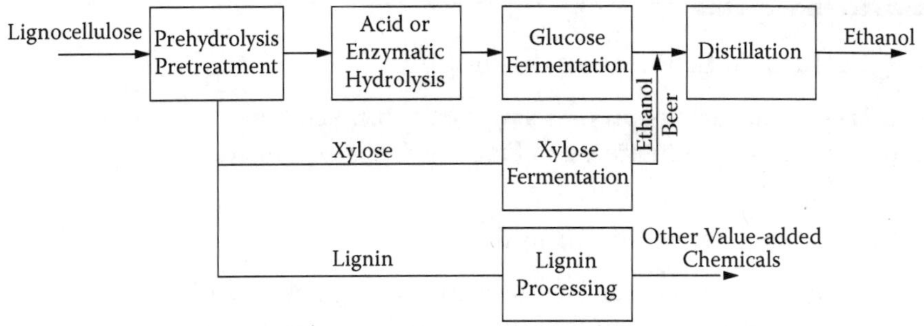

FIGURE 4.5
Conversion of lignocellulose to ethanol.

hydrolysis process that breaks the cellulose down into glucose. Enzymes, the biological catalysts, are highly specific, hence, the hydrolysis of cellulose to sugar does not further break down the sugars. Enzymatic processes are capable of achieving a yield approaching 100%. The glucose is then fermented to ethanol and combined with the ethanol from xylose fermentation. This dilute beer (i.e., dilute ethanol–water solution) is then concentrated to fuel-grade ethanol via distillation and further purification such as pressure swing adsorption (PSA).

The hemicellulose fraction, the second major component at around 25%, is primarily composed of xylan, which can be easily converted to the simple sugar xylose (or pentose). Xylose constitutes about 17% of woody angiosperms and accounts for a substantially higher percentage of herbaceous angiosperms. Therefore, xylose fermentation or conversion is essential for commercial bioconversion of lignocellulose into ethanol or other biochemicals. Xylose is more difficult than glucose to convert or ferment to ethanol, based on the current level of science and technology. From the process standpoint, it would be more beneficial to find or develop a more robust and optimal micro-organism that can ferment both glucose and xylose to ethanol in a single fermenter with high yield and selectivity. Methods have been identified using new strains of or metabolically engineered yeasts [23], bacteria, and processes combining enzymes and yeasts. Although none of these fermentation processes is yet ready for commercial use, considerable progress has been made.

Lignin, the third major component of lignocellulose (around 25%), is a large random phenolic polymer. In lignin processing, the polymer is broken down into fragments containing one or two phenolic rings. Extra oxygen and side chains are stripped from the molecules by catalytic methods and the resulting phenol groups are reacted with methanol to produce methyl aryl ethers. Methyl aryl ethers, or arylmethylethers, are high-value octane enhancers that can be blended with gasoline.

4.5 Cellulosic Ethanol Technology

In this section, various process stages of typical cellulose ethanol fermentation technology, as illustrated in Figure 4.5, are explained.

4.5.1 Acid or Chemical Hydrolysis

Acid hydrolysis of cellulosic materials has long been practiced and is relatively well understood. Among the important specific factors in chemical hydrolysis are surface-to-volume ratio, acid concentration, temperature, and time. The surface-to-volume ratio is especially important in that it also determines the magnitude of the yield of glucose. Therefore, smaller particle size results in better hydrolysis, in terms of the extent and rate of reaction [41]. With respect to the liquid-to-solids ratio, a higher ratio leads to a faster reaction. A trade-off must be made between the optimum ratio and economic feasibility because the increase in the cost of equipment parallels the increase in the ratio of liquid to solids. For chemical hydrolysis, a liquid/solids ratio of 10:1 seems to be most suitable [41].

In a typical system for chemically hydrolyzing cellulosic waste, the waste is milled to fine particle sizes. The milled material is immersed in a weak acid (0.2 to 10%), the temperature of the suspension is elevated to 180 to 230°C, and a moderate pressure is applied. Eventually, the hydrolyzable cellulose is transformed into sugar. However, this reaction has no effect on the lignin which is also present. The yield of glucose varies, depending upon the nature of the raw waste. For example, 84–86 wt% of kraft paper or 38–53 wt% of the ground refuse may be recovered as sugar. The sugar yield increases with the acid concentration as well as the elevation of temperature. A suitable concentration of acid (H_2SO_4) is about 0.5% of the charge.

A two-stage, low-temperature, and ambient-pressure acid hydrolysis process that utilizes separate unit operations to convert the hemicellulose and cellulose to fermentable sugars was developed [42] and tested by the Tennessee Valley Authority (TVA) and the U. S. Department of Energy (DOE). Laboratory and bench-scale evaluations showed more than 90% recovery and conversion efficiencies of sugar from corn stover. Sugar product concentrations of more than 10% glucose and 10% xylose were achieved. The inhibitor levels in the sugar solutions never exceeded 0.02 g/100 ml, which is far below the level shown to inhibit fermentation. An experimental pilot plant was designed and built in 1984. The acid hydrolysis pilot plant provided fermentable sugars to a 38 L/h fermentation and distillation facility built in 1980. The results of their studies are summarized as follows.

- Corn stover ground to 2.5 cm was adequate for the hydrolysis of hemicellulose.

- The time required for optimum hydrolysis in 10% acid at 100°C was 2 hours.
- Overall xylose yields of 86 and 93% were obtained in a bench-scale study at 1- and 3-hr reaction times, respectively.
- Recycled leachate, dilute acid, and prehydrolysis acid solutions were stable during storage for several days.
- Vacuum drying was adequate in the acid concentration step.
- Cellulose hydrolysis was successfully accomplished by cooking stover containing 66 to 78% acid for six hours at 100°C. Yields of 75 to 99% cellulose conversion to glucose were obtained in the laboratory studies.
- Fiberglass-reinforced plastics of vinyl ester resin were used for construction of process vessels and piping.

4.5.1.1 Process Description

The process involves two-stage sulfuric acid hydrolysis, relatively low temperature, and a cellulose prehydrolysis treatment with concentrated acid. Figure 4.6 is a schematic flow diagram of the TVA process. Corn stover is ground and mixed with the dilute sulfuric acid (about 10% by weight). The

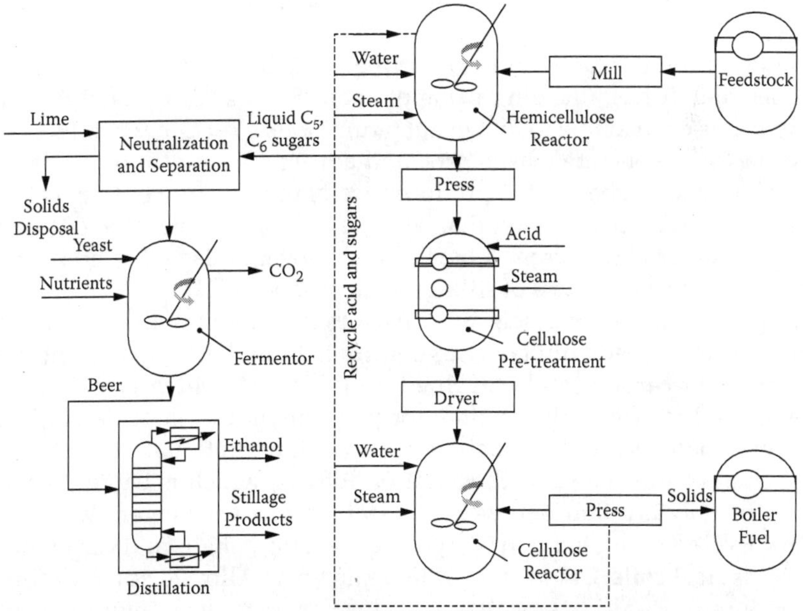

FIGURE 4.6
Low-temperature low-pressure two-stage acid hydrolysis concept for conversion of nonwoody feedstock to ethanol. (Modified from Farina, Barrier, and Forsythe, 1988. Fuel alcohol production from agricultural lignocellulosic feedstocks, *Energy Sources*, 10: 231–237.)

hemicellulose fraction of the stover is converted to pentose (xylose) sugars by heating the solution to 100°C for two hours in the first hydrolysis reactor. Raw corn stover contains, on a dry basis, about 40% cellulose, 25% hemicellulose, and 25% lignin. Sulfuric acid for the hydrolysis reaction is provided by recycling the product stream from the second hydrolysis step, which contains the sulfuric acid and hexose sugars. The pentose and hexose sugars, which are primarily xylose and glucose, respectively, are leached from the reactor with warm water. The sugar-rich leachate is then neutralized with lime (calcined limestone, CaO or calcined dolomite, $CaO \bullet MgO$), filtered to remove precipitated materials, and fermented to produce ethanol [43].

Residue stover from the first hydrolysis step (hemicellulose conversion) is dewatered and prepared for the second hydrolysis step (cellulose conversion) by soaking (prehydrolysis treatment step) in sulfuric acid (about 20–30% concentration) from one to two hours. The residue is then screened, mechanically dewatered, and vacuum dried to increase the acid concentration to 75–80% in the liquid phase before entering the cellulose reactor. The second hydrolysis reactor operates at 100°C and requires a time of four hours. The reactor product is filtered to remove solids (primarily lignin and unreacted cellulose). Because the second hydrolysis reactor product stream contains about 10% acid, it is used in the first hydrolysis step to supply the acid required for hemicellulose hydrolysis. Residue from the reactor is washed to recover the remaining sulfuric acid and the sugar not removed in the filtration step.

Lignin is the unreacted fraction of the feedstock that can be burned as a boiler fuel. It has a heating value of about 5,270 kcal/kg (or, 9,486 BTU/lb), which is comparable to that of subbituminous coal. Other products such as surfactants, concrete plasticizers, and adhesives can also be made from lignin. Stillage can be used to produce several products, including methane. Preliminary research showed that 30 liters of biogas containing 60% methane gas was produced from a liter of corn stover stillage. For each liter of ethanol produced, 10 liters of stillage were produced [42].

All process piping, vessels, and reactors in contact with corrosive sulfuric acid were made of fiberglass-reinforced vinyl ester resin. The dryer was made of carbon steel and lined with Kynar®, which is a trademark of Arkema Inc. (formerly Atofina) for poly(vinylidene fluoride), or PVDF. Conveyor belts were also made of acid-resistant material. Mild steel agitator shafts were coated with Kynar or Teflon®, which is DuPont's trademark for polytetrafluoroethylene, or PTFE. Heat exchangers were made with CPVC (chlorinated poly(vinyl chloride)) pipe shells and Carpenter 20 stainless steel coils. Carpenter 20, also known as Alloy 20, is a nickel-iron-chromium austenitic alloy that was developed for maximum corrosion resistance to acid attack, in particular sulfuric acid attack. Pumps were made with nonmetallic compound Teflon lining, or Carpenter 20 stainless steel. The two filter press units had plates made of polypropylene (PP) [42].

4.5.2 Enzymatic Hydrolysis

As a fermentable carbohydrate, cellulose differs from other carbohydrates generally used as a substrate for fermentation. Cellulose is insoluble and is polymerized as 1-4, β-glucosidic linkage. Each cellulose molecule is an unbranched polymer of 15 to 10,000 D-glucose units. Hydrolysis of crystalline cellulose is a rate-controlling step in the conversion of biomass to ethanol, because aqueous enzyme solutions have difficulty acting on insoluble, impermeable, highly structured cellulose. Therefore, making soluble enzymes act on insoluble cellulose is one of the principal challenges in process development of cellulosic ethanol.

Cellulose needs to be efficiently solubilized such that an entry can be made into cellular metabolic pathways. Solubilization is brought about by enzymatic hydrolysis catalyzed by a *cellulase* system of certain bacteria and fungi. Cellulase is a class of enzymes produced primarily by fungi, bacteria, and protozoans, that catalyze the hydrolysis of cellulose, that is, *cellulolysis*, as described below.

4.5.2.1 Enzyme System

There are several different kinds of cellulases, and they differ mechanistically and structurally. Each cellulolytic microbial group has an enzyme system unique to it. The enzyme capabilities range from those with which only soluble derivatives of cellulose can be hydrolyzed to those with which a cellulose complex can be disrupted. Although it is a usual practice to refer to a mixture of compounds that have the ability to degrade cellulose as cellulase, it is actually composed of a number of distinctive enzymes. Based on the specific type of reaction catalyzed, the cellulases may be characterized into five general groups, namely,

1. *Endocellulase* cleaves internal bonds to disrupt the crystalline structure of cellulose and expose individual cellulose polysaccharide chains.
2. *Exocellulase* detaches two or four saccharide units from the ends of the exposed chains produced by endocellulase, resulting in disaccharides or tetrasaccharides, such as cellobiose. *Cellobiose* is a disaccharide with the formula $[HOCH_2CHO(CHOH)_3]_2O$. There are two principal types of exocellulases, or cellobiohydrolases (CBH): (a) *CBH-I* works processively from the reducing end of cellulose and (b) *CBH-II* works processively from the nonreducing end of cellulose. In this description, processivity is the ability of an enzyme to continue repetitively its catalytic function without dissociating from its substrate. By an active enzyme being held onto the surface of a solid substrate, the chance for reaction is significantly enhanced.

3. *Beta-glucosidase* or cellobiase hydrolyzes the exocellulase products, disaccharides and tetrasaccharides, into individual monosaccharides.
4. *Oxidative cellulases* depolymerize and break down cellulose molecules by radical reactions, as in the case with a cellobiose dehydrogenase (acceptor), which is an enzyme that catalyzes the chemical reaction of

   $$\text{cellobiose} + \text{acceptor} \rightleftharpoons \text{cellobiono-1,5-lactone} + \text{reduced acceptor}$$

   by which cellobiose is dehydrogenated and the acceptor is reduced.

5. *Cellulose phosphorylases* depolymerize cellulose using phosphates instead of water.

In most cases, the enzyme complex breaks down cellulose to beta-glucose. This type of cellulase enzyme is produced mainly by symbiotic bacteria. *Symbiotic bacteria* are bacteria living in symbiosis (close and long-term interaction) with another organism or each other. Enzymes that hydrolyze hemicellulose are usually referred to as *hemicellulase* and are still commonly classified under cellulases. Enzymes that break down lignin are not classified as cellulase, strictly speaking. Along with diverse types of enzymes, it must be clearly pointed out that a principal challenge in hydrolytic degradation of biomass into fermentable sugars is how to make these different enzymes work together as a synergistic enzyme system. For example, cellulases and hemicellulases are secreted from a cell as free enzymes or extracellular cellulosomes (complexes of cellulolytic enzymes created by bacteria). The collective activity of these enzymes in a system is likely to be more active than, or at least quite different from, the individual activity of an isolated enzyme.

The enzymes described above can be classified into two types: progressive (also known as processive) and nonprogressive (or, nonprocessive) types. Progressive cellulase will continue to interact with a single polysaccharide strand, whereas nonprogressive cellulase will interact once, disengage, and then engage another polysaccharide strand.

Based on the enzymatic capability, cellulase is characterized into two groups, namely, C_1 enzyme or factor and C_X enzyme or factor [41]. The C_1 factor is regarded as an "affinity" or prehydrolysis factor that transforms highly ordered (crystalline) cellulose, (i.e., cotton fibers or Avicel) into linear and hydroglucose chains. The C_1 factor has little effect on soluble derivatives. Raw cotton is composed of 91% pure cellulose. As such, it serves as an essential precursor to the action of the C_X factor. The C_X (hydrolytic) factor breaks down the linear chains into soluble carbohydrates, usually cellobiose (a disaccharide) and glucose (a monosaccharide).

Microbes rich in C_1 are more useful in the production of glucose from the cellulose. Moreover, because the C_1 phase proceeds more slowly than the subsequent step, it is the rate controlling step. Among the many microbes, *Trichoderma reesei* surpasses all others in the possession of C_1 complex. *Trichoderma reesei* is an industrially important cellulolytic filamentous fungus and is capable of secreting large amounts of cellulases and hemicellulases [44]. Recent advances in cellulase enzymology, cellulose hydrolysis (cellulolysis), strain enhancement, molecular cloning, and process design and engineering are bringing *T. reesei* cellulases closer to being a commercially viable option for cellulose hydrolysis [45]. The site of action of cellulolytic enzymes is important in the design of hydrolytic systems (C_X factor). If the enzyme is within the cell mass, the material to be reacted must diffuse into the cell mass. Therefore, the enzymatic hydrolysis of cellulose usually takes place extracellularly, where the enzyme is diffused from the cell mass into the external medium.

Another important factor in the enzymatic reaction is whether the enzyme is adaptive or constitutive. A *constitutive enzyme* is present in a cell at all times. *Adaptive enzymes* are found only in the presence of a given substance, and the synthesis of the enzyme is triggered by an inducing agent. Most of the fungal cellulases are adaptive [15, 41].

Cellobiose is an inducing agent with respect to *Trichoderma reesei*. In fact, depending on the circumstances, cellobiose can be either an inhibitor or an inducing agent. It is inhibitory when its concentration exceeds 0.5 to 1.0%. Cellobiose is an intermediate product and is generally present in concentrations low enough to permit it to serve as a continuous inducer [46].

A milestone achievement (2004) accomplished by the National Renewable Energy Laboratory in collaboration with Genencor International and Novozyme Biotech is of significance in making effective cellulase enzymes at substantially reduced costs, as mentioned in an earlier section.

4.5.3 Enzymatic Processes

All enzymatic processes basically consist of four major steps that may be combined in a variety of ways: pretreatment, enzyme production, hydrolysis, and fermentation, as represented in Figure 4.7.

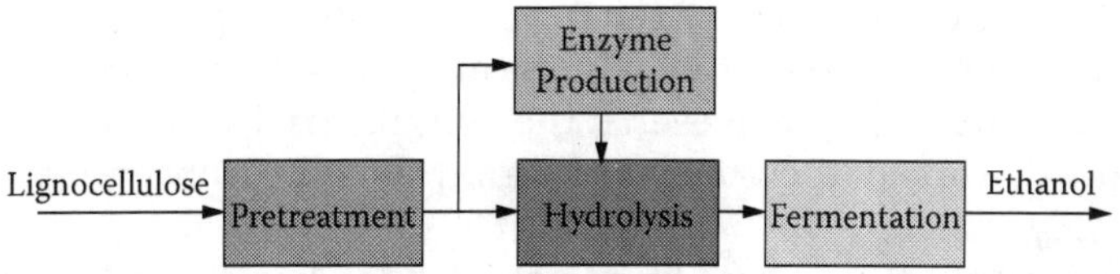

FIGURE 4.7
Fungal enzyme hydrolysis process. (Modified from Wright, 1988. Ethanol from biomass by enzymatic hydrolysis, *Chem. Eng. Prog.*, 84: 62–74.)

4.5.3.1 Pretreatment

It has long been recognized that some form of pretreatment is necessary to achieve reasonable rates and yields in the enzymatic hydrolysis of biomass. Pretreatment has generally been practiced to reduce the crystallinity of cellulose, to lessen the average degree of polymerization of the cellulose and the lignin–hemicellulose sheath that surrounds the cellulose, and to alleviate the lack of available surface area for the enzymes to attack. A typical pretreatment system consists of size reduction, pressure sealing, heating, reaction, pressure release, surface area increase, and hydrolyzate/solids separation [47].

Mechanical pretreatments such as intensive ball milling and roll milling have been investigated as a means of increasing the surface area, but they require exorbitant amounts of energy. The efficiency of the chemical process can be understood by considering the interaction between the enzymes and the substrate. The hydrolysis of cellulose into sugars and other oligomers is a solid phase reaction in which the enzymes must bind to the surface to catalyze the reaction. Cellulase enzymes are large proteins, with molecular weights ranging from 30,000 to 60,000 and are thought to be ellipsoidal with major and minor dimensions of 30 to 200 A°. The internal surface area of wood is very large, however, only about 20% of the pore volume is accessible to cellulase-sized molecules. By breaking down the tight hemicellulose–lignin matrix, hemicellulose or lignin can be separated and the accessible volume can be greatly increased. This removal of material greatly enhances the enzymatic digestibility.

The hemicellulose–lignin sheath can be disrupted by either acidic or basic catalysts. Basic catalysts simultaneously remove both lignin and hemicellulose, but suffer large consumption of the base through neutralization by ash and acid groups in the hemicellulose. In recent years, attention has been focused on the acidic catalysts. They can be mineral acids or organic acids generated in situ by autohydrolysis of hemicellulose.

Various types of pretreatments are used for biomass conversion. The pretreatments that have been studied in recent years are steam explosion autohydrolysis, wet oxidation, organosolv, and rapid steam hydrolysis (RASH). The major objective of most pretreatments is to increase the susceptibility of cellulose and lignocellulose material to acid and enzymatic hydrolysis. Enzymatic hydrolysis is a very sensitive indicator of lignin depolymerization and cellulose accessibility. Cellulase enzyme systems react very slowly with untreated material; however, if the lignin barrier around the plant cell is partially disrupted, then the rates of enzymatic hydrolysis are increased dramatically.

Most pretreatment approaches are not intended to actually hydrolyze cellulose to soluble sugars, but rather to generate a pretreated cellulosic residue that is more readily hydrolyzable by cellulase enzymes than native biomass. Dilute acid hydrolysis processes are currently being proposed for several near-term commercialization efforts until lower-cost commercial cellulase preparations become available. Such dilute acid hydrolysis processes typically result in no more than 60% yields of glucose from cellulose.

4.5.3.1.1 Autohydrolysis Steam Explosion

A typical autohydrolysis process [48] uses compressed liquid hot water at a temperature of about 200°C under a pressure that is higher than the saturation pressure, thus keeping the hot water in liquid phase, to hydrolyze hemicellulose in minutes. Hemicellulose recovery is usually high, and unlike the acid-catalyzed process, no catalyst is needed. The process is represented as shown in Figure 4.8. Very high temperature processes may lead to significant pyrolysis, which produces inhibitory compounds. The ratio of the rate of hemicellulose hydrolysis to that of sugar degradation (more pyrolytic in nature) is greater at higher temperatures. Low-temperature processes have lower xylose yields and produce more degradation products than well-controlled, high-temperature processes that use small particles.

According to a study by Dekker and Wallis [49], pretreatment of bagasse by autohydrolysis at 200°C for 4 min and explosive defibration resulted in a 90% solubilization of the hemicellulose (a heteroxylan) and in the production of a pulp that was highly susceptible to hydrolysis by cellulases from *Trichoderma reesei*. Saccharification yields were 50% after 24 hours at 50°C (pH 5.0) in enzymatic digests containing 10% (w/v) bagasse pulps and 20 filter paper cellulase units (FPU), and their saccharification yield could be increased to 80% at 24 hours by the addition of exogenous β-glucosidase from *Aspergillus niger*.

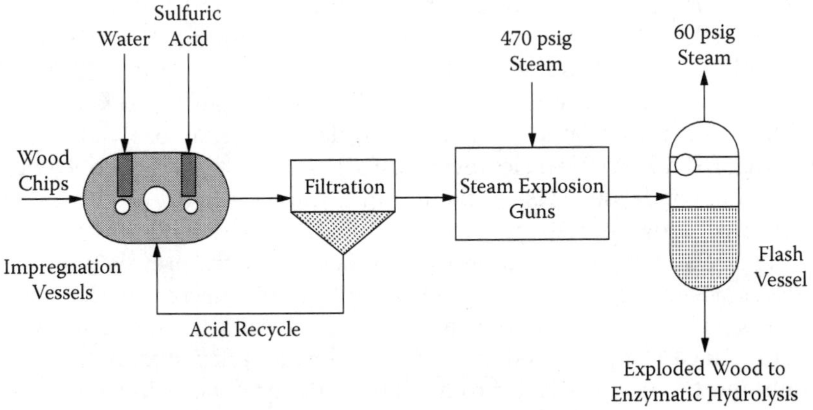

FIGURE 4.8
Steam explosion pretreatment process flow diagram. (Modified from Wright, 1988. Ethanol from biomass by enzymatic hydrolysis, *Chem. Eng. Prog.*, 84: 62–74.)

In general, xylose yields in autohydrolysis are low (30–50%). An autohydrolysis system is used as the pretreatment in separate hydrolysis and fermentation (SHF). The reaction conditions are 200°C for 10 minutes, with a xylose yield of 35%.

Steam consumption in autohydrolysis is strongly dependent upon the moisture content of the starting material. Wet feedstock requires considerably more energy because of the high heat capacity of retained water. An important advantage of autohydrolysis is that it breaks the lignin into relatively small fragments that can be easily solubilized in either base or organic solvents.

The steam explosion process [50] was first developed in 1925 for hardboard production and more recently was applied on aspen wood in the early 1980s. In a typical steam explosion process, cellulosic material is heated using high-pressure steam (20–50 atm, 210–290°C) for a short period (seconds to minutes). At the increased pretreatment pressure, water molecules diffuse into the inner microporous structure of the lignocellulose [47]. In this process, some steam condenses under high pressure, thereby wetting the material. The wetted material is then driven out of a reactor (i.e., ejected from a reactor) through a small nozzle by a pressure difference. Due to a rapid decrease in the pressure, the material is ejected through the discharge valve. The term "explosion" is used due to the process characteristics of ejection driven by a sudden large pressure drop of steam.

4.5.3.1.2 Dilute Acid Prehydrolysis

Lower temperature operation with reduced sugar degradation is achieved by adding a small amount of mineral acid to the pretreatment process. The acid increases reaction rates at a given temperature and the ratio of hydrolysis rate to the degradation rate is also increased.

A compromise between the reaction temperature and the reaction time exists for acid-catalyzed reactions. As for autohydrolysis, however, conditions explored range from several hours at 100°C to 10 seconds at 200°C with a sulfuric acid concentration of 0.5 to 4.0%. Acid catalysts have also been used in steam explosion systems with similar results. Xylose yields generally range from 70 to 95%. However, sulfuric acid processes produce lignin that is more condensed (52% of the lignin extractable in dilute NaOH) than that produced by an autohydrolysis system. Sulfur dioxide has also been investigated as a catalyst to improve the efficiency of the pretreatments. Use of excess water increases energy consumption and decreases the concentration of xylose in the hydrolyzate, thus decreasing the concentration of ethanol that can be produced in the xylose fermentation step. In a study by Ojumu and Ogunkunle [51], production of glucose was achieved in batch reactors from hydrolysis of lignocellulose under extremely low acid (ELA) concentration and high-temperature condition by pretreating the sawdust by autohydrolysis *ab initio*. The maximum glucose yield obtained was reported to be 70% for the pretreated sawdust at 210°C in the eighteenth minute of the experiment. This value is about 1.4 times the maximum glucose level obtained from the untreated sawdust under the nominally same condition [51].

The acid hydrolysis process has a long history of over 100 years. As an alternative to dilute acid hydrolysis, concentrated acid-based hydrolysis processes are also conceivable and available. However, these types of processes are generally more expensive to operate and render handling difficulties [52]. Sulfuric acid is the most common choice of catalyst; however, other mineral acids such as hydrochloric, nitric, and trifluoroacetic acids (CF_3COOH) have also been used.

4.5.3.1.3 Organosolv Pretreatment

The *organosolv process* is a pulping technique that uses an organic solvent to solubilize lignin and hemicellulose. The process was first developed as an environmentally benign alternative to kraft pulping. Its main advantages include the production of high-quality lignin for added values and easy recovery and recycling of solvents used in the process, thereby alleviating environmental stress on the water stream.

In this type of pretreatment of lignocellulose, an organic solvent (ethanol, butanol, or methanol) is added to the pretreatment reaction to dissolve and remove the lignin fraction. In the pretreatment reactor, the internal lignin and hemicellulose bonds (refer to Figure 4.1) are broken and both fractions are solubilized, whereas the cellulose remains as a solid. After leaving the reactor, the organic fraction is removed by evaporation (distillation) in the liquid phase. The lignin then precipitates and can be removed by filtration or centrifugation. Thus, this process cleanly separates the feedstock into a solid cellulose residue, a solid lignin that has undergone a few condensation reactions, and a liquid stream containing xylon, as shown in Figure 4.9.

The organosolv process is usually carried out at an elevated temperature of 140–230°C under pressure. High temperature is somewhat dictated by the desired bond cleavage reactions involving the liberation of lignin, and the high pressure is needed to keep the solvent process operation in the liquid phase. Ethanol has been regarded as a preferred solvent for organosolv due to its low price, availability, and easy solvent recovery. Butanol, also, has shown promise because of its superior capability of high lignin yield and immiscibility with water which make solvent recovery simple without energy-intensive distillation. Although butanol's effectiveness is quite appealing, its cost is considered to be somewhat prohibitive. As explained, a principal concern in these processes is the complete recovery of the solvent, which affects the overall process economics; as such, process engineering and optimization become important factors in process economics.

Results have shown that there are some reactions occurring during the organosolv process that strongly affect the enzymatic rate [53]. These reactions could be due to the physical or chemical changes in lignin or cellulose.

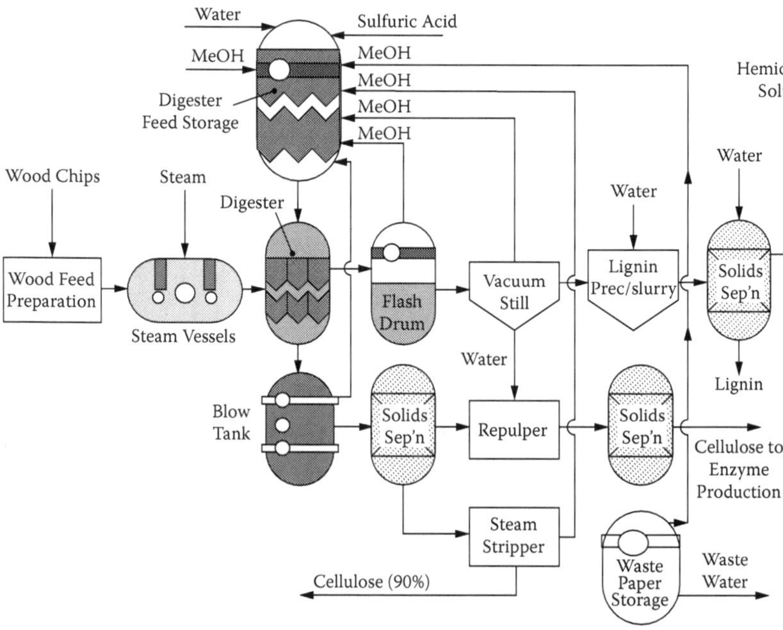

FIGURE 4.9
Organosolv pretreatment process. (Modified from Wright, 1988. Ethanol from biomass by enzymatic hydrolysis, *Chem. Eng. Prog.*, 84: 62–74.)

In general, organosolv processes have higher xylose yields than the other processes because of the influence of organic solvent on hydrolysis kinetics. In a recent study, Pan et al. [54] applied the ethanol organosolv pretreatment to lodgepole pine killed by mountain beetle and achieved 97% conversion to glucose. They recovered 79% of the lignin using the conditions of 170°C, 1.1 wt% H_2SO_4, and 65 vol% ethanol for 60 minutes.

4.5.3.1.4 Combined RASH and Organosolv Pretreatment

Attempts have been made to improve overall process efficiency by combining the two individual pretreatments of rapid steam hydrolysis and organosolv. Rughani and McGinnis [53] have studied the effect of a combined RASH–organosolv process upon the rate of enzymatic hydrolysis and the yield of solubilized lignin and hemicellulose. A schematic diagram of the process is shown in Figure 4.10. For the organosolv pretreatment, the steam generator is disconnected and the condensate valve closed. The rest of the reactor setup is similar to the typical RASH procedure.

The organosolv processes at low temperature are generally ineffective in removing lignin, as explained earlier; however, combining the two processes leads to increased solubilization of lignin and hemicellulose. RASH temperature is the major factor in maximizing the percentage of cellulose in the

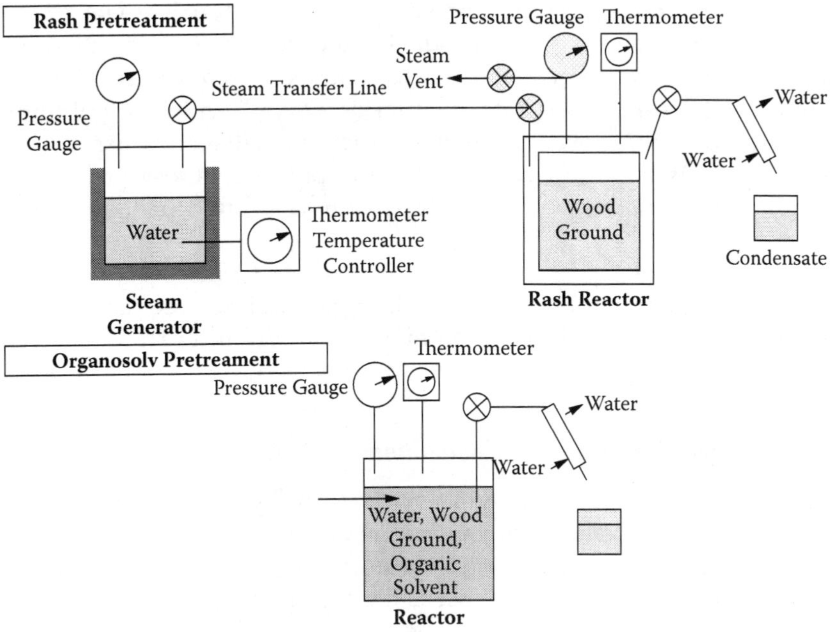

FIGURE 4.10
A combined RASH and organosolv pretreatment scheme. (Modified from Rughani and McGinnis, 1989. Combined rapid steam hydrolysis and organosolv pretreatment of mixed southern hardwoods, *Biotechnol. Bioeng.*, 33: 681–686,)

final product. The maximum yield of solubilized lignin was obtained at a temperature of 240°C for RASH and 160°C for the organosolv process.

4.5.3.1.5 Ionic Liquid Pretreatment

Ionic liquids are a relatively new class of solvents that have recently gained popularity as environmentally friendly alternatives to organic solvents. An ionic liquid is a salt composed of anions and cations that are poorly co-ordinated and which has a melting point typically below 100°C. Ionic liquids are also referred to as liquid electrolytes, ionic melts, liquid salts, ionic glasses, and the like. There are thousands of substances that fall into this category. Ionic liquids have been demonstrated as very efficient solvents in the fields of hydrogenation, esterification, nanomaterial synthesis, biocatalysis, and selective extraction of aromatics [55, 56].

The development of a novel biomass pretreatment technology using ionic liquids has only recently been initiated. The first demonstration of an ionic liquid as a cellulose solvent under relatively mild processing conditions was reported in 2002 by Swatloski [56]. In experiments using a range of anions and 1-butyl-3-methylimidazolium cations, some ionic liquids were able to completely dissolve microcrystalline cellulose, and the cellulose was recovered through the addition of an antisolvent such as water or ethanol. Moreover, the recovered product could be regenerated into a wide range of shapes and morphologies. The most effective cellulose solvents were the ionic liquids that contain chloride anions. An important finding associated with this novel pretreatment method is that enzymes can more efficiently hydrolyze into glucose the amorphous cellulose produced by ionic liquids than the microcrystalline cellulose found in lignocellulose naturally [55, 57].

Ionic liquids are an exciting area of new scientific discovery and inherently possess many processing merits in lignocellulose pretreatment. More in-depth R&D work needs to be conducted, however, before a commercially viable process can be fully developed and exploited.

4.5.3.2 Enzyme Production and Inhibition

The enzyme of interest is cellulase, needed for the hydrolysis of the cellulose, that is, cellulolysis. Cellulase is a multicomponent enzyme system consisting of endo-β-1,4-glycanases, exo-β-1,4-glucan glucohydrolases, and exo-β-1,4-glucan cellobiohydrolase. Cellobiose is the dominant product of this system but is highly inhibitory to the enzymes and is not usable by most organisms. Cellobiase hydrolyzes cellobiose to glucose, which is much less inhibitory and highly fermentable. Many fungi produce this cellobiase and most of the work that is presently being conducted is on *Trichoderma reesei* (*viride*). The cellulase produced by *T. reesei* is much less inhibited than other cellulases that have a major advantage for industrial purposes [58].

The type of inhibition exhibited by cellulases is the subject of much debate in research. Although most of the researchers favor competitive inhibition [59–64], some cellulases are noncompetitively [46, 62, 65, 66] or uncompetitively inhibited [60]. Uncompetitive inhibition takes place when an enzyme inhibitor binds only to the complex formed between the enzyme and the substrate, whereas noncompetitive inhibition takes place when an enzyme inhibitor and the substrate may both be bound to the enzyme at any given time. On substrates such as Solka Floc® (purified cellulose), wheat straw, and bagasse (biomass remaining after sugarcane stalks are crushed to extract their juice), *Trichoderma reesei* produced enzyme is competitively inhibited by glucose and cellobiose. On the other hand, some enzymes are noncompetitively inhibited by cellobiose using other substrates such as rice straw and Avicel®. *Avicel* is a registered trade name for microcrystalline cellulose that has been partially hydrolyzed with acid and reduced to a fine powder; it is used as a fat replacer. *Trichoderma viride* is uncompetitively inhibited by glucose in a cotton waste substrate [60].

Many mutants have been produced following *Trichoderma reesei*. The most prominent among these is the Rut C-30 [67], the first mutant with β-glucosidase production [43]. Other advantages of the strain include its hyperproducing properties and the fact that it is carbolite-repression resistant. The term *hyperproduction* means excessive production.

Cellulases from thermophilic bacteria have also been extensively examined. Among these, *Clostridium thermocellum* is perhaps the most extensively characterized organism. *C. thermocellum* is an anaerobic, thermophilic, cellulolytic, and ethanogenic bacterium capable of directly converting cellulosic substrate into ethanol. The enzymes isolated from thermophilic bacteria may have superior thermostability and hence will have longer half-lives at high temperatures. Although this is not always the case, cellulases isolated from *Clostridium thermocellum* have high specific activities [68], especially against crystalline forms of cellulose that have proven to be resistant to other cellulase preparations.

Enzyme production with *Trichoderma reesei* is difficult because cellulase production discontinues in the presence of easily metabolizable substrates. Thus, most production work has been carried out on insoluble carbon sources such as steam-exploded biomass or Solka-Floc [69]. Solka-Floc is composed of beta-1, 4-glucan units, is white, odorless, and flavorless, and has varying particle sizes [70]. In such systems, the rate of growth and cellulase production is limited because the fungi must secrete the cellulase and carry out slow enzymatic hydrolysis of the solid to obtain the necessary carbon. Average productivities have been approximately 100 IU/L/hr. [Hydrolytic activity of cellulose is generally in terms of international filter unit (IU). This is a unit defined in terms of the amount of sugar produced per unit time from a strip of Whatman filter paper.] The filter paper unit is a measure of the combined activities of all three enzymes on the substrate. High productivities have been reported with *Trichoderma reesei* mutant in a fed-batch system using lactose as a carbon source and steam-exploded aspen as an inducer. Although lactose is not available in sufficient quantities to supply a large ethanol industry, this does suggest that it may be possible to develop strains that can produce cellulases with soluble carbon sources such as xylose and glucose.

Productivity increases dramatically reduce the size and cost of the fermenters used to produce the enzyme. More rapid fermentation technologies would also decrease the risk of contamination and might allow for less expensive construction. Alternatively, using a soluble substrate may allow simplification of fermenter design or allow the design of a continuous enzyme production system. Low-cost but efficient enzymes for lignocellulosic ethanol technology must be developed in order to reduce the operational cost and improve the productivity of the process.

4.5.3.3 Cellulose Hydrolysis

4.5.3.3.1 Cellulase Enzyme Adsorption

The enzymatic hydrolysis of cellulose proceeds by adsorption of cellulase enzyme on the lignacious residue as well as the cellulose fraction. The adsorption on the lignacious residue is also interesting from the viewpoint of enzyme recovery after the reaction and recycling it for use on the fresh substrate. Obviously, the recovery efficiency is reduced by the adsorption of enzyme on lignacious residue, because a large fraction of the total operating cost is due to the production of enzyme. The capacity of lignacious residue to adsorb the enzyme is influenced by the pretreatment conditions, therefore the pretreatment should be evaluated, in part, by how much enzyme adsorbs on the lignacious residue at the end of hydrolysis as well as its effect on the rate and extent of the hydrolysis reaction.

The adsorption of cellulase on cellulose and lignacious residue has been investigated by Ooshima, Burns, and Converse [71] using cellulase from *Trichoderma reesei* and hardwood pretreated by dilute sulfuric acid with explosive decomposition. The cellulase was found to adsorb on the lignacious residue as well as on the cellulose during hydrolysis of the pretreated wood. A decrease in enzyme recovery in the liquid phase with an increase in the substrate concentration has been reported due to the adsorption on the lignacious residue. The enzyme adsorption capacity of the lignacious residue decreases as the pretreatment temperature is increased, whereas the capacity of the cellulose increases with higher temperature. The reduction of the enzyme adsorbed on the lignacious residue as the pretreatment temperature increases is essential for improving the ultimate recovery of the enzyme as well as enhancing the enzyme hydrolysis rate and extent. Lu et al. (2002) conducted an experimental investigation on cellulase adsorption and evaluated the enzyme recycle during the hydrolysis of SO_2-catalyzed steam-exploded Douglas fir and posttreated steam-exploded Douglas fir substrates [72]. After hot alkali peroxide posttreatment, the rates and yield of hydrolysis attained from the posttreated Douglas fir were significantly higher, even at lower enzyme loadings, than those obtained with the corresponding steam-exploded Douglas fir. This work suggests that enzyme recovery and reuse during the hydrolysis of posttreated softwood substrates could result in less need for the addition of fresh enzyme during softwood-based bioconversion processes [72].

An enzymatic hydrolysis process involving solid lignocellulosic materials can be designed in many ways. The common denominators are that the substrates and the enzyme are fed into the process, and the product stream (sugar solution), along with a solid residue, leaves it at various points. The residue contains adsorbed enzymes that are lost when the residue is removed from the system.

In order to ensure that the enzymatic hydrolysis process is economically efficient, a certain degree of enzyme recovery is essential. Both the soluble enzymes and the enzyme adsorbed onto the substrate residue must be reutilized. It is expected that the loss of enzyme is influenced by the selection of the stages at which the enzymes in solution and adsorbed enzymes are recirculated and the point where the residue is removed from the system.

Vallander and Erikkson [46] defined an enzyme loss function L, assuming that no loss occurs through filtration:

$$L = \frac{\textit{amount of enzyme lost through removal of residue}}{\textit{amount of enzyme at the start of hydrolysis}}$$

They developed a number of theoretical models to conclude that an increased enzyme adsorption leads to an increased enzyme loss. The enzyme loss decreases if the solid residue is removed late in the process. Both the adsorbed and dissolved enzymes should be reintroduced at the starting point of the process. This is particularly important for the dissolved enzymes. Washing of the entire residue is likely to result in significantly lower recovery of adsorbed enzymes than if a major part (60% or more) of the residue with adsorbed enzymes is recirculated. An uninterrupted hydrolysis over a given time period leads to a lower degree of saccharification than when hydrolyzate is withdrawn several times. Saccharification is also favored if the residue is removed at a late stage. Experimental investigations of the theoretical hydrolysis models have recovered more than 70% of the enzymes [46].

4.5.3.3.2 Mechanism of Hydrolysis

The overall hydrolysis is based on the synergistic action of three distinct cellulase enzymes and is dependent on the concentration ratio and the adsorption ratio of the component enzymes: endo-β-glucanases, exo-β-glucanases, and β-glucosidases. Endo-β-glucanases attack the interior of the cellulose polymer in a random fashion [43], exposing new chain ends. Because this enzyme catalyzes a solid phase reaction, it adsorbs strongly but reversibly to the microcrystalline cellulose (also known as Avicel). The strength of the adsorption is greater at lower temperatures. This enzyme is necessary for the hydrolysis of crystalline substrates. The hydrolysis of cellulose results in a considerable accumulation of reducing sugars, mainly cellobiose, because the extracellular cellulase complex does not possess cellobiose activity. Sugars that contain aldehyde groups that are oxidized to carboxylic acids are classified as reducing sugars.

Exo-β-glucanases remove cellobiose units (which are disaccharides with the formula $([HOCH_2CHO(CHOH)_3]_2O)$ from the nonreducing ends of cellulose chains. This is also a solid-phase reaction, and the exo-β-glucanases adsorb strongly on both crystalline and amorphous substrates. The mechanism of the reaction is complicated because there are two distinct forms of both endo- and exo-enzymes, each with a different type of synergism with the other members of the complex. As these enzymes continue to split off cellobiose units, the concentration of cellobiose in solution may increase. The action of exo-β-glucanases may be severely inhibited or even stopped by the accumulation of cellobiose in the solution.

The cellobiose is hydrolyzed to glucose by the action of β-glucosidase. *Glucosidase* is any enzyme that catalyzes hydrolysis of glucoside. β-Glucosidase catalyzes the hydrolysis of terminal, nonreducing beta-D-glucose residues with release of beta-D-glucose. The effect of β-glucosidase on the ability of the cellulase complex to degrade Avicel has been investigated by Kadam and Demain [73]. They determined the substrate specificity of the β-glucosidase and demonstrated that its addition to the cellulase complex enhances the hydrolysis of Avicel, specifically by removing the accumulated cellobiose. A thermostable β-glucosidase form, *clostridium thermocellum,* which is expressed in *Escherichia coli,* was used to determine the substrate specificity of the enzyme. The hydrolysis of cellobiose to glucose is a liquid-phase reaction and β-glucosidase adsorbs either quickly or not at all on cellulosic substrates. β-Glucosidase's action can be slowed or halted by the inhibitive action of glucose accumulated in the solution. The accumulation may also induce the entire hydrolysis to a halt as inhibition of the β-glucosidase results in a buildup of cellobiose, which in turn inhibits the action of exo-glucanases. The hydrolysis of the cellulosic materials depends on the presence of all three enzymes in proper amounts. If any one of these enzymes is present in less than the required amount, the other enzymes will be inhibited or lack the necessary substrates upon which to act.

The hydrolysis rate generally increases with increasing temperature. However, because the catalytic activity of an enzyme is also related to its shape, the deformation of the enzyme at high temperature can inactivate or destroy the enzyme. To strike a balance between increased activity and increased deactivation, it is preferable to run fungal enzymatic hydrolysis at approximately 40–50°C.

Although enzymatic hydrolysis is preferably carried out at a low temperature of 40–50°C, dilute acid hydrolysis is carried out at a substantially higher temperature. Researchers at the National Renewable Energy Laboratory (NREL) reported results for a dilute acid hydrolysis of softwoods in which the conditions of the reactors were as follows [74].

1. Stage 1: 0.7% sulfuric acid, 190°C, and a 3-min residence time
2. Stage 2: 0.4% sulfuric acid, 215°C, and a 3-min residence time

Their bench-scale tests also confirmed the potential to achieve yields of 89% for mannose, 82% for galactose, and 50% for glucose, respectively. Fermentation with *Saccharomyces cerevisiae* achieved ethanol conversion of 90% of the theoretical yield [75].

4.5.3.4 Fermentation

Cellulose hydrolysis and fermentation can be achieved by two different process schemes, depending upon where the stage of fermentation is actually carried out in the process sequence: (1) separate hydrolysis and fermentation, SHF, or (2) simultaneous saccharification and fermentation, SSF. The acronyms "SHF" and "SSF" are very commonly used in the field.

4.5.3.4.1 *Separate Hydrolysis and Fermentation (SHF)*

In SHF, the hydrolysis is carried out in one vessel and the hydrolyzate is then fermented in a second reactor. The most expensive items in the overall process cost are the cost of feedstock, enzyme production, hydrolysis, and utilities. The feedstock and utility costs are high because only about 73% of the cellulose is converted to ethanol in 48 hr, and the remainder of the cellulose, hemicellulose, and lignin are burned or gasified. Enzyme production is a costly step due to the large amount of the enzyme used in an attempt to overcome the end-product inhibition as well as its slow reaction rate. The hydrolysis step is also expensive due to the large capital and operating costs associated with large size tanks and agitators. The most important parameters are the hydrolysis section yield, the product quality, and the required enzyme loading, all of which are interrelated. Yields are typically higher in more dilute systems where inhibition of enzymes by glucose and cellobiose is minimized. Increasing the amount of enzyme loading can help to overcome inhibition and increase the yield and concentration, although it undoubtedly increases the overall cost. Increased reaction times also make higher yields and concentrations.

Cellulase enzymes from different organisms can result in markedly different performances. Figure 4.11 shows the effect of yield at constant solid and enzyme loading and the performance of different enzyme loadings. Increase in enzyme loading beyond a particular point has turned out to be of no use. It would be economical to operate at a minimum enzyme loading level. Or, the enzyme could be recycled by appropriate methods. As the cellulose is hydrolyzed, the endo- and exoglucanase components are released back into the solution. Because of their affinity for cellulose, these enzymes can be recovered and reused by contacting the hydrolyzate with fresh feed. The amount of recovery is limited because of β-glucosidase, which does not adsorb on the feed. Some of the enzyme remains attached to the lignin and unreacted cellulose; in addition, enzymes are thermally denatured during hydrolysis. A major difficulty in this type of process is maintaining sterility; otherwise, the process system would be contaminated. The power consumed in agitation is also significant and affects the economics of this process [43]. Even though the effect of yield on the selling price of ethanol in the figure was based on more classical ethanol production processes, it does explain the importance of yield on the final product cost.

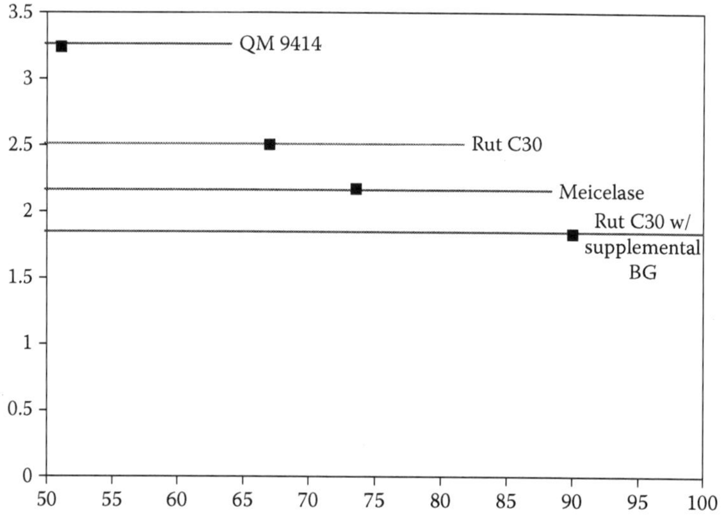

FIGURE 4.11
Effect of yield on selling price of ethanol. In the drawing, BG denotes β-glucanase. (Modified from Wright, 1988. Ethanol from biomass by enzymatic hydrolysis, *Chem. Eng. Prog.*, 84: 62–74.)

Um and Hanley [69] carried out an experimental study on high-solid enzymatic hydrolysis and fermentation of Solka Floc into ethanol. To lower the ethanol distillation cost of fermentation broths, a high initial glucose concentration is desired. However, an increase in glucose concentration typically reduces the ethanol yield due to the decreased mass and heat transfer rate. To overcome the incompatible temperatures between the enzymatic hydrolysis (50°C) and fermentation (30°C), saccharification, followed by fermentation (SFF), was employed with relatively high solid concentrations (10% to 20%) using a portion loading method. Glucose and ethanol were produced from Solka Floc, which was first digested by enzymes at 50°C for 48 hours, followed by fermentation. In this process, commercial enzymes were used in combination with a recombinant strain of *Zymomonas mobilis*. The highest ethanol yields of 83.6%, 73.4%, and 21.8%, based on the theoretical amount of glucose, were obtained with substrate concentrations of 10%, 15%, and 20%, respectively. These values also correspond to 80.5%, 68.6%, and 19.1%, based on the theoretical amount of the cell biomass and soluble glucose present after 48 hours of SFF. In addition to the substrate concentration effects, they also investigated the effects of reactor configurations [69].

As a classic study of the mechanism of the enzymatic hydrolysis of cellulose, Fan, Lee, and Beardmore investigated the effects of major structural features of cellulose on enzymatic hydrolysis. They found that the hydrolysis rate is mainly dependent upon the fine structural order of cellulose which can be best represented by the crystallinity rather than the simple surface area [76].

4.5.3.4.2 Simultaneous Saccharification and Fermentation (SSF)

The operating cost of the SSF process is generally lower than that of SHF as long as the process integration is synergistically done. As the name of the process implies, both the hydrolysis and fermentation are carried out in the same vessel. In this process, yeast ferments the glucose to ethanol as soon as the glucose is produced, thus preventing the sugars from accumulating and causing end-product inhibition. Using the yeast, *Candida brassicae,* and the Genencor enzyme (by Genencor International), the yield is increased to 79% and the ethanol concentration produced is 3.7% [43, 77].

Even in SSF, cellobiose (the soluble disaccharide sugar) inhibition occurs to an appreciable extent. The enzyme loading for SSF is only 7 IU/g of cellulose, compared to 33 IU/g in SHF. The cost of energy and feedstock is somewhat reduced because of the improved yield, and the increased ethanol concentration significantly reduces the cost of distillation and utilities. The cost of the SSF process is slightly less than the combined cost of hydrolysis and fermentation in the SHF process. The decreasing factor of the reactor volume due to the higher concentration of ethanol offsets the increasing factor in the reactor size caused by the longer reaction times (seven days for SSF vs. two days for hydrolysis and two days for fermentation). Earlier studies showed that fermentation is the rate-controlling step and the enzymatic hydrolysis process is not. With recent advances and developments in recombinant yeast strains that are capable of effectively fermenting both glucose and xylose, these process-configurational considerations for commercial exploitation, as well as the determination of a rate-limiting step for the overall process technology, may have to change accordingly.

The hydrolysis is carried out at 37°C and increasing the temperature increases the reaction rate; however, the ceiling temperature is usually limited by the yeast cell viability. The concentration of ethanol is also a limiting factor. (This was tested by connecting a flash unit to the SSF reactor and removing the ethanol periodically. This technique showed productivities up to 44% higher.) Recycling the residual solids may also increase the process yield. However, the most important limitation in enzyme recycling comes from the presence of lignin, which is inert to the enzyme. High recycling rates increase the fraction of lignin present in the reactor and cause handling difficulties.

Two major types of enzyme recycling schemes have been proposed: one in which enzymes are recovered in the liquid phase and the other in which enzymes are recovered by recycling unreacted solids [43]. Systems of the first type have been proposed for SHF processes that operate at 50°C. These systems are favored at such a high temperature because increasing temperature increases the proportion of enzyme that remains in the liquid phase. Conversely, as the temperature is decreased, the amount of enzyme adsorbed on the solid increases. Therefore, at lower temperatures encountered in SSF processes, solids recycling becomes a more effective option.

4.5.3.4.3 Comparison between the SSF and SHF Processes

SSF systems offer large advantages over SHF processes thanks to their reduction in end-product inhibition of the cellulase enzyme complex. The SSF process shows a higher yield than SHF (88% vs. 73% in an earlier example) and greatly increases product concentrations (equivalent glucose concentration of 10% vs. 4.4%). The most significant advantage of the SSF process is the enzyme loading, which can be reduced from 33 to 7 IU/g cellulose; this cuts down the cost of ethanol appreciably. With constant development of low-cost enzymes, the comparative analysis of the two processes will inevitably be changing. A comparative study of the approximate costs of the two processes was reported in Wright's 1988 article [43]. The results show that, based on the estimated ethanol selling price from a production capacity of 25,000,000 gallons/year, SSF is found to be more cost-effective than SHF by a factor of 1:1.49; that is, $\$_{SHF}/\$_{SSF} = 1.49$. It should be clearly noted that the number quoted here is the ratio of the two prices, not the direct dollar value of the ethanol selling price. Furthermore, this study was also based on the enzymes and bioconversion technologies available in the mid-1980s, which are significantly different from the most advanced current technologies of the twenty-first century. However, this ethanol production cost comparison between two different process configurations provides an idea about the complexity of interrelated cost factors among the reaction rates, temperature, processing time, enzyme adsorption, enzyme loading and recoverability, product inhibition, and more.

For the very same process economic reasons, it is anticipated that a hybrid hydrolysis and fermentation (HHF) process configuration is going to be widely accepted as a process of choice for production of lignocellulosic fuel ethanol, which begins with a separate prehydrolysis step and ends with a simultaneous saccharification (hydrolysis) and fermentation step. In the first stage of hydrolysis, higher-temperature enzymatic cellular saccharification is taking place, whereas in the second stage of SSF, mesophilic (moderate-temperature) enzymatic hydrolysis and biomass sugar fermentation are taking place simultaneously. The optimized process configurational scheme would have to change if a specific enzyme, proven to be highly efficient and cost-effective, is found to be intolerant against certain inhibitors that are associated with any of these processing steps.

4.5.3.5 Xylose Fermentation

Inasmuch as xylose accounts for 30–60% of the fermentable sugars in hardwood and herbaceous biomass, the fermentation of xylose to ethanol becomes an important issue. The efficient fermentation of xylose and other hemicellulose constituents is essential for the development of an economically viable process to produce ethanol from lignocellulosic biomass. Needless to say, co-fermentation of both glucose and xylose with comparably high efficiency would be most ideally desirable. As discussed earlier, xylose fermentation using pentose yeasts has proven to be difficult due to several factors including the requirement for O_2 during ethanol production, the acetate toxicity, and the production of xylitol as by-product. Xylitol (or, xyletol) is a naturally occurring low-calorie sugar substitute with anticariogenic (preventing production of dental caries) properties.

Other approaches to xylose fermentation include the conversion of xylose to xylulose (a pentose sugar, part of carbohydrate metabolism, that is found in the urine of individuals with the condition pentosuria [78]) using xylose isomerase prior to fermentation by *Saccharomyces cerevisiae,* and the development of genetically engineered strains [79].

A method for integrating xylose fermentation into the overall process is illustrated in Figure 4.12. In this example, dilute acid hydrolysis was adopted as a pretreatment step. The liquid stream is neutralized to remove any mineral acids or organic acids liberated in the pretreatment process, and is then sent to xylose fermentation. Water is added before the fermentation, if necessary, so that organisms can make full use of the substrate without having the yield limited by end-product inhibition. The dilute ethanol stream from xylose fermentation is then used to provide the dilution water for the cellulose–lignin mixture entering SSF. Thus, the water that enters during the pretreatment process is used in both the xylose fermentation and the SSF process.

The conversion of xylose to ethanol by recombinant *E. coli* has been investigated in pH-controlled batch fermentations [80]. Relatively high concentrations of ethanol (56 g/L) were produced from xylose with good efficiencies.

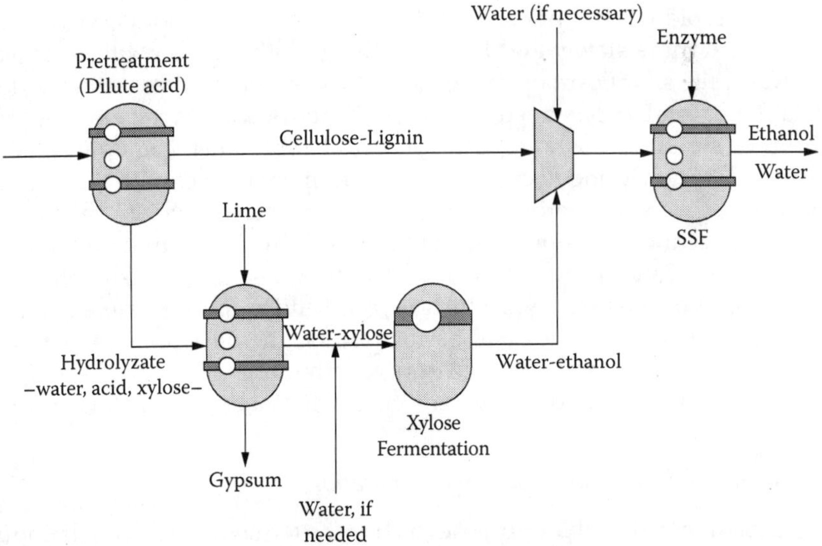

FIGURE 4.12
Integration of xylose fermentation and SSF. Modified from Wright. 1988. Ethanol from biomass by enzymatic hydrolysis, *Chem. Eng. Prog.*, 84: 62–74.)

In addition to xylose, all other sugar constituents of biomass, including glucose, mannose, arabinose, and galactose, can be efficiently converted to ethanol by recombinant *E. coli*. Neither oxygen nor strict maintenance of anaerobic conditions is required for ethanol production by *E. coli*. However, the addition of base to prevent excessive acidification is essential. Although less base was needed to maintain low pH conditions, poor ethanol yields and slower fermentations were observed below the pH of 6. Also the addition of metal ions, such as calcium, magnesium, and ferrous ions, stimulated ethanol production [80].

In general, xylose fermentation does not require precise temperature control, provided the broth temperature is maintained between 25 and 40°C. Xylose concentrations as high as 140 g/L have been positively tested to evaluate the extent to which this sugar inhibits growth and fermentation. Higher concentrations slow down growth and fermentation considerably. Ingram and coworkers [80–83] demonstrated that recombinant *Escherichia coli* expressing plasmid-borne *Zymomonas mobilis* genes for pyruvate decarboxylase (PDC) and alcohol dehydrogenase II (ADHII; adhB) can efficiently convert both hexose and pentose sugars to ethanol. Ethanologenic *E. coli* strains require simpler fermentation conditions, produce higher concentrations of ethanol, and are more efficient than pentose-fermenting yeasts for ethanol production from xylose and arabinose [84].

A study by Sedlak, Edenberg, and Ho [28] successfully developed genetically engineered *Saccharomyces* yeasts that can ferment both glucose and xylose simultaneously to ethanol. According to their experimental results, following rapid consumption of glucose in less than 10 hours, xylose was metabolized more slowly and less completely. Although the xylose conversion was quite significant by this genetically engineered yeast strain, xylose was still not totally consumed even after 30 hours. Ideally, xylose should be consumed simultaneously [26] with glucose at a similar efficiency and speed; however, this newly added capability of co-fermentation of both glucose and xylose has given new promise in the lignocellulosic ethanol technology leading to technological breakthroughs. They also found that ethanol was the most abundant product from glucose and xylose metabolism, but small amounts of the metabolic byproducts glycerol and xylitol also were obtained [28]. Certainly, later studies will be focused on the development or refinement of more efficient engineered strains, ethanol production with higher selectivity and speed, and optimized process engineering and flowsheeting.

4.5.3.6 Ethanol Extraction during Fermentation

In spite of the considerable efforts devoted to the fermentative alcohols, industrial applications have been delayed because of the high cost of production, which depends primarily on the energy input to the purification of dilute end-products, the low productivities of cultures, and the high cost of enzyme production. These issues are directly linked to inhibition phenomena.

Along with the conventional unit operations, liquid–liquid extraction with biocompatible organic solvents, distillation under vacuum, and selective adsorption on the solids have demonstrated the technical feasibility of the extractive fermentation concept. Lately, membrane separation processes that decrease biocompatibility constraints have been proposed. These include dialysis [85] and reverse osmosis [65]. More recently, the concept of supported liquid membranes has been reported. This method minimizes the amount of organic solvents involved and permits simultaneous realization of the extraction and recovery phases. Enhanced volumetric productivity and high substrate conversion yields have been reported [86] via the use of a porous Teflon® sheet (soaked with isotridecanol) as support for the extraction of ethanol during semi-continuous fermentation of *Saccharomyces bayanus*. This selective process results in ethanol purification and combines three operations: fermentation, extraction, and re-extraction (stripping) as schematically represented in Figure 4.13. As shown and suggested, novel process ideas can further accomplish maximized alcohol production, energy savings, and reduced cost in production.

4.5.4 Lignin Conversion

Lignin is produced in large quantities, approximately 250 billion pounds per year in the United States, as a by-product of the paper and pulp industry. Lignins are complex amorphous phenolic polymers that are not sugar-based,

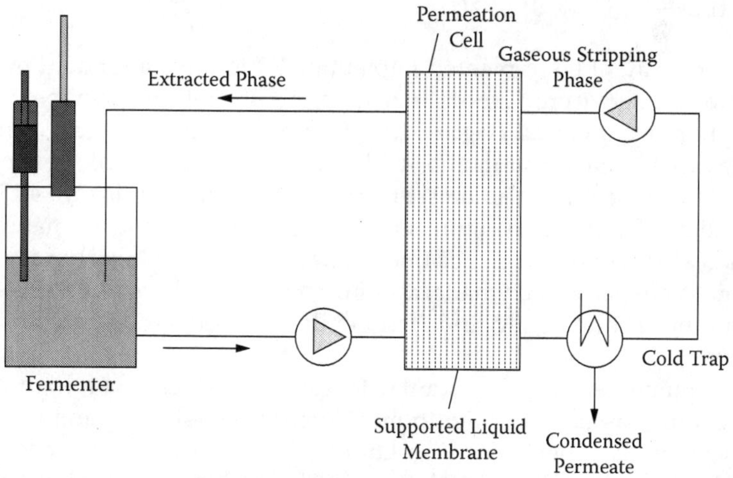

FIGURE 4.13
Extractive fermentation system: (1) fermenter; (2) permeation cell; (3) supported liquid membrane; (4) extracted phase; (5) gaseous stripping phase; (6) cold trap; (7) condensed permeate. (Modified from Christen, Minier, and Renon, 1990. Enhanced extraction by supported liquid membrane during fermentation, *Biotechnol. Bioeng.*, 36: 116–123.)

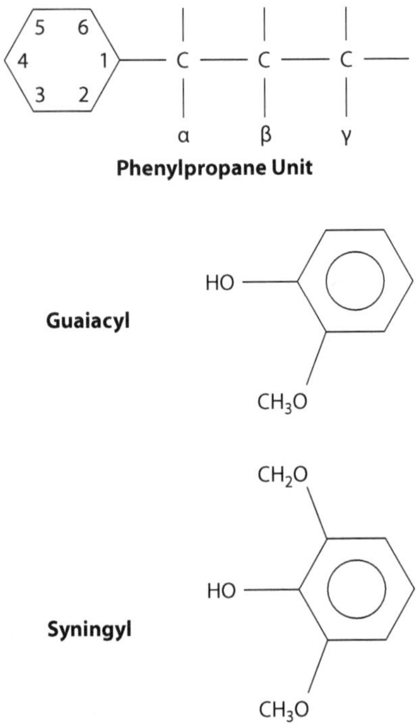

FIGURE 4.14
Monomer units in lignin. (Modified from Wright, 1988. Ethanol from biomass by enzymatic hydrolysis, *Chem. Eng. Prog.*, 84: 62–74.)

hence, they cannot be fermented into ethanol. Lignin is a random polymer made up of phenyl propane units, where the phenol unit may be either a guaiacyl or syringyl unit (Figure 4.14). These units are bonded together in many ways, the most common of which are α- or β-ether linkages. A variety of C–C linkages are also present, but are less common (Figure 4.15). The distribution of linkage in lignin is random because lignin formation is a free radical reaction that is not under enzymatic control. Lignin is highly resistant to chemical, enzymatic, and microbial hydrolysis due to extensive crosslinking. Therefore, lignin is frequently removed simply to gain access to cellulose.

Lignin monomer units are similar to gasoline, which has a high octane number; thus, breaking the lignin molecules into monomers and removing the oxygen makes them useful as liquid transportation fuels. The process for lignin conversion is mild hydrotreating to produce a mixture of phenolic and hydrocarbon materials, followed by reaction with methanol to produce methyl aryl ether. The first step usually consists of two principal parts: *hydrodeoxygenation* (removal of oxygen and oxygen-containing groups from the phenol rings) and *dealkylation* (removal of ethyl groups or large side chains from the rings). The major role of this stage is to carry out these reactions

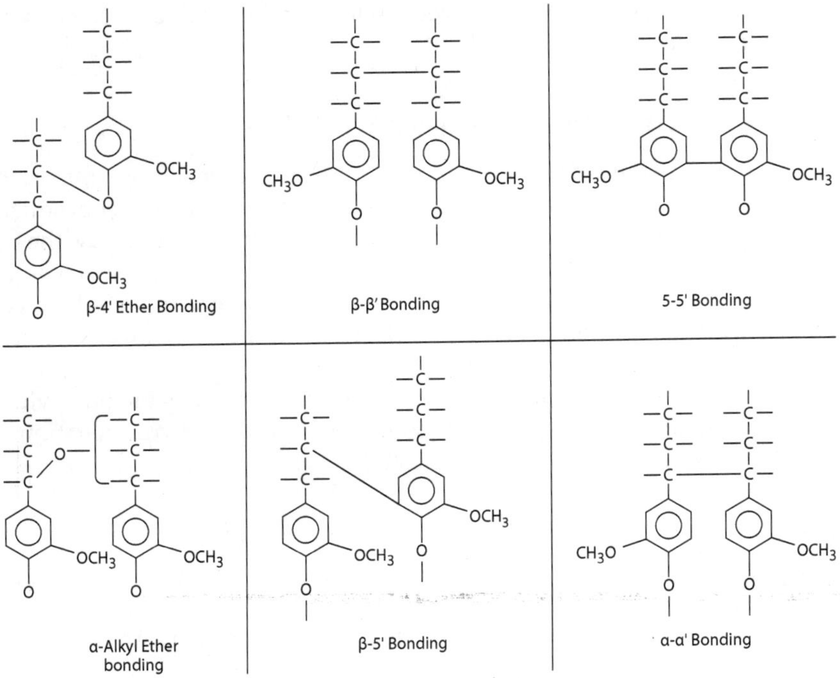

FIGURE 4.15
Ether and C–C bonds in lignin. (Modified from Wright, 1988. Ethanol from biomass by enzymatic hydrolysis, *Chem. Eng. Prog.*, 84: 62–74.)

to remove the unwanted chains without carrying the reaction too far; this would lead to excessive consumption of hydrogen and produce saturated hydrocarbons, which are not as effective as octane enhancers as are aromatic compounds. Furthermore, excessive consumption of hydrogen would represent additional cost for the conversion process. Catalysts to carry out these reactions have dual functions. Metals such as molybdenum and molybdenum/nickel catalyze the deoxygenation, and the acidic alumina support promotes the carbon–carbon bond cleavage.

Although lignin chemicals have applications in drilling muds, as binders for animal feed, and as the base for artificial vanilla, they have not been previously used as surfactants for oil recovery. According to Naee [87], lignin chemicals can be used in two ways in chemical floods for enhanced oil recovery. In one method, lignosulfonates are blended with tallow amines and conventional petroleum sulfonates to form a unique mixture that costs about 40% less to use than chemicals made solely from petroleum or petroleum-based products. In the second method, lignin is reacted with hydrogen or carbon monoxide to form a new class of chemicals called lignin phenols. Because they are soluble in organic solvents, but not in water, these phenols are good candidates for further conversion to produce chemicals that may be useful in enhanced oil recovery (EOR).

4.5.5 Coproducts of Cellulosic Ethanol Technology

In order to reduce the ethanol production cost from lignocellulosic materials, it is imperative to expand or develop the market for the process coproducts, by-products, or derivatives. Unlike the mature corn ethanol industry, the by-product (or coproduct) industry for the lignocellulosic ethanol industry is not very well defined or established yet. Potential coproducts include hemicellulose hydrolyzate (xylose), cellulose hydrolyzate (glucose of mixed sugars), cell mass, enzymes, soluble and insoluble lignins, lignin-derived chemicals and fuels, solid residues, and so on. Other valuable products include xylitol, which is a sugar alcohol sweetener and is produced by hydrogenation of xylose (an aldehyde) into a primary alcohol.

4.6 Energy Balance for Ethanol Production from Biomass

Biomass process development depends upon the economics of the conversion process, be it chemical, enzymatic, or a combination of both. A number of estimates have been computed based upon existing or potential technologies. One obvious factor is that, regardless of the process, transportation of the biomass material from its source to the site of conversion must be kept to an absolute minimum. Approximately 35% of the expected energy is consumed by transporting the substrate a distance of 15 miles [68]. This considerable expenditure of energy simply to transport the starting material dictates that any conversion plant be of moderate size and in close proximity to the production source of the starting material.

There are some objections to the production and use of ethanol as a fuel. Most important is the criticism that producing ethanol can consume more energy than the finished ethanol contains. The European analysis takes wheat as the feedstock and includes estimates of the energy involved to grow the wheat, transport it to the distillery, make the alcohol, and transport it to a refinery for blending with gasoline. It allows credit for by-products, such as animal feed from wheat, for savings on gasoline that come from replacing 5% with alcohol, and from the energy gained from the increase of 1.25 octane points.

As fully explained in Chapter 3, a more recent and very extensive assessment on the net energy value (NEV) of corn ethanol technology [88] using advanced process technology as well as more realistic industrial data [89, 90] decisively showed that the prevailing corn ethanol process in the United States generates a significantly positive net energy value. As for the cellulosic ethanol, a number of factors and issues including the feedstock diversity and availability, the use of nonfood crops, minimal or no use of fertilizers, non-use of arable land, and more complex but still evolving conversion process technology make such an assessment far more difficult and less meaningful. Yet, to confine debates on biomass fuels solely to the process energy balance would be misleading. Based on the merits of cellulosic biofuels as well as regional strengths, a number of cellulosic biofuel plants based on diverse process technologies are in operation or are under construction throughout the world [91].

Energy requirements to produce ethanol from different crops were evaluated by Da Silva et al. [92]. The industrial phase is always more energy intensive, consuming from 60 to 75% of the total energy. The energy expended in crop production includes all the forms of energy used in agricultural and industrial processing, except the solar energy that plants use for growth. The industrial stage, including extraction and hydrolysis, alcohol fermentation, and distillation, requires about 6.5 kg of steam per liter of alcohol. It is possible to furnish the total industrial energy requirements from the by-products of some of the crops. Thus, it is also informative to consider a simplified energy balance in which only agricultural energy is taken as input and only ethanol is taken as the output, the bagasse supplying energy for the industrial stage, for example.

Furthermore, technological data are often very difficult or nearly impossible to compare between different options, due to the wide variety of feedstock crops as starting lignocellulose. Therefore, the U.S. Department of Energy-sponsored projects chose corn stover as the model feedstock [93]. This selection is based on the fact that corn stover is the most abundant and concentrated biomass resource in the United States and its collection can leverage the existing corn ethanol infrastructure, including corn harvesting and ethanol production [93].

Unlike the cellulosic ethanol technology, sugarcane ethanol technology is far more straightforward and as such the energy balance evaluation is relatively straightforward. The energy balance results for ethanol production from sugarcane in Zimbabwe have shown that the energy ratio is 1.52 if all the major output is considered and 1.15 if ethanol is considered as the only output. The reported value of the net energy ratio for ethanol production from sugarcane in Brazil [92] is 2.41 and in Louisiana, United States [94], it is 1.85. The low ratio in Zimbabwe is due to (a) the large energy input in the agricultural phase, arising from a large fertilizer need, and (b) the large fossil-based fuel consumption in sugarcane processing. As shown in this comparative example, the energy balance results are dependent upon a large number of factors including process conversion technology, agricultural technology, climate and soil quality, logistical issues, and much more.

The NEV of cellulose ethanol from switchgrass was analyzed by Schmer et al. [95]. In this study, perennial herbaceous plants such as switchgrass were evaluated as cellulosic bioenergy crops. Two major concerns of their investigation were the net energy efficiency and economic feasibility of switchgrass and similar crops. This was a baseline study that represented the genetic material and agronomic technology available for switchgrass production in 2000 and 2001. Their study reported the following.

(a) The annual biomass yields of established fields averaged 5.2–11.1 Mg $\cdot$ ha^{-1} $\cdot$ y^{-1} with a resulting average estimated net energy yield (NEY) of 60 GJ $\cdot$ ha^{-1} $\cdot$ y^{-1}.
(b) Switchgrass produced 540% more renewable than nonrenewable energy consumed.
(c) Switchgrass monocultures managed for high yield produced 93% more biomass yield and an equivalent estimated NEY than previous estimates from human-made prairies that received low agricultural input.
(d) Estimated average greenhouse gas emissions from cellulosic ethanol derived from switchgrass were 94% lower than estimated GHG emissions from gasoline.

Generally speaking, the cost of production of ethanol decreases with an increase in capacity of the production facility, as is the case with most petrochemical industries. However, the minimum total cost corresponds to a point of inflection, at which point an increase in the production cost for every increase in the plant capacity is seen [42]. The possibility of the existence of an empirical relationship between plant size or output and the production cost has also been examined using various production functions and the computed *F* values at a 5% level of significance [96]. It is also imaginable that if the average distance of raw material transportation and acquisition becomes excessively long due to the increased plant capacity, then the production cost can be adversely affected by the plant size.

Xylose fermentation is being carried out by bacteria, fungi, yeast, enzyme–yeast systems, or genetically engineered micro-organisms. Advanced fermentation technology would reduce the cost by 25% or more in the case of herbaceous-type materials, as shown in the study by Schmer et al. [95]. Efforts are being made to achieve the yield of 100% and an increased ethanol concentration.

Lignin is another major component of biomass, and accounts for its large energy content because it has a much higher energy per pound than carbohydrates. Because it is a phenolic polymer it cannot be fermented to sugar, and is instead converted to materials such as methyl aryl ethers, which are compatible with gasoline as a high-octane enhancer. The combination of the above processes has the potential to produce transportation fuels at a competitive price.

4.7 Process Economics and Strategic Direction

McAloon et al. [97] studied the cost of ethanol production from lignocellulosic materials in comparison to that from corn starch. As properly pointed out in their study, the cost comparison was made between the mature corn-ethanol industry and the emerging lignocellulosic-ethanol industry. Based on the fixed price of the year 2000, the cost of fuel ethanol production from lignocellulose processes was determined to be \$1.50/gal, whereas that from corn processes was \$0.88/gal [97]. Needless to say, the cost values determined in 2000 cannot be considered valid for the current year, due to significant changes during the period in infrastructural and raw material costs as well as variable operating costs. In order to make lignocellulosic biorefinery technology a success, the following must be resolved.

1. The lignocellulose feedstock collection and delivery system has to be established on an economically sound basis. Feedstock preparation also becomes an issue.
2. Each step of the process technology needs to be separately investigated for various options and the interactions and connectivity between the steps must be completely evaluated. Interactive effects between stages become very important, because one stage's product and by-products may function as the next stage's inhibitors.
3. A thorough database for a variety of different feedstock must be established. A different feedstock can be chosen as a model feedstock for different countries and regions, depending upon the local availability, logistical constraints, and infrastructural benefits. Furthermore, conversion technologies should be readily adaptable to other lignocellulosic feedstock and agricultural residues [93].
4. Large-scale demonstration is crucially important for commercial operational experience as well as to minimize the risk involved in scale-up efforts. In addition, such an operation on a large scale helps demonstrate environmental life-cycle analysis whose results are more meaningful.
5. Low-cost but highly efficient enzymes for the technology must be developed in order to reduce operational cost and improve productivity. Current efforts by the National Renewable Energy Laboratory, Genencor International, and Novozymes Biotech are very significant and noteworthy in this regard. Further advances will make the full-scale commercialization of cellulosic ethanol closer and more economically feasible.

References

1. Saha, B.C. 2004. Lignocellulose biodegradation and applications in biotechnology. In B.C. Saha and K. Hayashi (Eds.), *Lignocellulose Biodegradation*, pp. 2–34.
2. Cohen, D. 2004. *Form and Distribution of Trace Elelments in Biomsass for Power Generation*, Tech. Rep. 48, July. Australia: QCat Technology Transfer Center.
3. Odian, G. 2004. *Principles of Polymerization.* Hoboken, NJ: Wiley-Interscience.
4. Bettiga, M., Bengtsson, O., Hahn-Hagerdal, B., and Gorwa-Grauslund, M.F. 2009. Arabinose and xylose fermentation by recombinant *Saccharomyces cerevisiae* expressing a fungal pentose utilization pathway, *Microbial Cell Factories,* 8: 1–12.
5. Lin, Y., He, P., Wang, Q., Lu, D., Li, Z., Wu, C., and Jiang, N. 2010. The alcohol dehydrogenase system in the xylose-fermenting yeast *Candida maltosa, PloS One,* 5: 1–9.
6. Jeffries, T.W. and Shi, N.Q. 1999. Genetic engineering for improved xylose fermentation by yeasts, *Adv. Biochem. Eng. Biotechnol.,* 65: 117–161.
7. La Rovere, E.L. 2004, The Brazilian ethanol program: Biofuels for transport. Presented at *International Conference for Renewable Energies,* http://www.renewables2004.de/ppt/Presentation4-SessionIVB(11-12.30h)-LaRovere.pdf.
8. Lichts, F.O. 2011. *Industry Statistics: 2010 World Fuel Ethanol Production.* Renewable Fuels Association. Available at: http://www.ethanolrfa.org/pages/statistics#E.
9. Federal Energy Regulatory Commission. 2005. Energy Policy Act (EPAct) of 2005. http://www.ferc.gov/legal/fed-sta/ene-pol-act.asp (August 2011).
10. Klasson, K.T., Ackerson, M.D., Clausen, E.C., and Gaddy, J.L. 1992. Bioconversion of synthesis gas into liquid or gaseous fuels, *Enzyme Microbial Technol.,* 14: 602–608.
11. Ellis, G. 1937. *Chemistry of Petroleum Derivatives.* New York: Reinhold.
12. Judice, C.A. and Pirkle, L.E. 1963. U.S. Patent 3,095,458.
13. Lewis, W.K. 1936. U.S. Patent 2,045,785, June 30.
14. Miller, S.A. 1969. *Ethylene and Its Industrial Derivatives.* London: Ernest Benn.
15. Bailey, J.E. and Ollis, D.F. 1986. *Biochemical Engineering Fundamentals, 2nd Ed.* New York: McGraw-Hill.
16. Demirbas, A. 2009. Political, economic and environmental impacts of biofuels: A review, *Appl. Energy*, 86: S108–S117.
17. Goldemberg, J. and Guardabassi, P. 2010. The potential for first-generation ethanol production from sugarcane, *Sustainable Chem. Green Chem.,* 4: 17–24.
18. Park, S.C. and Baratti, J. 1991. Batch fermentation kinetics of sugar beet molasses by *Zymomonas mobilis, Biotechnol. Bioeng.,* 38: 304–313.
19. Willey, J.M., Sherwood, L.M., and Woolverton, C. 2011. *Prescott's Microbiology, 8th Ed.* New York: McGraw-Hill.
20. Viikari, L. 1988. Carbohydrate metabolism in Zymomonas, *Crit. Rev. Biotechnol.,* 7: 237–261.
21. Kalnenieks, U., Galinina, N., Toma, M.M., and Poole, R.K. 2000. Cyanide inhibits respiration yet stimulatesaerobic growth of Zymomonas mobilis, *Microbiology,* 146: 1259–1266.

22. Cazetta, M.L., Celligoi, M.A.P.C., Buzato, J.B., and Scarmino, I.S. 2007. Fermentation of molasses by Zymomonas mobilis: Effects of temperature and sugar concentration on ethanol production, *Bioresource Technol.*, 98: 2824–2828.
23. Eliasson, A., Christensson, C., Wahlbom, C.F., and Hahn-Hägerdal, B. 2000. Anaerobic xylose fermentation by recombinant *Saccharomyces cerevisiae* carrying XYL1, XYL2, and XKS1 in mineral medium chemostat cultures, *Appl. Environ. Microbiol.*, 66: 3381–3386.
24. Zhang, M. Zymomonas mobilis, special topics session, microbial pentose metabolism. Presented at *25th Symposium on Biotechnology for Fuels and Chemicals.* http://www1.e ere.energy.gov/biomass/pdfs/34264.pdf.
25. Tao, H. 2001. Engineering a homo-ethanol pathway in Escherichia coli: Increased glicolytic flux and levels of expression of glycolytic genes during xylose fermentation, *J. Bacteriol.*, 183: 2979–2988.
26. U.S. DOE Office of Science and Office of Energy Efficiency and Renewable Energy, 2006. *Breaking the Biological Barriers to Cellulosic Ethanol: A Joint Research Agenda. A Research Roadmap Resulting from the Biomass to Biofuels Workshop,* December 7–9, 2005, Rockville, MD, Tech. Rep. DOE/SC-0095, June.
27. Matsushika, A., Inoue, H., Kodaki T., and Sawayama, S. 2009. Ethanol production from xylose in engineered *Saccharomyces cerevisiae* strains: Current state and perspectives, *Appl. Microbiol. Biotechnol.*, 84: 37–53.
28. Sedlak, M., Edenberg, H.J., and Ho, N.W.Y. 2003. DNA microarray analysis of the expression of the genes encoding the major enzymes in ethanol production during glucose and xylose co-fermentation by metabolically engineered *Saccharomyces* yeast, *Enzyme Microbial Technol.*, 33: 19–28.
29. Bonnardeaux, J. 2007. *Potential Uses for Distillers Grains.* South Perth, Australia: Department of Agriculture and Food, Government of Western Australia.
30. U.S. Congress Office of Technology Assessment, 1980. *Energy from Biological Process,* U.S. Government Printing Office, Washington, D.C., Tech. Rep. 2, pp. 142–177.
31. Sneller, T. and Durant, D. 2008. The impact of ethanol production on food, feed, and fuel, *Food, Feed, Fuel,* Issue Brief by *Ethanol Across America.*
32. Novozymes. 2011. Cellulosic biofuel greatly benefits society. http://www.bioenergy.novozymes.com/cellulosic-ethanol/advantages-of-cellulosic-ethanol/; (August).
33. Szamant, H. 1978. Big push for a biomass bonanza, *Chem. Week,* 122: 40.
34. Zhu, J.Y., Pan, X.J., Wang, G.S., and Gleisner, R. 2009. Sulfite pretreatment (SPORL) for robust enzymatic saccharification of spruce and red pine, *Bioresource Technol.*, 100: 2411–2418.
35. Eggeman, T. and Atiyeh, C. 2010. The role for biofuels, *Chem. Eng. Prog.*, 106 (March): 36–38.
36. The White House. 2011. Growing America's fuel: An innovation to achieving the biofuels target. www.whitehouse.gov/sites/default/files/rss_viewer/growing_americas_fuels.pdf. October.
37. National Renewable Energy Laboratory. 2011, Reducing enzyme costs increases market potential of biofuels. *NREL/Innovation* http://www.nrel.gov/innovation/pdfs/47572.pdf. December.
38. U.S. EPA. 2011. Wastes - non-hazardous wastes - municipal solid wastes. http://www.epa.gov/osw/nonhaz/municipal/index.htm

39. Goldstein, I.S. 1990. Department of Wood and Paper Science, North Carolina University, *C&EN*, pp. 68, September 10.
40. C&EN. 1990. Major polymeric components of plant materials, *Chem. Eng. News* (September 10).
41. Diaz, L.F., Savage, G.M., and Golueke, C.G. 1984. Critical review of energy recovery from solid wastes, *CRC Critic. Rev. Environ. Control,* 14: 285–288.
42. Farina, G.E., Barrier, J.W., and Forsythe, M.L. 1988. Fuel alcohol production from agricultural lignocellulosic feedstocks, *Energy Sources,* 10: 231–237.
43. Wright, J.D. 1988. Ethanol from biomass by enzymatic hydrolysis, *Chem. Eng. Prog.,* 84: 62–74.
44. Kumar, R., Singh, S., and Singh, O.V. 2008. Bioconversion of lignocellulosic biomass: Biochemical and molecular perspectives, *J. Ind. Microbiol. Biotechnol,* 35: 377–391.
45. Viikari, L., Alapuranen, M., Puranen, T., Vehmaanperä, J., and Siika-Aho, M. 2007. Thermostable enzymes in lignocellulose hydrolysis, *Adv. Biochem. Eng. Biotechnol.,* 108: 121–145.
46. Vallander, L. and Erikkson, K. 1991. Enzymatic hydrolysis of lignocellulosic materials: I. Models for the hydrolysis process - A theoretical study, *Biotechnol. Bioeng.,* 38: 135–138.
47. Cort, J.B., Pschorn P., and Stromberg, B. 2010. Minimize scale-up risk, *Chem. Eng. Prog.,* 106 (March): 3–49.
48. Carvalheiro, F., Duarte L.C., and Girio, F.M. 2008. Hemicellulose biorefineries: A review on biomass pretreatments, *J. Sci. Indust. Res.,* 67: 849–864.
49. Dekker, R.F.H. and Wallis, A.F.A. 1983. Enzymic saccharification of sugarcane bagasse pretreated by autohydrolysis–steam explosion, *Biotechnol. Bioeng.,* 25: 3027–3048.
50. Mason, W.H. 1926. Process and apparatus for disintegrating wood and the like, U.S. Patent No. 1,578,609, March 30.
51. Ojumu, T.V. and Ogunkunle, O.A. 2005. Production of glucose from lignocellulosics under extremely low acid and high temperature in batch process - Authohydrolysis approach, *J. Appl. Sci.* 5: 15–17.
52. Galbe, M. and Zacchi, G. 2002. A review of the production of ethanolfrom softwood, *Appl. Microbial Biotechnol.,* 59: 618–628.
53. Rughani, L. and McGinnis, G.D. 1989. Combined rapid steam hydrolysis and organosolv pretreatment of mixed southern hardwoods, *Biotechnol. Bioeng.,* 33: 681–686.
54. Pan, X.J., Xie, D., Yu, R.W., Lam, D., and Saddler, J.N. 2007. Pretreatment of lodgepole pine killed by mountain pine beetle using the ethanol organosolv process: Fractionation and process optimization, *Ind. Eng. Chem. Res.,* 46: 2609–2617.
55. Simmons, B.A., Singh, S., Holmes B.M., and Blanch, H.W. 2010. Ionic liquid pretreatment, *Chem. Eng. Prog.,* 106(March): 50–55.
56. Swatloski, R.P. 2002. Dissolution of cellulose with ionic liquids, *J. Amer. Chem. Soc.,* 124: 4974–4975.
57. Dadi, A.P. 2007. Mitigation of cellulose recalcitrance to enzymatic hydrolysis by ionic liquid pretreatment, *Appl. Biochem. Biotechnol.,* 137:407–421.
58. Holtzapple, M.T., Cognata, M., Shu, Y., and Hendrickson, C. 1991. Inhibition of *Trichoderma reesei* cellulase by sugars and solvents, *Biotechnol. Bioeng.,* 38: 296–303.

59. Blotkamp, P.J., Takagi, M., Pemberton, M.S., and Emert, G.H. 1978. Biochemical engineering: Renewable sources of energy and chemical feedstocks. In J.M. Nystrom and S.M. Barnett (Eds.), *AIChE Symposium Series*, New York: AIChE.
60. Beltrame, P.L., Carniti, P., Focher, B., Marzetti, A., and Sarto, V. 1984. Enzymatic hydrolysis of cellulosic materials: A kinetic study, *Biotechnol. Bioeng.*, 26: 1233–1238.
61. Ohmine, K., Ooshima, H., and Harano, Y. 1983. Kinetic study on enzymatic hydrolysis of cellulose by cellulase from *Trichoderma viride*, *Biotechnol. Bioeng.*, 25: 2041–2053.
62. Okazaki, M. and Young, M. 1978. Kinetics of enzymatic hydrolysis of cellulose: Analytic description of mechanistic model, *Biotechnol. Bioeng.*, 20: 637–663.
63. Ryu. D.Y. and Lee, S.B. 1986. Enzymatic hydrolysis of cellulose: Determination of kinetic parameters, *Chem. Eng. Commun.*, 45: 119–134.
64. Gonzales, G., Caminal, G., de Mas, C., and Santin, J.L. 1989. *J. Chem. Tech. Biotechnol.*, 44: 275.
65. Garcia, A., Lannotti, E.L., and Fischer, J.L. 1986. Butanol fermentation liquor production and separation by reverse osmosis, *Biotechnol. Bioeng.*, 28: 785–791.
66. Vallander, L. and Erikkson, K. 1991. Enzymatic hydrolysis of lignocellulosic materials: II. Experimental investigation of theoretical hydrolysis process models for an increased enzyme recovery, *Biotechnol. Bioeng.*, 38: 139–144.
67. Szengyel, Z., Zacchi, G., Varga, A., and Réczey, K. 2000. Cellulase production of Trichoderma reesei Rut C 30 using steam-pretreated spruce. Hydrolytic potential of cellulases on different substrates, *Appl. Biochem. Biotechnol.*, 84–86: 679–691.
68. Moses, V., Springham, D.G., and Cape, R.E. 1991. *Biotechnology - The Science and the Business*, Harwood Academic.
69. Um, B.H. and Hanley, T.R. 2008. High-solid enzymatic hydrolysis and fermentation of solka floc into ethanol, *J. Microbiol. Biotechnol.*, 18: 1257–1265.
70. International Fiber Corporation. 2011. Solka floc. http://www.ifcfiber.com/products/solkafloc.php, August.
71. Ooshima, H., Burns D.S., and Converse, A.O. 1990. Adsorption of cellulase from *Trichoderma reesei* on cellulose and lignacious residue in wood pretreated by dilute sulfuric acid with explosive decompression, *Biotechnol. Bioeng.*, 36: 446–452.
72. Lu, Y., Yang, B., Gregg, D., Saddler, J.N., and Mansfield, S.D. 2002. Cellulase adsorption and an evaluation of enzyme recycle during hydrolysis of steam-exploded softwood residues, *Appl. Biochem. Biotechnol.*, 98–100: 641–654.
73. Kadam, S. and Demain, A. 1989. Addition of cloned beta-glucosidase enhances the degradation of crystalline cellulose by the Clostridium thermocellum cellulase complex, *Biochem. Biophys. Res. Comm.*, 161: 706–711.
74. Torget R. 1996. *Milestone Completion Report: Process Economic Evaluation of the Total Hydrolysis Option for Producing Monomeric Sugars Using Hardwood Sawdust for the NREL Bioconversion Process for Ethanol Production*, Golden, CO: National Renewable Research Laboratory.
75. Nguyen, Q. 1998. *Milestone Completion Report: Evaluation of a Two-Stage Dilute Sulfuric Acid Hydrolysis Process*, Golden, CO: National Renewable Research Laboratory.

76. Fan, L.T., Lee, Y., and Beardmore, D.H. 1980. Mechanism of the enzymatic hydrolysis of cellulose: Effects of major structural features of cellulose on enzymatic hydrolysis, *Biotechnol. Bioeng.*, 22: 177–199.
77. Spindler, D.D., Wyman, C.E., Mohagheghi, A., and Grohmann, K. 1988. Thermotolerant yeast for simultaneous saccharification and fermentation of cellulose to ethanol, *Appl. Biochem. Biotech.*, 279–293.
78. Anonymous. 2004. *The American Heritage Stedman's Medical Dictionary*. New York: Houghton Mifflin.
79. Sarthy, A., McConaughy, L., Lobo, Z., Sundstorm, A., Furlong E., and Hall, B. 1987. Expression of the *Escherichia coli* xylose isomerase gene in *Saccharomyces cerevisiae*, *Appl. Environ. Microbiol.*, 53: 1996–2000.
80. Beall, D.S., Ohta, K., and Ingram, L.O. 1991. Parametric studies of ethanol production from xylose and other sugars by recombinant *Escherichia coli*, *Biotechnol. Bioeng.*, 38: 296–303.
81. Ohta, K., Beall, D.S., Mejia, J.P., Shanmugam, K.T., and Ingram, L.O. 1991. Genetic improvement of Escherichia coli for ethanol production: Chromosomal integration of Zymomonas mobilis genes encoding pyruvate decarboxylase and alcohol dehydrogenase Il, *Appl. Environ. Microbiol.*, 57: 893–900.
82. Ingram, L.O. and Conway, T. 1988. Expression of different levels of ethanologenic enzymes from Zymomonas mobilis in recombinant strains of Escherichia coli, *Appl. Environ. Microbiol.*, 54: 404.
83. Alterthum, F. and Ingram, L.O. 1989. Efficient ethanol production from glucose, lactose, and xylose by recombinant Escherichia coli, *Appl. Environ. Microbiol.*, 55: 1943–1948.
84. Skoog, K. and Hahn-Hagerdal, B. 1988. Xylose fermentation, *Enzyme Microbiol. Technol.*, 10: 66–80.
85. Kyung, K.H. and Gerhardt, P. 1984. Continuous production of ethanol by yeast immobilized in membrane-contained fermenter, *Biotechnol. Bioeng.*, 26: 252.
86. Christen, P., Minier, M., and Renon, H. 1990. Enhanced extraction by supported liquid membrane during fermentation, *Biotechnol. Bioeng.*, 36: 116–123.
87. Naee, D.G. 1990. ACS Press Conference, 200th National Meeting of the ACS, August 1990, pp. 17, September 10.
88. Shapouri, H., Duffield, J.A., and Wang, M. 2004. The energy balance of corn ethanol: An update, U.S. Department of Agriculture, October.
89. Mueller, S. and Copenhaver, K. 2010. News from corn ethanol: Energy use, co-products, and land use, *Near-Term Opportunities for Biorefineries Symposium*, October 11–12, Champaign, IL.
90. Mueller, S. 2008. Detailed report: 2008 national dry mill corn ethanol industry. University of Illinois-Chicago. http://ethanolrfa.3cdn.net/2e04acb7ed88d08d21_99m6idfc1.pdf.
91. de Groot, P. and Hall, D. 1986. Power from the farmers, *New Scientist*, 50–55.
92. Da Silva, J.G., Serra, G.E., Moreira, J.R., Concalves J.C., and Goldenberg, J. 1978. Energy balance for ethyl alcohol production from crops, *Science*, 201: 903–906.
93. US DOE, Office of Energy Efficiency and Renewable Energy, 2002. Bioethanol process based on enzymatic cellulose hydrolysis, *NREL/PO-510-31788*, June.
94. Hopkinson, C.S. and Davy, J.W. 1980. Net energy analysis of alcohol production from sugarcane, *Science*, 207: 302–304.

95. Schmer, M.R., Vogel, K.P., Mitchell, R.B., and Perrin, R.K. 2008. Net energy of cellulosic ethanol from switchgrass, *Proc. Nat. Acad. Sci. USA,* 105 (January 7): 464–469.
96. Gladius, L. 1984. Some aspects of the production of ethanol from sugar cane residues in Zimbabwe, *Solar Energy,* 33: 379–382.
97. McAloon, A., Taylor, F., Yee, W., Ibsen K., and Wooley, R. 2000. *Determining the Cost of Producing Ethanol from Corn Starch and Lignocellulosic Feedstocks,* Tech. Rep. NREL/TP-580-28893. Golden, CO: National Renewable Energy Laboratory.

5

Fast Pyrolysis and Gasification of Biomass

5.1 Biomass and Its Utilization

5.1.1 Definition of the Term *Biomass*

The term *biomass* has been an important part of legislation enacted by Congress for many decades and has evolved over time, resulting in a variety of differing and sometimes conflicting definitions [1]. These definitions are critical to all parties engaged in the research, development, finance, and application of biomass to produce energy. The term biomass is more generally defined as "different materials of biological origin that can be used as a primary source of energy" [2–5]. Alternately, biomass is defined as "plant materials and animal wastes used especially as a source of fuel" [6]. These biomass definitions contain the generalized statements for the origins of the materials or their intended uses and applications, however, the definitions are not meant to provide sufficient and necessary conditions for certain specific material to be classified, or qualified, as biomass.

Riedy and Stone (2010) explained the evolving nature of biomass definitions and analyzed its trend in biomass-related legislation [1]. Based on the common definition that biomass is biologically originated matter that can be converted into energy, more readily conceivable and common examples of biomass include food crops, nonfood crops for energy generation, crop residues, woody materials and by-products, animal waste, and residues of biological fuel-processing operations. Over the past decades, however, the term biomass has grown to encompass algae and algae-processing residues, municipal solid waste (MSW), yard waste, and food waste. The term still remains highly flexible and open to divergent interpretations, including specific inclusions and special exceptions, often based on a number of factors involving technoeconomic considerations, technological advances, and new scientific findings, renewability and sustainability issues, environmental and climate change concerns, ecological issues, strategic directions of local and federal governments, regional economic strengths and weaknesses, and more. Simply put, biomass is a very broad term and encompasses a wide variety of matters. This is why the term biomass itself has been a part of modern legislation promulgated by the U.S. Congress. Legislation can have many purposes: to regulate, to authorize, to provide funds and incentives, to sanction, to grant, to proscribe, to declare, or to restrict. Using a globally generic definition of biomass in specific legislation would be not only grossly insufficient and inappropriate, but also potentially conflicting and controversial [1].

This book deals with conversion of biomass into biofuels and bioenergy, generally speaking. For the same reason described above, the individual chapters of this book are subdivided, based on the specific types of biomass and their associated transformation technologies, into the technologically categorized topics of corn ethanol, cellulosic ethanol, biodiesel, algae biodiesel, waste-to-energy, biomass pyrolysis and gasification, and so on.

The discussion of biomass definitions lately has centered around the issues involving: (i) the types of forestry products considered eligible biomass sources, (ii) the lands where biomass removal can occur, specifically Federal and Indian lands, and (iii) the kinds and types of waste that qualify as biomass, specifically municipal solid waste and construction and debris (C&D) [1].

5.1.2 Renewability and Sustainability of Biomass Feedstock

An aspect that is quite attractive in biomass utilization is its renewability that ultimately guarantees nondepletion of the resource. Considering all plants and plant-derived materials, all energy is originally captured, transformed, and stored via a natural process of photosynthesis. Strictly following the aforementioned definitions of biomass, it can be safely said that energy from biomass has been exploited by humans for a very long time in all geographical regions of the world. The combustion, or incineration, of biological substances such as woody materials and plant oils has long been exploited to provide warmth, illumination, and energy for cooking. It has been estimated that, in the late 1700s, approximately two-thirds of the volume of wood removed from the American forest was for energy generation [3]. Because wood was one of the only renewable energy sources readily exploitable at the time, its use continued to grow until the mid-to-late 1800s, when petroleum was discovered and town gas infrastructure based on coal gasification was introduced. It was reported that during the 1800s, single households consumed an average of 70 to 145 m3 of wood annually for heating and cooking [7, 8]. A small percentage of rural communities in the United States still use biomass for these purposes. Countries including Finland use the direct combustion of wood for a nontrivial percentage of their total energy consumption [9]. Furthermore, Finland has spent significant R&D efforts in biomass utilization programs and has successfully developed a number of advanced biomass conversion technologies. Finland and the United States are not the only countries that use biomass consumption for supplementing their total energy usage. In fact, the percentage of biomass energy of the total energy consumption for a country is far greater in African nations and many other developing countries.

An assessment by the World Energy Council (WEC) [10] reported that the 1990 biomass usage in all forms accounted for 1,070 MTOE, which is approximately 12% of global energy consumption of 8,811 MTOE assessed for the same year. MTOE stands for metric tonnes of oil equivalent. In 2010, about 16% of global energy consumption came from renewables, of which about 10% was contributed from traditional biomass, which was mainly used for heating, 3.4% from hydroelectricity, and 2.8% from so-called "new renewables" which included small hydro, modern biomass, wind, solar, geothermal, and biofuels [11]. The last category of new renewables has been growing very rapidly based on the development of advanced technologies as well as the global fear of depletion of conventional petroleum fuel.

On a larger scale, biomass is currently the primary fuel in the residential sector in many developing countries. Their biomass resources may be in the form of wood, charcoal, crop waste, or animal waste. For these countries, the most critical function of biomass fuel is for cooking, with the other principal uses being lighting and heating. The dependence on biomass for the critical energy supply for these countries is generally decreasing, whereas that for industrialized countries is more strategically targeted for new generation biomass energy. According to REN21 [11], the top five nations in terms of existing biomass power capacity in 2011 are the United States, Brazil, Germany, China, and Sweden in order of one to five. The top two nations of this list are also the top two nations in bioethanol production, not coincidentally. In other words, the biomass power category has so far been propelled and dominated mostly by the bioethanol transportation fuel sector.

5.1.3 Woody Biomass and Its Utilization

Of diverse biomass resources, woody biomass is of particular interest for biomass energy. Woody biomass is used to produce bioenergy and a variety of biobased products including lumber, composites, pulp and paper, wooden furniture, building components, round wood, ethanol, methanol, and chemicals, and energy feedstock including firewood.

The U.S. national Energy Policy Act (EPAct) of 2005 recognized the importance of a diverse portfolio of domestic energy. This policy outlined 13 recommendations designed to increase America's use of renewable and alternative energy. One of these recommendations directed the Secretaries of the Interior and Energy to re-evaluate access limitations to federal lands in order to increase renewable energy production, such as biomass, wind, geothermal, and solar. The Departments of Agriculture and Interior are jointly implementing the National Fire Plan (NFP), the President's Healthy Forests Initiative, the Healthy Forest Restoration Act, and the Tribal Forest Protection Act of 2004 to address the risk of catastrophic wildland fires, reduce their impact on communities, assure firefighting capabilities for the future, and improve forest and rangeland health on federal lands by thinning biomass density. The NFP includes five key points: (1) firefighting preparedness, (2) rehabilitation and restoration of burned areas, (3) reduction of hazardous fuels, (4) community assistance, and (5) accountability [12].

On June 18, 2003, the U.S. Departments of Energy, Interior, and Agriculture jointly announced an initiative to encourage the use of woody biomass from forest and rangeland restoration and hazardous fuels treatment projects. The three departments signed a Memorandum of Understanding (MOU) on Policy Principles for Woody Biomass Utilization for Restoration and Fuel Treatment on Forests, Woodlands, and Rangelands, supporting woody biomass utilization as a recommended option to help reduce or offset the cost and increase the quality of the restoration or hazardous fuel reduction treatments [12, 13].

One of the gateway process technologies for bioenergy generation from woody biomass is gasification, whose resultant product is biomass synthesis gas, also known as biomass syngas or biomass gas. The biomass syngas is similar in nature and composition to coal-based or natural gas-based syngas, whereas differences are largely originated from the source-specific properties. Similarly to the syngas generated via coal gasification or natural gas reformation, biomass syngas is also rich in hydrogen, carbon monoxide, carbon dioxide, and methane and as such this syngas can be used as building block chemicals [14] for a variety of synthetic fuels and petrochemicals and also as a feedstock for electric power generation.

5.1.4 Thermal and Thermochemical Conversion of Biomass

There are five thermal or thermochemical approaches that are commonly used to convert biomass into an alternative fuel/energy: direct combustion, gasification, liquefaction, pyrolysis, and partial oxidation. These five modes of conversion are also applicable to the conventional utilization of coal. When biomass is heated under oxygen-deficient conditions, it generates bio-oil and bio gas that consists primarily of carbon dioxide, methane, and hydrogen. When biomass is gasified at a higher temperature with an appropriate gasifying medium, synthesis gas is produced, which has similar compositions to that of coal syngas. This syngas can be directly burned or further processed for other gaseous or liquid fuel products. In this sense, thermal or thermochemical conversion of biomass is very similar to that of coal [5]. Figure 5.1 shows a variety of process options of biomass treatment and utilization as well as their resultant alternative fuel products.

Of a variety of thermochemical conversion options of biomass, this chapter is mainly focused on fast pyrolysis and gasification of biomass due to their technological significance, and other options of thermochemical conversion are discussed whenever deemed relevant.

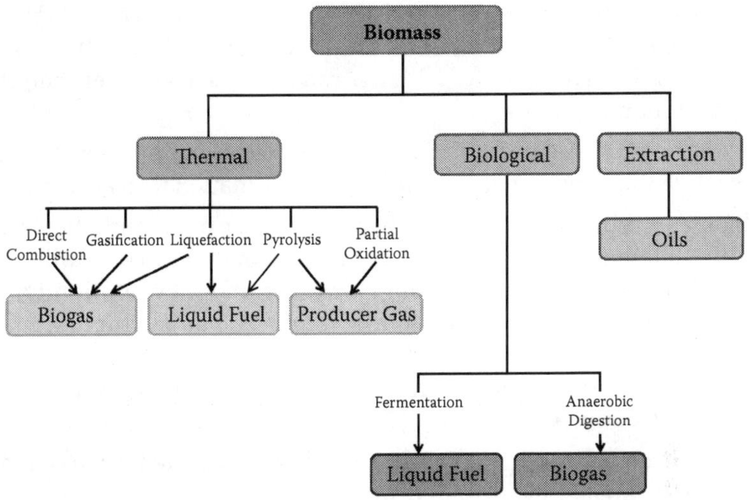

FIGURE 5.1
Conversion of biomass to alternative fuels. (Modified from Lee, Speight, and Loyalka, 2007. *Handbook of Alternative Fuel Technologies*. Boca Raton, FL: CRC Press.)

5.2 Analysis and Composition of Biomass

5.2.1 Similarities and Differences between Biomass and Coal as Feedstock

Even though biomass gasification has long been practiced on a variety of scales with and without modern scientific understanding, the subject itself has greatly benefited from the coal science and technology which has been studied far more in depth [15]. Most of the scientific tools developed for and applicable to coal technology are more or less relevant to biomass utilization technology and they include analytical methods, solids handling technology, chemical reaction pathways, reactor designs and configurations, process integration, waste heat recovery and energy integration, gas cleanup, product separation, safety practice and measures, and much more. However, the compositional differences between coal and biomass feedstock as well as their impact on processing technologies must be clearly understood for full and beneficial exploitation of the advances and innovations made in coal processing technology. The differences between biomass and coal feedstock are summarized below.

1. The hydrogen content in biomass is significantly higher than that of coal. Coal is a very mature product of a lengthy and slow coalification process whose principal chemical reaction is carbonization, whereas biomass is not. The coal rank basically indicates the degree

of carbonization progressed and a higher rank coal is a petrologically older coal. Therefore, the H/C ratio of coal is much lower than that of biomass. The H/C ratio of a higher rank coal is also lower than that of a lower rank coal.

2. A higher H/C ratio of biomass feedstock makes it generally more reactive than coal for conventional transformational treatments.
3. Biomass typically has much higher moisture content than coal. This statement is applicable to both forms of moisture: equilibrium and chemically bound. Among various ranks of coal, lignite, the lowest rank coal which is also the youngest, has the highest moisture content. Therefore, of all ranks of coal, lignite may be considered the closest to biomass in terms of both proximate analysis and ultimate analysis.
4. Biomass contains a significantly higher oxygen content than coal due to its oxygenated molecular structures of carbohydrates (or polysaccharides), cellulose, glycerides, fatty acids, and the like. Weathered coals (i.e., coal exposed to atmosphere after mining) show an increased level of oxygen content over that of freshly mined coal. However, the oxygen content of weathered coal is still far lower than that of typical biomass. Due to the high oxygen content of biomass fuel, its heating value is substantially lower than that of coal.
5. Coal contains 0.5–8 wt% sulfur (S), whereas biomass has little or no sulfur content. Coal with lower than 1 wt% sulfur may be classified as low-sulfur coal, whereas coal with higher than 3 wt% sulfur may be considered high-sulfur coal. No such designation or classification is needed for typical biomass. In coal syngas sulfurous compounds, if not removed, affect the downstream processing severely by poisoning the catalysts and also by starting corrosion on metallic parts and equipment. In this regard, biomass is considered a sulfur-free raw material. Sulfurous compounds in coal syngas typically include H_2S, carbonyl sulfide (COS), and mercaptans (R-SH), whose prevailing abundance depends largely on the gasifying environment as well as the feed coal composition. Furthermore, coal sulfur is subdivided largely into three different forms: pyritic sulfur, organic sulfur, and sulfatic sulfur [14]. However, such a subcategorization for forms of sulfur is unnecessary for biomass.
6. Alkali metals such as sodium (Na) and potassium (K) as well as low-boiling heavy metals such as lead (Pb) and cadmium (Cd) are typically present in raw biomass syngas [16]. This is not as severe for coal syngas and the trace element problems with coal syngas are more source-specific. Due to the trace mineral components in the biomass syngas, downstream processing of biomass syngas, in particular catalytic processing, requires rather comprehensive purification

pretreatment of feed syngas or use of robust and poison- and fouling-resistant catalysts.

7. Fuel analysis of both coal and biomass is represented by proximate analysis and ultimate analysis. Proximate and ultimate analyses of a variety of biomass samples found in the literature are presented in Tables 5.1 and 5.2, respectively.
8. Due to the high abundance of moisture, high oxygen content, and noncombustible impurities in biomass, the heating value of biomass is typically much lower than that of coal. The energy density of biomass feedstock on a volume basis is therefore substantially inferior to that of coal.
9. Biomass has substantially higher volatile matter (VM) content than coal, although it has much lower fixed carbon (FC) content than coal. Therefore, a large amount of hydrocarbon species can be extracted/obtained from biomass simply via devolatilization or pyrolysis, whereas devolatilization or pyrolysis of coal generates a high amount of char.

TABLE 5.1

Proximate Analysis of Biomass Species (Dry Basis)

	Fixed Carbon, %	Volatile Matter, %	Ash, %	HHV, kJ/kg (measured)	Data Source
Cotton shell briquette	17.1	77.8	5.1	19,055	Parikh, Channiwala, and Ghosal, 2005 [18]
Block wood	14.6	83.3	2.1	18,261	Parikh, Channiwala, and Ghosal, 2005 [18]
Plywood	21.8	74.2	4.0	19,720	Parikh, Channiwala, and Ghosal, 2005 [18]
Corn stover	17.6	78.7	3.7	17,800	Demirbas, 1997 [19]
Corn cob	18.5	80.1	1.4	18,770	Demirbas, 1997 [19]
Wood chips	23.5	76.4	0.1	19,916	Jenkins, 1980 [20]
Hazelnut shell	28.3	69.3	1.4	19,300	Demirbas, 1997 [19]
Redwood	19.9	79.7	0.4	20,720	Demirbas, 1997 [21]
Softwood	28.1	70.0	1.7	~20,000	Demirbas, 1997 [19]
Eucalyptus	21.3	75.4	3.3	18,640	Parikh, Channiwala, and Ghosal, 2005 [18]
Douglas fir	25.8	73,0	1.2	22,098	Tillman, 1978 [22]
Walnut	20.8	78.5	0.7	19,967	Parikh, Channiwala, and Ghosal, 2005 [18]
Wheat straw	23.5	63.0	13.5	~17,000	Parikh, Channiwala, and Ghosal, 2005 [18]
Rice straw (ground)	16.2	68.3	15.5	15,614	Parikh, Channiwala, and Ghosal, 2005 [18]

Note: All percentages are in dry weight percents.

TABLE 5.2

Ultimate Analysis of Biomass Species (Dry Basis)

	Elemental Analysis (Dry wt%)					
	C	H	O	N	S	Date Source
Cotton shell briquette	–	–	–	–	–	Parikh, Channiwala, and Ghosal, 2005 [18]
Block wood	46.9	6.07	43.99	0.95	0	Parikh, Channiwala, and Ghosal, 2005 [18]
Plywood	–	–	–	–		Parikh, Channiwala, and Ghosal, 2005 [18]
Corn stover	–	–	–	–	–	Demirbas, 1997 [19]
Corn cob	46.58	5.87	45.46	0.47	0.01	Demirbas, 1997 [19]
Wood chips	48.1	5.99	45.74	0.08	0	Jenkins, 1980 [20]
Hazelnut shell	52.9	5.6	42.7	1.4	–	Demirbas, 1997 [19]
Redwood	50.64	5.98	42.88	0.05	0.03	Jenkins and Ebeling, 1985 [21]
Softwood	52.1	6.1	41	0.2	–	Demirbas, 1997 [19]
Eucalyptus	46.04	5.82	44.49	0.3	0	Parikh, Channiwala, and Ghosal, 2005 [18]
Douglas fir	56.2	5.9	36.7	0	0	Tillman, 1978 [22]
Walnut	48.2	6.25	43.24	1.61	–	Parikh, Channiwala, and Ghosal, 2005 [18]
Wheat straw	45.5	5.1	34.1	1.8	–	Parikh, Channiwala, and Ghosal, 2005 [18]
Rice straw (ground)	–	–	–	–	–	Parikh, Channiwala, and Ghosal, 2005 [18]

Note: All percentages are in dry weight percents.

10. Biomass is generally composed of softer organic materials and its grindability or pulverizability is poor using common size reduction equipment. Considering the irregular shapes and nonuniform compositions of untreated biomass components, cost-effective size reduction for manageable transportation as well as continuous reactor feeding often becomes a technological challenge. Pretreatment of biomass feedstock is usually required for industrialized utilization.
11. Both biomass gasification and coal gasification encounter varying degrees of tar formation during thermal/chemical transformation, however, the severity of tar formation is typically more significant with biomass gasification. Although tar is collectively a carcinogenic species, it condenses at reduced temperatures, thereby blocking and clogging pipelines and valves as well as fouling process equipment and parts.

5.2.2 Analysis of Biomass

As mentioned in the previous section, the proximate and ultimate analyses of specific biomass material provide very valuable information about the biomass feedstock. This compositional information provides the science and engineering information needed to identify or determine the fuel heating value, ash amount projected, maximum achievable gasification and liquefaction efficiency, moisture content of feedstock, predicted behavior of the feedstock in a processing environment, and much more. The proximate analysis is a procedure for determination, by prescribed methods, of moisture (MO), volatile matter (VM), fixed carbon (FC), and ash. The amount of fixed carbon is determined by difference. The term proximate analysis involves neither determination of quantitative amounts of chemical elements nor determination other than those categorically named or prescribed. The group of analyses involved in proximate analysis is defined in ASTM D 3172. On the other hand, the ultimate analysis is a procedure of the determination of the elemental composition of the organic portion of carbonaceous materials, as well as the total ash and moisture. The ultimate analysis is also called elemental analysis. And this analysis is also determined by prescribed methods.

An extensive tabulation of both proximate and ultimate analysis data on over 200 biomass species was presented in Channiwala's PhD dissertation (1992) [17]. Some representative values of proximate and ultimate analyses of a variety of biomass species, as found from the literature sources, are presented in Tables 5.1 and 5.2, respectively. For comparison, the classification of coal and typical analysis is also shown in Table 5.3.

Comparing between the analyses of coal and biomass, the following generalized statements can be made.

1. Biomass has a very high oxygen (O) content, which is the second most abundant atomic species present in biomass and is nearly as much as the carbon (C) content. However, the oxygen content in coal is much lower than the carbon content and this trend is even more noticeable with higher rank coals. The higher the rank of a coal, the lower its oxygen content is. It may be said that deoxygenation (i.e., oxygen rejection) was an important part of a petrochemical process of coalification or carbonization.
2. Due to the higher oxygen content in biomass, the heating value of biomass is much lower than that of coal. Bio-oil derived from biomass also has a high oxygen content, which makes the oil more corrosive to metallic parts and piping. Therefore, efficient use of biomass as fuel or fuel precursor will involve a certain level of oxygen rejection (i.e., deoxygenation) in its process scheme.
3. The H/C ratio of biomass is substantially higher than that of coal. The reactivity of biomass is generally higher than that of coal and its processability is also better than coal's.

TABLE 5.3

Coal Classification and Analysis

	Average Analysis – Dry and Ash-Free (Daf) Basis						
	Volatile Matter (wt. %)	Hydrogen (wt. %)	Carbon (wt. %)	Oxygen (wt. %)	Heating Value (kJ/kg)	$\frac{C}{H}$	$\frac{C+H}{O}$
Anthracite							
Meta-	1.8	2.0	94.4	2.0	34,425	46.0	50.8
Anthracite	5.2	2.9	91.0	2.3	35,000	33.6	42.4
Semi-	9.9	3.9	91.0	2.8	35,725	23.4	31.3
Bituminous							
Low-volatile	19.1	4.7	89.9	2.6	36,260	19.2	37.5
Med-volatile	26.9	5.2	88.4	4.2	35,925	16.9	25.1
High-volat. A	38.8	5.5	83.0	7.3	34,655	15.0	13.8
High-volat. B	43.6	5.6	80.7	10.8	33,330	14.4	8.1
High-volat. C	44.6	4.4	77.7	13.5	31,910	14.2	6.2
Subbituminous							
Subbitum. A	44.7	5.3	76.0	16.4	30,680	14.3	5.0
Subbitum. B	42.7	5.2	76.1	16.6	30,400	14.7	5.0
Subbitum. C	44.2	5.1	73.9	19.2	29,050	14.6	4.2
Lignite							
Lignite A	46.7	4.9	71.2	21.9	28,305	14.5	3.6

Source: Lee, *Alternative Fuels*. Philadelphia: Taylor & Francis, 1996.

4. Among various ranks of coal, lignite is the closest to biomass in a number of properties including its high moisture content, high oxygen content, low carbon content, and low heating value. As such, lignite has often been considered as a co-fed companion fuel with biomass.

The standardized analysis of biomass fuel is conducted following the ASTM standards and Table 5.4 shows the list of these codes for specific analyses.

5.2.3 Thermochemical Conversion of Biomass

Thermochemical treatment of biomass can convert biomass into solid, liquid, and gaseous fuel products whose compositional distribution is governed by the imposed process treatment conditions. The solid product is usually char (or biochar), the liquid product is bio-oil, and the gaseous product is biosyngas. The process also involves formation of the unwanted product of tar. The thermochemical conversion process involves heating of the biomass feedstock, which triggers a series of parallel and consecutive reactions including

TABLE 5.4

Standardized Testing Procedure for Biomass Fuels

Test	Standardized Procedure	Desired Units
C	ASTM D5373	Weight %
H	ASTM D5373	Weight %
N	ASTM D5373	Weight %
Cl	ASTM D3761	mg/kg
S	ASTM D4239	Weight %
Proximate	ASTM D3172	Weight %
Moisture	ASTM D2013	Weight %
Ash	ASTM D5142	Weight %
Heat of Combustion	ASTM D5865	kJ/kg or BTU/lb

devolatilization of volatile matter, pyrolytic decomposition of hydrocarbons and other carbonaceous matters, gas–solid type gasification reactions, coke and char formation, tar and its precursor formation, and more. Simply speaking, depending upon the processing temperature and reactor residence time, thermochemical treatment of biomass can be regrouped into three basic types of process treatment, namely carbonization, fast pyrolysis, and gasification. As shown in Table 5.5, the principal product, or intended product, of fast pyrolysis is a liquid fuel, whereas the desired product of gasification is a gaseous fuel [23].

Even though it is not listed in Table 5.5, indirect liquefaction via the biosyngas route is also a viable option for liquid fuel production, as well demonstrated in the fields of coal and natural gas syngas [5, 14, 24, 25]. As the name implies, indirect liquefaction goes through two stages of process treatment, viz., gasification followed by liquefaction by which liquid hydrocarbon fuels such as methanol, dimethylether (DME), higher alcohols, gasoline, diesel, and jet fuel are synthesized using the syngas produced by the gasification stage. In this case, biomass syngas is a thermochemical intermediate for the next stage synthesis of liquid fuel.

TABLE 5.5

Biomass Treatment Processes and Their Product Distribution

	Process Treatment			Product Compositions		
	Temperature L, M, H	Residence Time	Air or O_2 Y, N	Solid Char (%)	Liquid Bio-oil (%)	Gas or Syngas (%)
Carbonization	Low	Long	N	35	30	35
Fast Pyrolysis	Medium	Short	N	12	75	13
Gasification	High	Long	Y	10	5	85

5.2.4 Analysis of Biomass Feedstock and Product Compositions

Fast pyrolysis of biomass generates a wide variety of organic and inorganic chemical compounds and the product compositions vary significantly depending upon the types of feedstock as well as the process treatment conditions to which the biomass is subjected. Therefore, studies of process modeling and technoeconomic analysis are often carried out using model compounds carefully chosen for the specific process and typical feedstock [26]. The analysis of corn stover samples used by Mullen et al. [27] for their fast pyrolysis study is presented in Table 5.6, as an example of the compositional analysis of biomass feedstock. They carried out the fast pyrolysis in a bubbling fluidized bed of quartz sand at a temperature of 500°C.

Inorganic elemental composition of corn stover used for the aforementioned pyrolysis determined by x-ray fluorescence (XRF) is given in Table 5.7. Also compared in the same table are the XRF analysis data for corn cobs which were also tested for fast pyrolysis by Mullen et al. (2010) [27]. As can be seen, the elemental composition between corn cob and corn stover are quite different. The most abundant element in corn cob was K, whereas Si was the most abundant species in corn stover. High levels of K and P in both samples are expected, although the high levels of Ca, Mg, Al, Fe, and Mn in corn stover are noteworthy. Mineral matters in the biomass feedstock can reappear as contaminants or trace elements in bio-oils and biosyngas, which can potentially affect the catalytic activity of the downstream processing by fouling or poisoning.

The yield data of the USDA corn stover fast pyrolysis by Mullen et al. [27] is shown in Table 5.8. The pyrolysis product distribution in terms of the product phases was bio-oil 61.7%, biochar 17.0%, and noncondensable gas (NCG) 21.9%. As explained earlier and also summarized in Table 5.5, the principal product of fast pyrolysis of biomass, that is, corn stover in this example, is bio-oil.

TABLE 5.6

Ultimate and Proximate Analysis of Corn Stover

Ultimate Analysis (Dry Basis)		Proximate Analysis (Wet Basis)	
Element	**Mass %**	**Ingredient**	**Mass %**
Carbon (C)	46.60	Moisture	25.0
Hydrogen (H)	4.99	Volatile Matter	52.8
Oxygen (O)	40.05	Fixed Carbon	17.7
Nitrogen (N)	0.79	Ash	4.5
Sulfur (S)	0.22		
Ash	6		

Source: Mullen et al. 2010. Bio-oil and bio-char production from corn cobs and stover by fast pyrolysis, *Biomass Bioenergy*, 34: 67–74.

TABLE 5.7

Inorganic Elemental Compositions of Corn Cobs and Corn Stovers by XRF

Inorganic Element	Corn Cobs (In g/kg or 1,000 ppm)	Corn Stover (In g/kg or 1,000 ppm)
Si	5.33	27.9
Al	0.18	5.09
Fe	0.08	2.35
Ca	0.23	3.25
Mg	0.55	2.34
Na	0.10	0.23
K	10.38	4.44
Ti	0.003	0.37
Mn	0.01	0.98
P	1.11	2.15
Ba	0.11	0.02
Sr	0.002	0.005
S (inorganic)	0.14	0.05

Source: Mullen et al. 2010. Bio-oil and biochar production from corn cobs and stover by fast pyrolysis, *Biomass Bioenergy*, 34: 67–74.

From the product compositions, the following observations are deemed significant:

1. The gaseous effluent of fast pyrolysis has a heating value of only 6.0 MJ/kg. The gas composition is dominated by carbon oxides (CO and CO_2), followed by methane and hydrogen.
2. High levels of oxygen in the effluent gas show that the gaseous effluent served at least as an outlet for deoxygenation of biomass.
3. Bio-oil also showed a very high level of oxygen and its heating value was seriously affected. In order to enhance the fuel quality of bio-oil as well as to enhance the fast pyrolysis process, a systematic rejection of oxygen from the products' molecular structures (i.e., deoxygenation) would become crucially important.
4. Biochar showed a heating value nearly as high as that of bio-oil, even though it contained a high level of ash.
5. Biochar showed a high C/H ratio, which is indicative of its lack of volatile hydrocarbons. Thus, biochar is a useful by-product of the fast pyrolysis process of biomass.

TABLE 5.8

Product Analysis of Fast Pyrolysis of Corn Stover

Gaseous Compounds	Vol. %
CO_2	40.3
CO	51.6
H_2	2.0
CH_4	6.0
HHV (MJ/kg)	6.0
Bio-Oil Compounds	**Mass %**
C	53.97
H	6.92
N	1.18
S	<0.05
O	37.94
Ash	<0.09
HHV (MJ/kg)	24.3 (dry)
Biochar	**Mass %**
C	57.29
H	2.86
N	1.47
S	0.15
O	5.45
Ash	32.78
HHV (MJ/kg)	21.0

5.3 Chemistry of Biomass Gasification

Gasification is a conversion process that transforms macromolecular carbonaceous matters contained in fossil fuels and biological substances into simpler gaseous molecular products. The gaseous products from a gasification reaction are called synthesis gas, syngas, or producer's gas. Depending upon the original carbonaceous feedstock, the syngas may be further labeled as coal syngas, natural gas (NG) syngas, or biomass syngas. Gasification of biomass usually takes place at an elevated temperature with the aid of a gasifying medium, which may be regarded as a gaseous reactant. Because gasification involves both heat and chemical(s) that induce concurrent thermal decomposition and chemical reactions, the gasification process is classified as a *thermochemical conversion* process.

5.3.1 Chemical Reactions Taking Place during Biomass Gasification

Typical biomass gasification takes place in the presence of injected air (or oxygen) and steam under high pressure at an elevated temperature, $T > 850°C$. In this regard, typical biomass gasification is very similar to advanced coal gasification process technologies [5, 28, 29]. The chemical reactions taking place in a biomass gasifier are very complex and they include: (1) pyrolytic decomposition of hydrocarbons and oxygenated organics such as carbohydrates (or saccharides) and cellulose, (2) further decomposition of fragmented hydrocarbons (of reduced molecular weights), (3) recombination of methylene and methyl radicals, (4) partial oxidation of hydrocarbons and oxygenates, (5) steam gasification of hydrocarbons and oxygenates, (6) water gas shift reaction, (7) formation of polycyclic aromatic hydrocarbons (PAHs) and potential coking precursors, (8) carbon dioxide gasification of carbonaceous materials, and more.

$$C_xH_y \rightarrow C_aH_b + C_cH_d + e \cdot H_2$$

$$C_xH_y \rightarrow C_fH_g + h \cdot CH_4 + j \cdot H_2$$

$$C_uH_vO_w \rightarrow C_kH_lO_{w1} + C_mH_nO_{w2} + p \cdot H_2O + q \cdot H_2 + r \cdot CO_2 + s \cdot CO$$

$$C_{x1}H_{y1} \rightarrow C_{f1}H_{g1} + K \cdot (\cdot CH_2 \cdot)$$

$$(\cdot CH_2 \cdot) + H_2 \rightarrow CH_4$$

$$(\cdot CH_2 \cdot) + (\cdot CH_2 \cdot) \rightarrow C_2H_4$$

$$C_xH_y + \left(\frac{x}{2} + \frac{y}{4}\right) \cdot O_2 \rightarrow x \cdot CO + \frac{y}{2} \cdot H_2O$$

$$C_uH_vO_w + \left(\frac{u}{2} + \frac{v}{4} - \frac{w}{2}\right) \cdot O_2 \rightarrow u \cdot CO + \frac{v}{2} \cdot H_2O$$

$$C_xH_y + x \cdot H_2 \rightarrow x \cdot CO + \left(x + \frac{y}{2}\right) \cdot H_2$$

$$C_uH_vO_w + (u - w) \cdot H_2O \rightarrow u \cdot CO + \left(\frac{v}{2} + u - w\right) \cdot H_2$$

$$CO + H_2O \leftrightarrow H_2 + CO_2$$

$$C_xH_y + x \cdot CO_2 \rightarrow 2x \cdot CO + \frac{y}{2} \cdot H_2$$

$$C_uH_vO_w + u \cdot CO_2 \rightarrow 2u \cdot CO + \left(\frac{v}{2} - w\right) \cdot H_2 + w \cdot H_2O$$

The first five reactions represent pyrolytic decomposition reactions of hydrocarbons and oxygenates, which provide some explanation for the formation of methane and lighter hydrocarbon species. The last five reactions explain the formation of carbon oxides and hydrogen, principal ingredients of biomass syngas. One can also notice that the above reactions are analogous to the four classical gasification reactions of carbon and concurrent water gas shift reaction as shown below [14].

$$C_s + H_2O \rightarrow CO + H_2$$

$$C_s + CO_2 \rightarrow 2 \cdot CO$$

$$C_s + 2 \cdot H_2 \rightarrow CH_4$$

$$C_s + O_2 \rightarrow CO/CO_2$$

$$CO + H_2O \leftrightarrow H_2 + CO_2$$

where C_s denotes carbon on the solid surface.

The reactions listed above are called steam gasification of carbon, Boudouard reaction, hydrogasification of carbon, partial oxidation of carbon, and water gas shift reaction, respectively. It should be clearly noted that the last three reactions, as written, are exothermic, whereas the first two are endothermic at their typical operating conditions. The water gas shift reaction can proceed either in the forward or reverse direction depending upon the temperature and imposed/developed reaction environment. The forward water gas shift reaction is mildly exothermic, whereas the reverse water gas shift reaction is mildly endothermic.

Chemical equilibrium constants for a wide range of temperatures for selected chemical reactions that are of significance to the gasification of carbon are listed in Table 5.9. Although the values are for reactions of carbon, neither of biomass nor of coal char, they still provide general ideas for reactions involving carbonaceous matters. As the C/H ratio of solid materials increases or the number of carbon atoms in a hydrocarbon molecule increases, their thermodynamic equilibrium values become closer to those of carbon reactions. Furthermore, if biomass is pretreated before any gasification reactions,

TABLE 5.9

Chemical Equilibrium Constants for Carbon Reactions

		ln K_p					
T, K	1/T, K^{-1}	$C + 1/2\, O_2 = CO$	$C + O_2 = CO_2$	$C + H_2O = CO + H_2$	$C + CO_2 = 2\,CO$	$CO + H_2O = CO_2 + H_2$	$C + 2H_2 = CH_4$
300	0.003333	23.93	68.67	−15.86	−20.81	4.95	8.82
400	0.0025	19.13	51.54	−10.11	−13.28	3.17	5.49
500	0.002	16.26	41.26	−6.63	−8.74	2.11	3.43
600	0.001667	14.34	34.40	−4.29	−5.72	1.43	2.00
700	0.001429	12.96	29.50	−2.62	−3.58	0.96	0.95
800	0.00125	11.93	25.83	−1.36	−1.97	0.61	0.15
900	0.001111	11.13	22.97	−0.37	−0.71	0.34	-0.49
1,000	0.001	10.48	20.68	0.42	0.28	0.14	−1.01
1,100	0.000909	9.94	18.80	1.06	1.08	−0.02	−1.43
1,200	0.000833	9.50	17.24	1.60	1.76	−0.16	−1.79
1,300	0.000769	9.12	15.92	2.06	2.32	−0.26	−2.1
1,400	0.000714	8.79	14.78	2.44	2.80	−0.36	−2.36

Source: Walker, P.L., Rusinko, F., and Austin, L.G. 1959, Gas reactions in carbon. In D.D. Eley, P.W. Selwood, and P.B. Weisz (Eds.), *Advances in Catalysis*, New York: Academic Press.

then the equilibrium values in the table would be more relevant and closer to the actual values.

The temperature where $K_p = 1$ (i.e., ln $K_p = 0$) has some extra significance, by indicating the general location of chemical equilibrium shift. The temperatures where $K_p = 1$ for steam gasification, Boudouard reaction, hydrogasification, and water gas shift reaction are 947 K (674°C), 970 K (697°C), 823 K (550°C), and 1,087 K (814°C), respectively. For example, it may be said that for the steam gasification of carbon to proceed in the forward direction, the gasification temperature must be higher than 674°C. Among the reactions listed in Table 5.9, the temperature-dependent variation of the equilibrium constant is the weakest for the water gas shift reaction, thus exhibiting an easily reversible nature of the chemical equilibrium for a very wide range of temperatures. This is the reason why the water gas shift reaction equilibrium becomes a player in nearly all syngas reaction systems under widely varying process conditions.

When the coal gasification reaction is explained or modeled, most technologists denote and simplify coal more or less as carbon, that is, C(s), based on the fact that the hydrogen content of coal is much lower than that for most hydrocarbons. However, such practice in the case of biomass gasification would become an oversimplification, inasmuch as the oxygen and hydrogen content in biomass feedstock are much higher than those of high rank coal. The abundance of oxygenated functional groups such as hydroxyl (–OH) groups in biomass makes most decomposition and transformation reactions proceed more easily.

5.3.1.1 Pyrolysis or Thermal Decomposition

Pyrolysis or thermal decomposition is the molecular breakdown of organic materials such as hydrocarbons via cleavages of chemical bonds at elevated temperatures without the involvement of oxygen or air. Typical chemical bond cleavages during pyrolysis involve the C–C and C–H bonds at most operating temperatures, whereas double bonds of C=C, C=O are substantially more difficult to break at most practical conditions. As can be imagined by the chemical bonds that are typically broken during pyrolysis, the pyrolysis reaction starts at a temperature as low as 150–200°C, where the intrinsic reaction rate is very slow and the extent of reaction is far from completion in any reasonable time. It should be clearly noted that this low temperature is not a typical operating temperature for pyrolysis, considering that the pyrolytic decomposition reactions are still present at this low temperature, even though not active at all. Most practical pyrolysis of hydrocarbons without using any catalyst is pursued at a temperature higher than 450°C. Pyrolysis involves the concurrent change of chemical compositions and physical phases, and the process is irreversible.

If high molecular weight hydrocarbons are pyrolyzed in an oxygen-deprived environment at a temperature of 450–650°C, lighter hydrocarbons (reduced carbon numbers and lower molecular weights), hydrogen, and solid char would be typically formed. Lighter hydrocarbons nearly always involve methane (CH_4), ethylene (C_2H_4), ethane (C_2H_6), and other fragmented hydrocarbons, of which methane is most dominant. Depending upon the pyrolysis conditions, liquid range hydrocarbons, $C_4 - C_{15}$, are also obtained. Char formation is believed to be via a route similar to the formation of polycyclic aromatic hydrocarbons (PAHs) which involves polymerization of highly reactive free radicals of fragmented hydrocarbons and unsaturated hydrocarbon species. Hydrogen formation during pyrolysis is via cleavage reactions of C–H chemical bonds of the original and intermediate hydrocarbon molecules and some of the hydrogen molecules formed in such a manner are recombined with methyl radicals and ethylene thus producing methane and ethane. As illustrated, pyrolysis of hydrocarbons can yield materials of all three different phases (i.e., solid, liquid, and gas) as its end products depending upon the treatment conditions. Furthermore, the actual number of chemical reactions involved and the number of final and intermediate chemical species are countless. Therefore, pyrolysis collectively represents a class of chemical reactions taking place as thermochemical decomposition. A more generalized chemical reaction equation for hydrocarbon pyrolysis may be written as follows:

$$C_aH_b \rightarrow c \cdot CH_4 + C_dH_e + C_fH_g + h \cdot H_2$$

$$a = c + d + f$$

$$b = 4c + e + g + 2h$$

$$C_uH_vO_w \rightarrow C_kH_lO_{w1} + C_mH_nO_{w2} + p \cdot H_2O + q \cdot H_2 + r \cdot CO_2 + s \cdot CO$$

$$u = k + m + r + s$$

$$v = l + n + 2p + 2q$$

$$w = w1 + w2 + p + 2r + s$$

In the first reaction expression, methane is explicitly written in the product side, because methane is always a dominant hydrocarbon product species of hydrocarbon pyrolysis.

Biomass chemical compounds are far more oxygenated than straight hydrocarbons and biomass also contains a high level of moisture. Therefore, thermal decomposition or pyrolysis of biomass also generates carbon oxides in addition to the aforementioned pyrolysis products of condensable hydrocarbons, methane, and hydrogen. If biomass is microbially degraded in anaerobic conditions, it generates a product gas rich in methane and carbon dioxide. This product gas is called *biogas,* or *landfill gas.* A process system developed for exploiting this biogas is called an *anaerobic digester,* which can produce methane rich gas from waste materials on a small-scale unit.

The scientific definition of pyrolysis presented in this section precludes oxygen involvement in its mechanistic reaction steps. This was necessary in defining pyrolysis as a thermochemical reaction by itself. However, it should be clearly noted that the actual pyrolysis reaction can also occur as a component reaction of many reactions simultaneously taking place in many different chemical process environments, including both oxidative as well as reducing environments. In such environments, pyrolytic decomposition reactions compete with other chemical reactions also occurring in the system and as such the reaction environment becomes that much more complex in terms of both the nature and the total number of simultaneous reactions taking place in the system.

Of course, it is also true that pyrolysis alone in the absence of oxygen can be targeted in certain process environments, as is the case with fast pyrolysis of biomass. Typical fast pyrolysis processes are operated at a temperature that is substantially lower than typical gasification temperatures of steam gasification, Boudouard reaction, hydrogasification, and partial oxidation. Hence, fast pyrolysis as a transformation process is more or less strictly a combination of devolatilization and pyrolytic decomposition reactions in an oxygen-deprived environment. Furthermore, it must be clearly understood that the principal targeted product of fast pyrolysis of biomass is liquid-phase bio-oil, not gaseous synthesis gas.

Biomass can be gasified without any gasifying agent additionally introduced into the reactor and this type of gasification is called *pyrogasification.*

Pyrolytic gasification or pyrogasification of biomass takes advantage of both pyrolysis and gasification and it can be carried out both catalytically [31, 32] and noncatalytically. In pyrogasification, no separate gasifying medium or oxygen (or air) is introduced, it is expected that a gasifying reactant such as steam has to be in situ provided from biomass pyrolysis. In pyrogasification, biomass pyrolysis also produces biochar and this biochar reacts with steam via steam gasification to generate product gas.

Pyrolytic gasification of wood using a stoichiometric nickel aluminate catalyst was carried out by Arauzo and coworkers [32] in a fluidized bed reactor and near-equilibrium yields of products were obtained above 650°C. Although they obtained 85–90% gas yields, tar production was not detected. They further tested the process using a modified nickel–magnesium aluminate stoichiometric catalyst and also an addition of potassium as a promoter. They found that magnesium addition to the catalyst crystalline lattice enhanced attrition resistance of the catalyst with a minor loss of gasification activity and an increased production of coke. However, they found little effect from an addition of potassium component. Catalyst fouling by carbon deposition, that is, surface coverage by coke, was shown and regeneration of magnesium-containing catalyst by carbon burn-off was demonstrated.

Asadullah et al. [31] comparatively evaluated the catalytic performances of $Rh/CeO_2/SiO_2$, steam reforming catalyst G-91, and dolomite for a number of different biomass gasification modes, including pyrolytic, CO_2, O_2, and steam. With respect to the biomass conversion to product gas and selectivity of useful gaseous species, $Rh/CeO_2/SiO_2$ has shown superior results in all gasification modes. In the pyrogasification case, about 79% of the carbon in biomass was converted to the product gas at 650°C. There was no tar detected in the effluent gas stream. The gasifier used for their experiment was a lab-scale continuously fed fluidized bed reactor.

5.3.1.2 Partial Oxidation

In chemical processes that generate synthesis gas from fossil or biomass feedstock, partial oxidation has been proven to be an important pivotal gasification reaction. The partial oxidation has several inherent merits, namely,

1. The reaction rate is very fast.
2. The reaction irreversibly proceeds over a very wide range of temperatures.
3. The reaction generates exothermic heat, which helps sustain the system's energy balance.
4. The reaction is universally and nearly equally efficient on all hydrocarbon molecules of widely different carbon numbers.

5. Partial oxidation of hydrocarbons generates hydrogen and carbon monoxide as principal product species, which are major components of synthesis gas product.
6. The partial oxidation reaction is an excellent companion reaction to many other chemical reactions, including steam gasification, steam reformation, and Boudouard reaction among others.
7. If partial oxidation is properly used in conjunction with other gaseous reactions, synergistic effects can result in (i) efficient process energy management including an autothermal operation, (ii) higher gas yield or higher gasification efficiency, (iii) higher conversion of carbon, (iv) tailormade gas composition or control of H_2/CO ratio in syngas, (v) reduction of char formation or resistance against coking, (vi) tar reduction, and more.

Most advanced coal gasification processes such as the Texaco gasifier and Shell gasifier utilize partial oxidation of coal as a principal reaction [5]. The reaction is often carried out in the copresence of steam, which gets involved in steam gasification as well as the water gas shift (WGS) reaction as:

$$C_{(s)} + \frac{1}{2}O_2 \rightarrow CO$$

$$C_{(s)} + H_2O \rightarrow CO + H_2$$

$$CO + H_2O \leftrightarrow CO_2 + H_2$$

If partial oxidation is poorly managed or improperly designed, an unnecessarily high extent of complete combustion of hydrocarbons can take place resulting in a large amount of carbon dioxide, thereby wasting the useful heating value of feedstock hydrocarbons as well as increasing greenhouse gas (GHG) formation even without subjecting the fuel to useful end-uses.

5.3.1.3 Steam Gasification

Steam reacts with carbonaceous materials including hydrocarbons, carbohydrates, oxygenates, natural gas, and even graphite at elevated temperatures and generates carbon monoxide and hydrogen. The stoichiometric chemical reactions in this class of reaction include:

$$C_{(s)} + H_2O_{(g)} = CO_{(g)} + H_{2\,(g)}$$

$$Coal + Steam = CO + H_2$$

$$CH_4 + H_2O = CO + 3 \cdot H_2$$

$$C_aH_b + a \cdot H_2O = a \cdot CO + \left(a + \frac{b}{2}\right) \cdot H_2$$

The first reaction represents steam gasification of carbon, whereas the second reaction is steam gasification of coal. The third reaction is steam reformation of methane (or, methane steam reformation, MSR), whereas the fourth reaction is known as reformation of hydrocarbon fuels. The chemical equilibrium favors the forward reaction of steam gasification of carbon, if the temperature of reaction exceeds 674°C, as explained in Section 5.3.1 and Table 5.9. This threshold temperature (and its vicinity) for forward reaction progress is nearly universally applicable to all hydrocarbon species including coal. As clearly shown, product hydrogen in these reactions at least partially originates from water (steam) molecules. Without separately going through a water-splitting reaction, this reaction efficiently extracts hydrogen out of water molecules and carbon atoms in the hydrocarbon molecules react with oxygen atoms from water molecules. As expected, the forward reactions, that is, steam gasification reactions as written, are highly endothermic at practical operating temperatures, requiring high energy input.

As mentioned earlier in the pyrolysis section, even when hydrocarbons are reformed or gasified by steam at elevated temperatures, thermochemical conversion due to pyrolysis is also taking place as a competing and parallel reaction to the steam gasification reaction. If a hydrocarbon feedstock, such as coal and biomass, is introduced into a reactor where steam gasification is desired at an elevated temperature, the resultant reaction usually proceeds as an apparent two-stage reaction, viz., appearing to be the pyrolysis reaction followed by steam gasification. Mathematically, this apparent two-stage reaction process can be explained by the result of superposition of two parallel reactions between one very fast reaction (pyrolysis) and one slow reaction (gasification). In this explanation, easily pyrolyzable components are rapidly broken down in the early period (e.g., a matter of a few seconds), whereas much slower gasification takes place more steadily over a much longer period of time (e.g., a matter of 0.5–3 hours). In coal gasification studies, some researchers interpreted this early stage pyrolysis result as an initial conversion [33].

Because biomass generally contains a high level of moisture, steam gasification reaction is nearly always present with or without a separate feed of steam into the reactor, except in the case of fast pyrolysis. In the fast pyrolysis of biomass, the typical temperature of operation is around 500°C, which is substantially lower than the steam gasification temperature, and therefore the biomass moisture is not involved in the steam gasification reaction.

5.3.1.4 Boudouard Reaction or Carbon Dioxide Gasification Reaction

Among the endothermic gasification reactions of hydrocarbons, the speed of carbon dioxide gasification reaction is the slowest at practical operating temperatures. Most advanced gasification technologies produce carbon dioxide as a component in their syngas products. The gasification using CO_2 has not been popularly attempted, due to its poorer thermal efficiency and inferior energetics compared to steam gasification. However, due to the growing concerns of greenhouse gas emissions as well as the roles of carbon dioxide as a major greenhouse gas, various technologies including the capture of CO_2, its reduction, utilization in carbon gasification, and conversion into other petrochemicals are actively pursued and developed. Gasification of biomass or coal coupled with CO_2 management is also an environmentally prudent option.

Complete combustion of biomass or fossil fuels generates carbon dioxide. Because carbon dioxide is chemically very stable, its reactivity is limited. Therefore, the conversion of carbon dioxide into far more reactive carbon monoxide is one of the technological options, whereas the direct conversion of carbon dioxide into hydrocarbons is another. The two types of reactions are categorized under the reduction of carbon dioxide, and finding energetically prudent pathways for CO_2 reduction is a challenge in modern fuel chemistry. The first group of chemical reactions includes the Boudouard reaction and the reverse water gas shift reaction:

$$C_{(s)} + CO_{2(g)} = 2CO_{(g)}$$

$$CO_{2(s)} + H_{2(g)} = CO_{(g)} + H_2O_{(g)}$$

As can be seen from Table 5.9, the temperature for $K_p > 1$ for the forward reactions as written to proceed for the Boudouard reaction and the reverse water gas shift (RWGS) reaction are 697°C and 814°C, respectively. Also, the RWGS reaction requires hydrogen as a reactant, which generally makes the process conversion costly.

Lee et al. [34] studied the kinetics of carbon dioxide gasification of various coal char samples for a temperature range between 800°C and 1,050°C using a unified intrinsic kinetic model and compared with the literature values obtained for various carbon, coal, and char samples. The Arrhenius activation energy values obtained for the carbon dioxide gasification for these samples are shown in Table 5.10 [34].

Obtained from independent investigations by various investigators on diverse carbonaceous materials, the activation energy values for the kinetic rate equations for carbon dioxide gasification are around 60 kcal/mol or 250 kJ/mol. This high activation energy is also indicative of the nature of chemical reaction which requires a high temperature reaction to attain a

TABLE 5.10
Activation Energy for CO_2 Gasification Reaction of Coal/Char/Graphite

Sample	Arrhenius Activation Energy, E		Investigators
	kcal/mol	kJ/mol	
Carbon	59–88	247–368	Walker et al., 1959 [30]
Anthracite, Coke	49–54	205–226	Von Federsdorff, 1963 [35]
Coke	68	285	Hottel et al., 1977 [36]
Graphite	87	364	Strange and Walker, 1976 [37]
Montana Rosebud char	60	251	Lee et al., 1984 [34]
Illinois No. 6 char	58	243	Lee et al., 1984 [34]
Hydrane No. 49 char	65	272	Lee et al., 1984 [34]

practically significant reaction rate. If we check the (E/RT) value for carbon dioxide gasification at 1,000°C, then the value becomes

$$\frac{E}{RT} = \frac{60,000}{1.987 * 1273} = 23.7$$

This (E/RT) value is within the range of the values for most industrially practiced petrochemical reactions, often used as a rule of thumb. The study [34] also established that the kinetic rate of the noncatalytic carbon dioxide gasification of coal char at practical operating conditions, such as 900°C and 250 psi, is substantial. The rate was found to be about two to four times slower than that of steam gasification at the same T and P conditions.

5.3.1.5 Hydrogasification

The term *hydrogasification* is the reaction between carbonaceous material and hydrogen, strictly speaking. However, an augmented definition of "hydrogasification" involves the reaction of carbonaceous material in a hydrogen-rich environment to generate methane as a principal product. The gasification reaction in certain steam environments, such as in the copresence of steam and hydrogen, often qualifies for this hydrogen-rich environment for methane generation. The latter is called *steam hydrogasification* [38]. However, the steam gasification whose principal goal is to produce syngas should still be referred to as "steam gasification," not simply as "hydrogasification." "Hydro-" as a prefix is used for "of water" or "of hydrogen," depending upon the situation. As far as the gasification of coal and biomass is concerned, the prefix "hydro-" means "of hydrogen," as in the case with "hydrocracking."

Carbonaceous materials undergo *hydrocracking* under high pressures of hydrogen at elevated temperatures. Hydrocracking generates lighter hydrocarbons as cleavage products from larger hydrocarbons. Although the hydrocracking reaction is chemically distinct from hydrogasification of carbon, the difference between the two becomes small when it is applied to coal or coal char whose molecular structure is deficient in hydrogen.

Unlike other gasification reactions involving steam and carbon dioxide, this gasification reaction is *exothermic*, that is, generating reaction heat, as

$$C_{(s)} + 2H_{2(g)} = CH_4 \quad \left(-\ H^0_{298}\right) = 74.8\, kJ/mol$$

Coal char hydrogasification can be regarded as two simultaneous reactions differing considerably in their reaction rates [33], as also mentioned in the earlier section on steam gasification. This statement of apparent two-stage reactions of pyrolysis and gasification is valid for biomass gasification as well. Due to the high moisture content in raw untreated biomass, biomass hydrogasification always involves steam hydrogasification, where all three principal modes of gasification—including pyrolysis, steam gasification, and hydrogasification—take place simultaneously. Of these reactions, pyrolysis is by far the fastest chemical reaction at the operating conditions.

Hydrogasification of carbons and biomass can be catalyzed for faster and more efficient reactions [39]. Many metallic ingredients have been shown to have catalytic effects on hydrogasification of coal char and carbon and these catalysts include aluminum chloride [40], iron-based catalysts [41], nickel-based catalysts [42], and calcium salt-promoted iron group catalysts [43].

An interesting study was carried out by Porada [44], in which hydrogasification and pyrolysis of basket willow (*Salix viminalis*), bituminous coal, and a 1:1 mixture of the two were compared. Their study employed a nonisothermal kinetics approach, in which the reaction temperature was increased at a constant rate of 3 K/min from ambient temperature to 1,200 K under the hydrogen pressure of 2.5 MPa. Of the test samples, the highest gas yields were obtained during hydrogasification of coal and the lowest yields were observed in the basket willow processing. It was also established that the conversion ratio to C_1–C_3 hydrocarbons from C under a relatively low H_2 pressure was approximately five times higher than the pyrolysis conducted in an inert atmosphere. This clearly explains that the beneficial role of hydrogen gas is very significant in gasification of biomass as well as coal.

5.3.1.6 Water Gas Shift Reaction

The water gas shift reaction plays an important role in manufacturing hydrogen, ammonia, methanol, and other chemicals. Nearly all synthesis gas reaction involves a water gas shift reaction in some manner. The WGS reaction is of a reversible kind, whose proceeding direction can be relatively easily reversed by changing the gaseous composition as well as varying the temperature of the reaction. As shown in Table 5.9, the temperature dependence of chemical reaction equilibrium constant (K_p) is the mildest of the important syngas reactions considered in the table. Furthermore, the temperature when K_p becomes unity is around 814°C, where most carbon gasification reactions begin to be kinetically active.

The water gas shift reaction has dual significance. The forward water gas shift reaction converts carbon monoxide and water into additional hydrogen and carbon dioxide. This reaction is utilized to enhance the hydrogen yield from raw or intermediate syngas, such as raw product of steam reformation of hydrocarbons. If the water gas shift reaction is exploited in its reverse direction, that is, as a reverse water gas shift reaction, carbon dioxide can be reduced to carbon monoxide that is far more reactive than carbon dioxide. This enables further chemical conversion of carbon monoxide into other useful petrochemicals, instead of direct conversion of carbon dioxide, which is much more difficult as a task. The catalytic reverse water gas shift reaction could be very useful as a reduction method of carbon dioxide.

In many industrial reactions, the water gas shift reaction is a companion reaction to the principally desired reaction in the main stage, as evidenced in methanol synthesis and in steam reformation of methane. Whenever deemed appropriate, WGS is also carried out as a secondary stage reaction to result in additional conversion of water gas into hydrogen, as desired by a fuel reformer to generate hydrogen for proton exchange membrane (PEM) fuel cell application. Because the forward water gas shift is an exothermic reaction, low temperature thermodynamically favors higher CO conversion, and its intrinsic kinetic rate without an aid of an effective catalyst is inevitably low at low temperatures. Therefore, most water gas shift reaction is carried out catalytically at a low temperature such as 180–240°C. This type of catalyst is called a *low-temperature shift* (*LTS*) catalyst, which has long been used industrially. One of the LTS catalyst formulations is coprecipitated Cu/ZnO/Al_2O_3 catalyst, whose formulation is better known for the low-pressure methanol synthesis catalyst [24].

In biomass gasification for generation of biosyngas, the water gas shift reaction plays an important role, inasmuch as it can produce additional hydrogen and it can also be used to control the ratio of H_2:CO in the synthesis gas composition. The WGS reaction not only enhances the targeted gas composition with a higher selectivity, but also prepares a syngas more suitable for the next-stage conversion by adjusting it for an optimal syngas composition.

5.3.2 Biosyngas

Depending upon the compositions of resultant gaseous products, syngas may be classified as (i) balanced syngas whose H_2:CO molar ratio is close to 2:1, (ii) unbalanced syngas whose H_2:CO molar ratio is substantially lower than 2:1, (iii) CO_2-rich syngas whose CO_2 molar concentration exceeds 10%, and so on. The balanced syngas is also referred to as hydrogen-rich syngas, whereas unbalanced syngas is called CO-rich syngas. The gaseous product can also be classified based on its heating value (HV) as (i) high BTU gas, (ii) medium BTU gas, and (iii) low BTU gas. Alternately, syngas may be classified based on its origin as: (i) natural gas-derived syngas, (ii) coal-derived syngas, (iii) biomass syngas, and (iv) coke oven gas. These descriptions are commonly used for all types of syngas derived from a variety of feedstock including natural gas, coal, and biomass. The first category of classification has been used widely in the synthesis of clean liquid fuels such as methanol and dimethylether (DME) synthesis, whereas the second category has come from the classical design of coal gasifiers [5].

For further clarification of the terminology, *biogas* usually stands for a gas produced by anaerobic digestion of organic materials, and is largely comprised of methane (about 65% or higher) and carbon dioxide. Therefore, the term "biogas" is not interchangeable with the term "biomass syngas." The methane-rich biogas is a high BTU gas and is also called "marsh gas," "landfill gas," or "swamp gas." As the name implies, swamp gas is produced by the same anaerobic processes (where oxygen is absent and unavailable for biological conversion) that take place during the underwater decomposition of organic matter in wetlands. *Anaerobic digestion* is a series of biochemical processes in which micro-organisms break down biodegradable material in the absence or serious deficiency of oxygen. The biochemical processes carried out by micro-organisms consist of four principal stages, viz., (i) hydrolysis, (ii) acidogenesis, (iii) acetogenesis, and (iv) methanogenesis. Therefore, the biogas obtained from anaerobic digesters is not classified as a thermochemical intermediate of biomass conversion.

5.3.3 Tar Formation

Tar is neither a chemical name for certain molecular species, nor a clearly defined terminology in materials. Tar has been operationally defined in gasification work as the material in the product stream that is condensable in the gasifier or in downstream processing steps or subsequent conversion devices and parts [45]. This physical definition is inevitably dependent upon the types of processes, the nature of the feedstock, and specifically applicable treatment conditions. Producer gases from both biomass gasification and coal gasification contain tars. The generalized composition of tars is mostly aromatic and the average molecular weight is fairly high. Even with the very same biomass feed, the amount and nature of tars formed are different depending upon the gasifiers used and process conditions employed. Similarly, the same gasifier would generate different amounts and types of tars depending upon the feedstock properties and compositions. Therefore, successful implementation of efficient gasification technology depends on the effective control of tar formation reactions as well as the efficient removal or conversion of tar from the produced gas.

A number of investigators studied various aspects of tar in terms of its formation, maturation scheme, properties, molecular species, and relationship

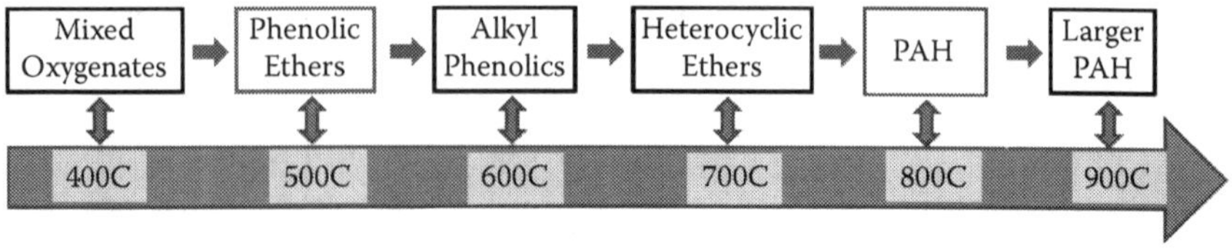

FIGURE 5.2
Tar maturation scheme proposed by Elliott. (Modified from Milne and Evans, 1998, Biomass gasifier "tars"; Their nature, formation, and conversion. U.S. Department of Energy, National Energy Technology Laboratory, and Elliott, 1988, *Relation of Reaction Time and Temperature to Chemical Composition of Pyrolysis Oils.* Washington, DC: American Chemical Society.)

between the tar yield and the reaction temperature. Elliott extensively reviewed the composition of biomass pyrolysis and gasifier tars from various gasification processes and proposed a tar maturation scheme [46], as shown in Figure 5.2.

Nickel-based catalysts are known to be effective in biomass gasification for tar reduction to produce synthesis gases, because of their relatively lower cost and good catalytic effects. Several different types of nickel-based catalysts for biomass gasification were reviewed with respect to tar reduction efficiency by Wu and Williams [47].

Several methods for the sampling and analysis of tar have been developed. Most of these methods are based on the condensation of tar in a liquid phase or adsorption of tar on a solid material. The collected samples are subsequently analyzed gravimetrically or by using a gas chromatograph (GC). The SPA (solid-phase absorption) method was originally developed by KTH, Sweden, and according to the SPA method a gas sample is passed through an amino-sorbent which collects all tar compounds [48]. The ensuing step is to use different solvents to collect aromatic and phenolic compounds separately. These compounds are then analyzed on a gas chromatograph and positive identification of the condensed material is achieved by a gas chromatograph-mass spectrometer (GC-MS). The tar amount determined by the GC analysis is called *GC-detectable* tar. Tar can also be analyzed gravimetrically and the gravimetrically determined tar is called *gravimetric tar.* Gravimetric tar is evaporation/distillation residue from particle-free solution(s) determined by gravimetric analysis. Both GC-detectable and gravimetric tar are reported in mg/m^3. Both chromatographic and gravimetric determination of tar is based on the European Technical Specification, TC BT/TF 143 WI CSC 03002.4: 2004 (E), developed by Technical Committee CEN/BT/TF 143 "Measurement of Organic Contaminants (Tar) in Biomass Producer Gas" [49]. This technical specification is applicable to sampling and analysis of tars and particles in the concentration range between 1 mg/m^3 to 300 g/m^3 at all relevant sampling port conditions (0–900°C and 0.6–60 bars). The application of the technical specification allows determination of four different analytical values [49]:

- The concentration of gravimetric tars in mg/m^3
- The sum of the concentrations GC-detectable tars in mg/m^3
- The concentration of individual organic compounds in mg/m^3
- The concentration of particles in mg/m^3

5.4 Fast Pyrolysis of Biomass

Pyrolysis of biomass is an important process option either as a pretreatment for gasification (i.e., the first stage of two-stage gasification) or as an independent process such as fast pyrolysis. The former is usually aimed at producing biosyngas, whereas the latter is intended to produce a liquid fuel product. Typical biomass pyrolysis takes place actively at around 500°C and produces a liquid product via fast cooling (shorter than two seconds) of volatile pyrolytic products. The liquid product produced is called bio-oil or pyrolysis oil. Bio-oil is considered greenhouse gas neutral, because it only puts back into the atmosphere what was initially removed by the plant during its lifetime [50]. Bio-oil is nearly sulfur-free.

As can be seen from the prevailing pyrolysis temperature of ~500°C, pyrolysis of biomass as a process treatment is quite similar to oil shale pyrolysis [51] and coal pyrolysis [14], wherein hydrocarbon species are devolatilized and thermally cracked. Bio-oil production via biomass pyrolysis is typically carried out via flash-pyrolysis or fast pyrolysis. The biomass fast pyrolysis process [23, 52, 53] is a thermochemical conversion process that converts biomass feedstock into gaseous, solid, and liquid products via heating of biomass in the absence of oxygen or air. The principal product of typical fast pyrolysis of biomass is a liquid product of bio-oil. As the name of the process implies, the process is intended to take place very fast in a matter of a couple of seconds or shorter; that is, $\tau > 2$ s. As such, heat and mass transfer conditions in the reactor become crucially important in both design and operation of fast pyrolysis of biomass. A variety of reactor designs has been proposed and tested on pilot scales, and they include traditional fluidized bed reactors, circulating fluidized bed reactors, rotating cone reactors, vacuum reactors, ablative tubes, and more. Table 5.11 shows a partial list of operational pilot-scale biomass pyrolysis units.

A typical biomass fast pyrolysis process involves several stages of operation including biomass feed drying, comminuting, fast pyrolysis, separation

TABLE 5.11

Pilot-scale Fast Pyrolysis of Biomass

Process	Type of Reactor	Capacity (kg/h)	References
Dynamotive Energy Systems (Canada)	Bubbling fluidized bed	400	Dynamotive Energy Systems, 2011; Bain. 2004 [50, 54]
Union Fenosa (Spain)	Bubbling fluidized bed	200	Ringer, Putsche, and Scahill, 2006 [55]
Wellman Process Eng. Ltd. (UK)	Fluidized bed	250	Conversion and Resource Evolution (CARE), Ltd., 1998–2002 [56]
Resource Technology (RTI)	Fluidized bed	20	Scott, 1999 [57]
Red Arrow /Ensyn	Circulating fluid bed	1,000	Czernik and Bridgwater, 2004 [53]
VTT /Ensyn (Finland)	Circulating fluid bed	20	VTT Technical Research Centre of Finland, 2002 [58]
ENEL /Ensyn (Italy)	Circulating Transported bed	625 kg/h	Gradassi, 2002 [59]
BTG/KARA (The Netherlands)	Rotating cone	200 kg/h	Biomass Technology Group (BTG), 2011 [60]
Pyrovac	Vacuum, stirred bed	3500 kg/h	Ray, 2000 [61]
Enervision (Norway)	Ablative tube		Bridgwater, 1999 [62]
Fortum Oy (Finland)	Own	350 kg/h	U.S. Department of Energy–Energy Efficiency and Renewable Energy–Biomass Program, 2011 [63]

of char, and liquid recovery. Fast pyrolysis of biomass has several distinct merits as an alternative fuel process technology and they are as follows:

- The principal product of biomass fast pyrolysis is bio-oil, a liquid product. As such, storage and transportation of the product are easy.
- The process takes place very quickly in a matter of 0.5–2 seconds and as such the reactor residence time is very short.
- The process technology is very widely and universally applicable to a variety of biomass feedstock.
- The process chemistry is simple and straightforward.
- The process equipment needed is relatively simple and not complex.
- Due to the high reactor throughput and simple chemistry, the capital cost is not high.
- Small-scale process feasibility has been demonstrated. However, process economics for a small-scale power generation (<5 MW_e) based on fast pyrolysis of biomass is substantially less favorable than that of a larger scale (>10 MW_e) system.

However, the drawbacks of fast pyrolysis of biomass include the following:

- The moisture level of the feed biomass needs to be controlled below 10% or even lower. Otherwise, the feed water and reaction-produced water will end up in the final liquid oil product.
- The biomass feedstock needs to go through size reduction and preconditioning. The feed material has to be in particulate form in order to minimize the heat and mass transfer resistance; preferably ~2 mm for bubbling bed and ~6 mm for circulating fluidized bed (CFB).
- The quality of bio-oil product is generally of poor quality. Bio-oil has a high oxygen content which makes the oil more corrosive and unstable, in addition to possessing a lower heating value. Bio-oil also contains metallic compounds and nitrogen species, which can foul and deactivate most of the fuel upgrading catalysts. The upgrading process requires a large amount of hydrogen and becomes costly.
- The process requires a very fast heating rate, which is costly in both operation and capital investment.
- The overall energy efficiency of the process is not high, by itself.
- Large-scale process operation may be subjected to significant logistical burdens of feedstock collection, storage, and pretreatment.

The first biocrude in the United States was produced at a 30 kg/hr scale [64] at operating conditions of approximately 500°C and a residence time of one second, that is, at a typical fast pyrolysis condition. Since then, Canadian researchers have converted woody biomass into fuel via pyrolysis in a 200 kg/hr pilot plant [65]. The fuel oil substitute was produced (on 1,000 ton/day dry basis) at approximately $3.4/GJ, based on the 1990 fixed price (1990 U.S. dollars). At the time, the cost for light fuel oil was $4.0–4.6/GJ [66], thus indicating that the pyrolyzed biomass fuel was a more economical alternative. However, the skepticism of high transportation costs of the biomass to the pyrolyzer outweighing potential profits limited R&D funding for the following years. This was when the petroleum-based liquid fuel price was much lower than that in the twenty-first century. To circumvent this biomass transportation and logistical problem, the Energy Resources Company (ERCO) in Massachusetts developed a mobile pyrolysis prototype for the U.S. EPA.

ERCO's unit was designed to accept biomass with 10% moisture content at a rate of 100 tons/day. At this rate, the system had a minimal net energy efficiency of 70% and produced gaseous, liquid, and char end products. The process, which was initially started using an outside fuel source, became completely self-sufficient shortly after its startup. This was achieved by implementing a cogeneration system to convert the pyrolysis gas into the electricity required for operation. A small fraction of the pyrolysis gas is also

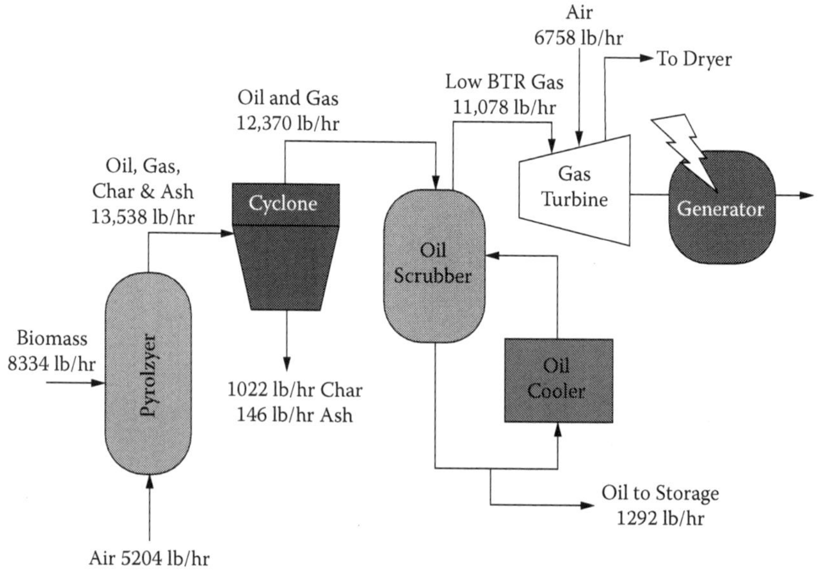

FIGURE 5.3
A schematic and material balance for ERCO's Mobile Pyrolysis Unit.

used to dry the entering feedstock to the required 10% moisture. A simplified version of ERCO's mobile unit is shown in Figure 5.3 [67].

The end-products are pyrolysis oil and pyrolytic char, both of which are more economical to transport than the original biomass feedstock. The average heating values for the pyrolysis oil and char are 10,000 BTU/lb and 12,000 BTU/lb, respectively [67]. The pyrolysis gas, which has a nominal heating value of 150 BTU/scf, is not considered an end product because it is directly used in the cogeneration system. If classified based on the coal syngas criterion, typical pyrolysis gas of biomass would be classified as a low-BTU gas whose usual criterion for the heating value is less than 300 BTU/scf [5]. The mobility, self-sufficiency, and profitability of the system lifted some of the hesitancy of funding research on the pyrolysis of biomass. In addition, ERCO's success led to additional investigation of "dual," or cogeneration systems, which produce both useful heat and electric power, that is, combined heat and power (CHP).

In a fluidized bed fast pyrolysis reactor, fine particles (2–3 mm size) of biomass are introduced into a bed fluidized by a gas, which is usually a recirculated product gas [68]. High heat and mass transfer rates result in rapid heating of biomass particles via convective heat. Char attrition and bio-oil contamination may take place and char carbon could appear in the bio-oil product. Heat can be supplied externally to the bed or by a hot fluidizing gas. In most designs, reactor heat is usually provided by heat exchanger tubes through which hot gas from char combustion flows. The reactor effluent

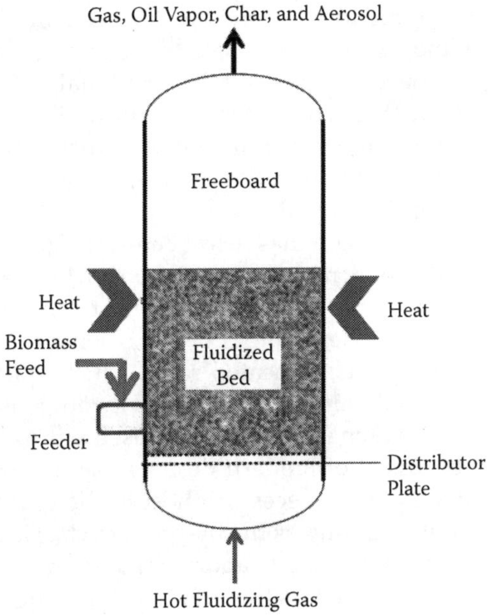

FIGURE 5.4
Bubbling fluidized bed reactor for fast pyrolysis of biomass.

leaves from the top of the reactor as noncondensable gas, char, bio-oil vapor, and aerosol. The fluid bed may contain, in addition to biomass particles, fluidizing media such as hot sand particles for enhanced heat transfer. *Aerosol* is generally defined as a suspension of solid and liquid particles in a gas. The term aerosol includes both the particles and the suspending gas and the particle size may range from 0.002 to larger than 100 µm [49]. There is some dilution of the products due to the use of fluidizing gas, which makes it more difficult to condense and then separate the bio-oil vapor from the gas exiting the condensers. The general operational principle is quite similar to that of the traditional gas–solid fluidized bed reactor [69]. The fluidized bed fast pyrolysis process has been pilot tested by several companies including Dynamotive [54], Wellman [56], and Agri-Therm [70], as listed in Table 5.11. Special versions of fluidized bed reactors popularly used in biomass fast pyrolysis are bubbling fluidized bed (BFB) and circulating fluidized bed (CFB) reactors. A principal difference between the two types is that the former has a freeboard space above the fluid bed (as shown in Figure 5.4), whereas the latter achieves a full entrainment of the particle–fluid mixture in the reactor.

Bubbling fluidized bed reactors have long been used in chemical and petroleum processing. An earlier version of bubbling fluidized sand bed reactor was utilized by the WFPP (Waterloo Fast Pyrolysis Process) [71, 72]. Larger units based on the BFB are a 200 kg/hr system by Union Fenosa (Spain) and a

400 kg/hr system by DynaMotive (Canada). Both systems were based on the WFPP developed at the University of Waterloo (Canada) and designed by its spin-off company Resource Transforms International (RTI) in Canada [55]. Bubbling fluidized beds (BFBs) occur when the incoming carrier gas velocity is sufficiently above the minimum fluidization velocity to cause the formation of bubblelike structures within the particulate bed. In such a condition, the bed appears more or less bubbling.

A variety of different designs has also been introduced. The particulate or granular bed in the BFB may be composed of biomass particles only without any inert media and all the process heat can be supplied by hot fluidizing gas. Alternately, the granular bed can contain a hot heat-transfer solid medium such as indirectly heated sand, which enhances heat transfer efficiency and generally allows for a larger throughput for the process. An earlier version of the DynaMotive system used natural gas to heat their pilot-scale reactor, but most of modern designs use the exothermic heat of char combustion to supply the necessary heat to the pyrolysis reactor. The BFB is, in principle, self-cleaning as the by-product char is carried out of the reactor by the product gases and oil vapors [55]. For this process feature, the density of char will have to be less than that of fluidizing media so that the by-product char will literally "float" on top of the bed. The allowable particle size range for this type of reactor is quite narrow and should be carefully managed. Furthermore, the bio-oil produced has a carbon contamination possibility, inasmuch as the oil vapor has to pass through the char-rich layer on its way out of the bed [55]. The gas flow rate for this reactor will have to be determined based on the dimensions of fluidizing media and the desired residence time of gas and oil vapor in the freeboard section of the reactor, which is above the bed. The residence time is generally between 0.5 and 2.0 seconds. The BFB is currently most popularly used for both fast pyrolysis and gasification processes of biomass. In order to achieve a short residence time for volatiles, a shallow bed depth, a high gas flow rate, or usually both are utilized [57, 73]. A high gas-to-biomass feed ratio is adopted for necessary fluidization and a short residence time, which in turn results in product dilution and lowers the thermal efficiency of the process.

Circulating fluid bed reactors and derivative types of reactor design are frequently utilized for fast pyrolysis of biomass. The basic concept of CFB reactors involves efficient and rapid heat transfer in a convective mode and short residence times for both biomass particles and product vapors. Fine biomass particles are introduced into a circulating fluidized bed of hot sand. Hot sand and biomass particles move together with the transport gas which is usually a recirculated product gas. In reactor engineering, this transport gas is usually referred to as a fluidizing gas. In practice, the residence times for solid biomass particles are not uniform and only a little longer than the volatiles. Therefore, solid recycling of partially reacted feed would become necessary or very fine particle size of biomass would have to be used. Most CFB reactors are dilute phase units and their heat transfer rates are high,

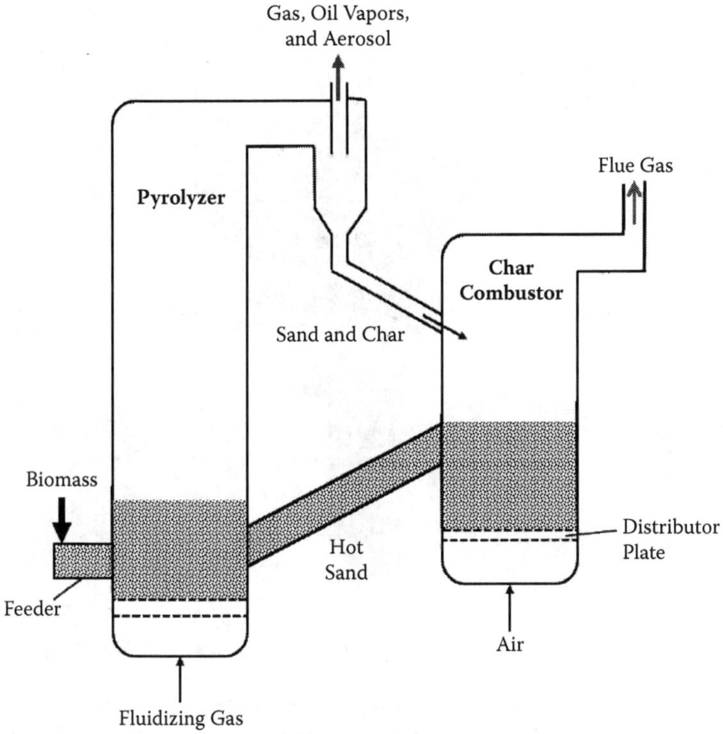

FIGURE 5.5
Circulating fluidized bed reactor for fast pyrolysis of biomass.

but not as high as particularly desired, because the mode of heat transfer is gas–solid convective heat transfer [57, 74]. If a twin-bed reactor system is used, that is, the first for fast pyrolysis and the second as a char combustor to reheat the circulating solids, as shown in Figure 5.5, there is a strong possibility for ash carry-over to the pyrolysis reactor and ash buildup in circulating solids [57].

The ash attrition and char carry-over problem could also be high and if not controlled properly, some level of contamination of bio-oil products is also possible. One of the main advantages of the CFB is the possibility to achieve a short and controllable residence time for char [75]. The alkaline compounds in biomass ash are known to possess a catalytic effect for cracking organic molecules contained in volatile vapors, thereby potentially lowering the volatile bio-oil yield. Red Arrow and VTT processes are based on circulating fluid bed reactors.

A rotating cone fast pyrolysis system (Figure 5.6) operates based on the idea of intensive mixing between biomass particles and hot sand particles, thereby providing good heat and mass transfer. This type of reactor requires very fine to fine biomass particle size. This process does not

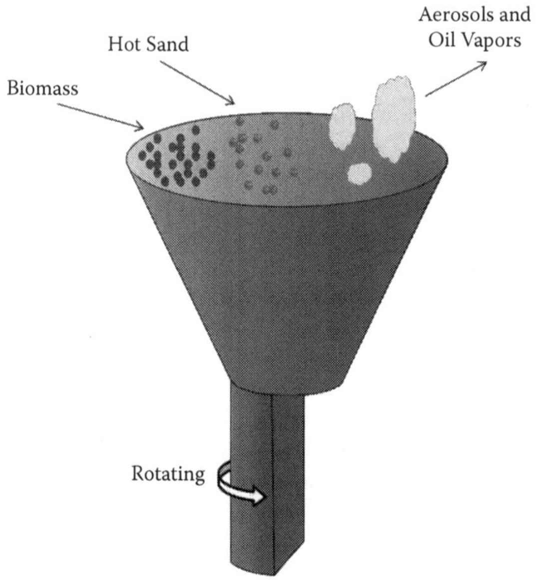

FIGURE 5.6
A schematic of rotating cone reactor for fast pyrolysis of biomass.

require any carrier gas for its operation and therefore the reactor size can be made compact. The BTG's fast pyrolysis process developed by Biomass Technology Group (the Netherlands) is based on a modified rotating cone reactor [60].

In this reactor, efficient heat transfer between the hot sand particles and biomass particles is accomplished, and a good portion of the process heat is retained in the hot sand particles. A wide variety of different biomass feedstock can be processed in the pyrolysis process which is operated at 450–600°C. Before entering the reactor, the biomass feedstock must be reduced in size to a particulate form finer than 6 mm, and its moisture content to below 10 wt%. Sufficient excess heat is normally available from the pyrolysis plant to dry the feed biomass from 40–50 wt% moisture to below 10 wt%. A schematic of rotating cone reactor is shown in Figure 5.6.

From the BTG process, up to 75 wt% pyrolysis oil and 25 wt% char and gas are produced as primary products [60]. Because no "inert" carrier gas is used in this process, no additional gas heating is required and the pyrolysis products are undiluted vapor. This undiluted vapor flow allows the downstream equipment to be of a minimum size. In a condenser, the oil vapor product is rapidly cooled yielding the oil product and some permanent (noncondensable) gases. In only a few seconds of process treatment, the biomass is transformed into pyrolysis oil. Biomass char and hot sand used in the reactor are recycled to a combustor, where char is combusted to reheat the sand. After reheating the sand by char combustion heat, the sand is recirculated back

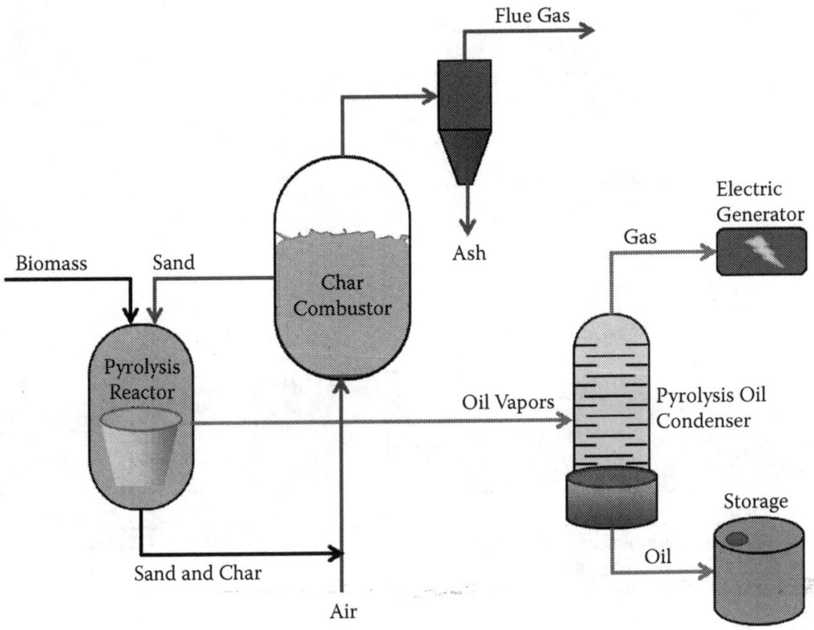

FIGURE 5.7
A schematic of BTG process for fast pyrolysis of biomass.

to the reactor. The permanent gases (or noncondensable gas) can be utilized in a gas engine to generate electricity or simply flared off. In principle, no external utilities are required for operation of the process, that is, energy-wise self-sufficient. A schematic of a BTG process for fast flash of biomass is shown in Figure 5.7.

The vacuum pyrolysis process operates under low pressure (vacuum or atmospheric) and has principal process merits in its processability of larger biomass particles as well as its short residence time for volatiles. These biomass particles are taken out using a vacuum pump from the reactor regardless of the particles' residence time.

Due to the lack of convective gas flow inside the reactor, however, the heat and mass transfer rates are slower than those for the fluidized bed reactors, hence requiring a longer residence time for biomass particles in the vacuum reactor. A longer biomass residence time in turn makes the reactor and equipment size inevitably larger. Biomass in a vacuum reactor moves downward by gravity and rotating scrapers through the multiple hearth pyrolyzer with the temperature increasing from about 200°C to 400°C, as shown in Figure 5.8. Pyrovac's pyrocycling^{TM} process is based on vacuum pyrolysis technology [61]. According to Pyrovac, their process technology has the following features:

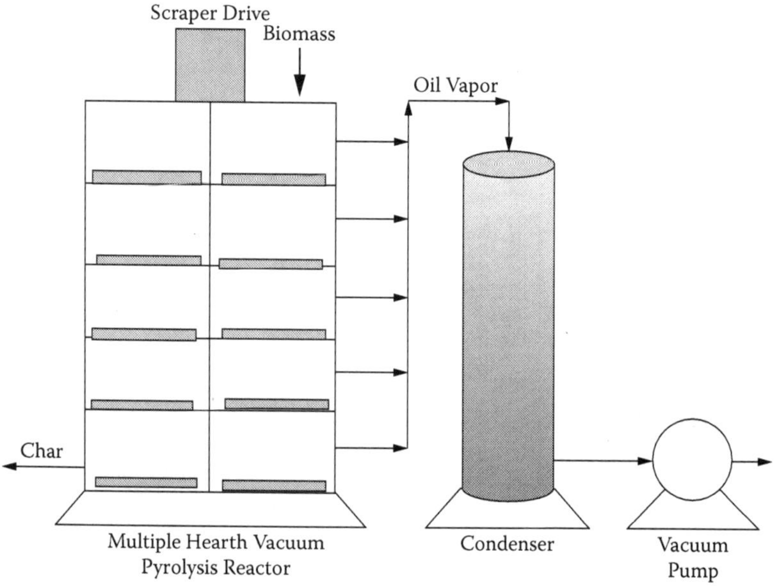

FIGURE 5.8
A schematic of vacuum pyrolysis reactor for biomass fast flash.

- There is no need to pulverize or grind the feed biomass materials. Particles of up to 20 mm can be fed without any difficulty.
- The process system can be operated under torrefaction mode (<300°C) or pyrolysis mode (>450°C). The first mode generates biochar as a principal product, whereas the latter mode produces bio-oil.
- The process can be operated under vacuum or atmospheric conditions to enhance the production of either bio-oil or biochar. The process operation as well as its product portfolio is more versatile compared to competing process technologies.
- The process adopts a moving and stirred bed reactor.
- Molten salt heat carrier at 575°C is in indirect contact with biomass feedstock, thus aiding in efficient heat transfer.
- There are two heating plates with internal raking systems.
- The process uses two condensing towers. The first tower mainly collects heavy bio-oil and contains little water and acids. The second tower mainly recovers the aqueous acidic phase.

The ablative pyrolysis process is based on the heat transfer taking place when a biomass particle slides over a hot surface, as shown in Figure 5.9. High pressure applied to biomass particles on a hot reactor wall or surface to provide good contact is achieved by centrifugal or mechanical motion [68].

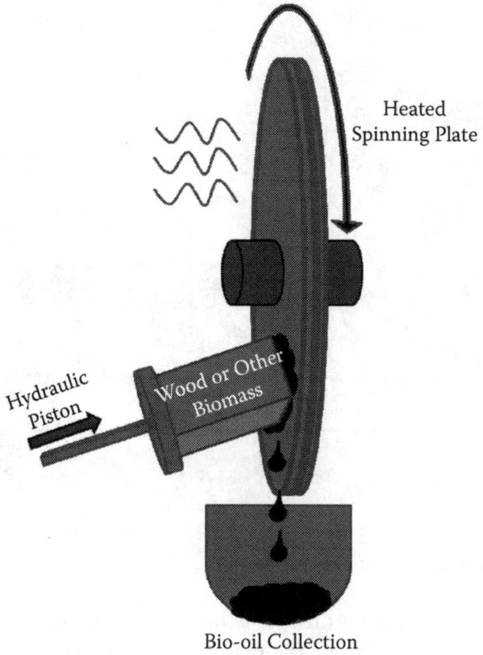

FIGURE 5.9
A schematic of ablative pyrolysis reactor for fast pyrolysis of biomass.

This type of reactor does not require a small particle size and can handle large particles without difficulty. The reactor does not require any carrier gas or sweep gas. The reactors of cyclonic type have difficulty in achieving sufficiently long residence times for the biomass particles that are required to allow a high degree of conversion.

Therefore, it is usually necessary to recycle partially reacted solids back to the reactor, as is the case with a circulating fluidized bed reactor. A high degree of char attrition also takes place and tends to contaminate the product bio-oil with a high level of carry-over carbon [57]. Some variations of the ablative pyrolysis process include the cone type and plate type for hot surfaces. These processes are mechanically more complex and difficult to scale up, because the moving parts are subjected to the high temperatures required for pyrolysis. Furthermore, the loss of thermal energy from the ablative process in general could be high, inasmuch as the hot surface needs to be at a substantially higher temperature than the desired pyrolysis temperature.

Fast pyrolysis of biomass can be achieved in an auger-type reactor or auger reactor, in which an auger or an advancing screw assembly drives the biomass and hot sand through the reactor barrel. The operational principle is very similar to that of a polymer twin-screw extruder, as shown in Figure 5.10. This type of reactor achieves a very good mixing of materials

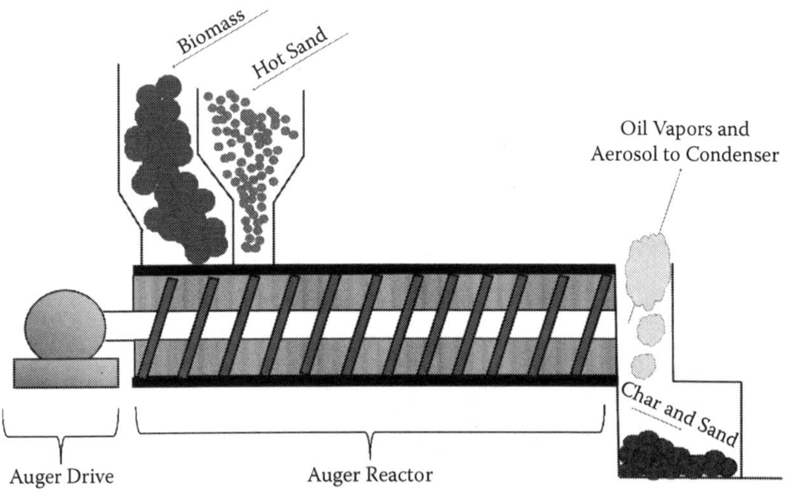

FIGURE 5.10
An auger type reactor for biomass fast pyrolysis.

in the reactor, thereby enhancing the heat and mass transfer efficiency. In an auger-type reactor for biomass fast flash, the usual mechanism of heat transfer is via direct heat transfer between hot sand and biomass particles.

At the outlet of the reactor, solids including hot sands, char, and ash are separately recovered from the oil vapors and aerosol. Hot sand is reheated using char combustion heat and recirculated back to the reactor, and the oil vapor and aerosol are sent to the condenser for bio-oil recovery. The process does not require a carrier gas, but the biomass particle size is preferentially small for smooth operation. The design of auger reactors has received extensive benefits from the industrial practice and design experience of the polymer and twin-screw extruder industries [76]. Therefore, this type of process is relatively straightforward to design and fabricate and is deemed to be suitable for small-scale production.

5.5 Biomass Gasification Processes

The first biomass gasification system investigated at the pilot scale was a fluidized bed that incorporated dry ash-free (daf) corn stover as the gasifier feed. Corn stover has been selected as the feed to the gasifier since 1977 [77], even when U.S. corn production was less than half of the 2010 production of 312 million metric tons [78]. The amount of corn stover that can be sustainably collected and made available in 2003 was estimated to be 80–100 million dry metric tons/yr [79]. Of this total, potential long-term demand for corn

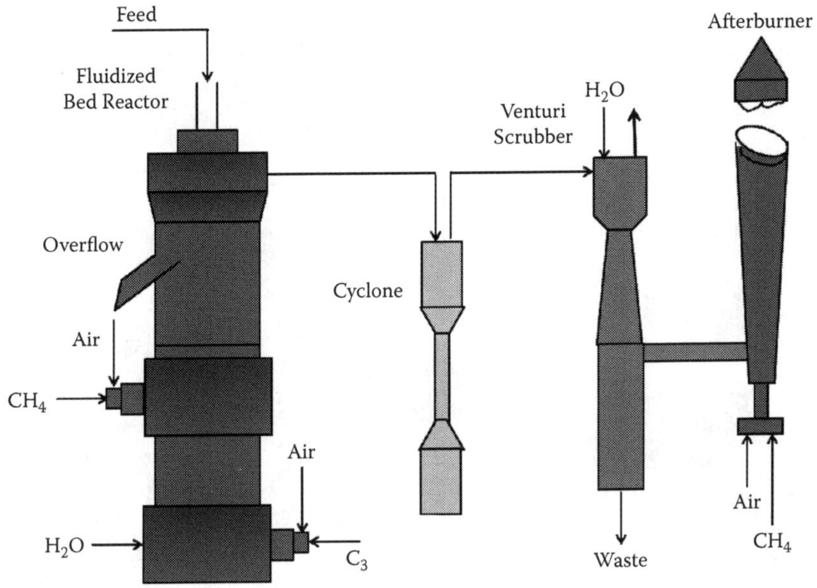

FIGURE 5.11
KSU's pilot-scale fluidized bed gasifier of corn stover.

stover by nonfermentative applications such as biomass gasification in the United States was estimated to be about 20 million dry tons/yr.

An early pilot-scale system designed and operated at Kansas State University, shown in Figure 5.11, has a 45.5 kg bed capacity [80]. Fluidizing gas and heat for biomass gasification were supplied by the combustion of propane in the presence of air. The particulates and char were removed using a high temperature cyclone. A Venturi scrubber was then used to separate the volatile matter into noncondensable gas, a tar-oil fraction, and an aqueous waste fraction. Raman et al. [80] conducted a series of tests with temperatures ranging from 840 to 1,020 K. The optimal gas production was obtained using a feed rate of 27 kg/hr and a temperature of 930 K. At these conditions, 0.25 x 106 BTU/hr of gas was produced. This was enough to operate a 25-hp internal combustion engine operating at 25% efficiency [80].

Another one of the extensively studied gasification systems for biomass conversion is Sweden's VEGA gasification system. Skydkraft AB, a Swedish power company, decided in June 1991 to build a cogeneration power plant in Värnamo, Sweden to demonstrate the integrated gasification combined cycle (IGCC) technology. Bioflow, Ltd. was formed in 1992 as a joint venture between Skydkraft and Alstrom to develop the pressurized air-blown circulating fluidized bed gasifier technology for biomass [81]. The biomass integrated gasification combined cycle (BIGCC) was commissioned in 1993 and fully completed in 1995. VEGA is a biomass-fuel based IGCC system that combines heat and power for a district heating system. It generated 6.0

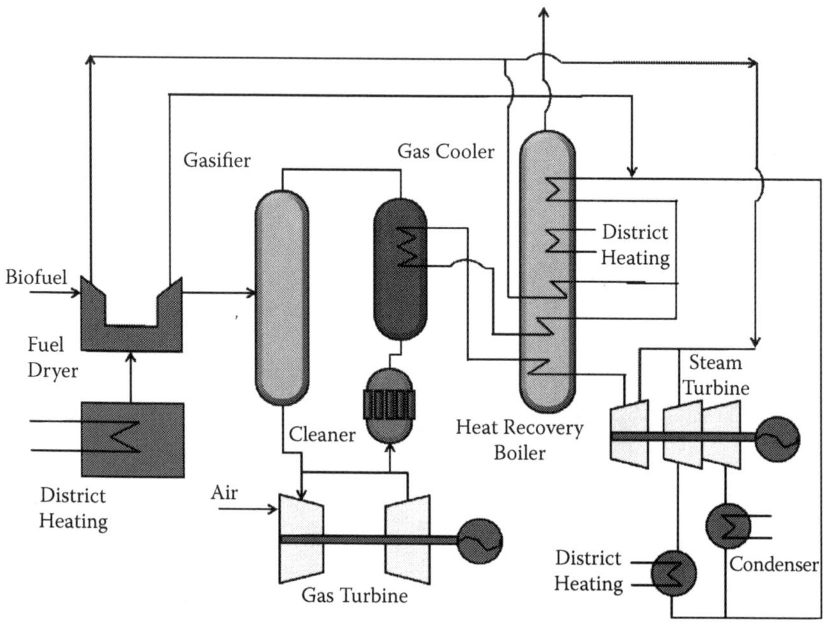

FIGURE 5.12
A schematic of VEGA process for biomass gasification.

MW_e and 9.0 MW_{th} for district heating of the city of Värnamo, Sweden. This was the first complete BIGCC for combined heat and power from biomass feedstock. As shown in Figure 5.12, the moisture of the entering biomass feedstock is removed via a "biofuel dryer" to decrease gaseous emissions [81]. The dried biomass is then converted into a "biofuel" in a combined cycle gasifier. The resulting gas is cooled before entering the heat recovery boiler and distribution to district heating. The gasifier is known as a bioflow gasifier or bioflow pressurized circulating fluidized-bed gasifier.

Biomass gasification and power generation technology has long been developed with significant technological advances in Finland, where about 20% of total energy consumption derives from biomass. This high percentage of biomass energy utilization is mainly due to the recycling of biowaste produced as a by-product of the forest industry. VTT Energy and Condens Oy of Finland developed a new type of fixed bed biomass gasifier [58], whose configuration is based on forced feed flow that allows the use of low bulk-density fibrous biomass feedstock. This gasifier is a combination of updraft and cocurrent gasification technologies, where gasifying medium and solid feed move upward through the gasifying section of the reactor. In 1999–2001, a 500 kW_{th} pilot plant was operated in a test facility and very positive test results were obtained. Some of the principal features of the technology include [58, 82]:

- Fuel feeding is not based on natural gravity alone.
- The process is suitable for various biomass residues and waste-derived fuels.
- The process achieves high carbon conversion and generates low tar content.
- The process can be scaled up to above 8 MW, unlike its predecessor technology of Bioneer gasifiers.
- There was no problem with leaking feeding system or blocking gas lines.
- The VTT successfully demonstrated a variety of feedstock including
  - Forest wood residues chips (moisture level of 10–55wt%)
  - Sawdust and wood shavings
  - Crushed bark with maximum moisture of 58%
  - Demolition wood
  - Residue from plywood and furniture industry
  - Recycled fuel manufactured from household waste
  - Sewage sludge in conjunction with other fuels

Condens Oy is offering this technology for a wide range of fuel feedstock. The Kokemäki (Finland) CHP plant of 1.8 MWe/3.9 MWth based on this technology was started up in 2005 [82].

The most common method of gasifying biomass is using an air-blown circulating fluidized bed gasifier with a catalytic reformer, even though there are many different variations. Most fluidized bed gasification processes use closed-coupled combustion with very little or no intermediate gas cleaning [83]. This type of process is typically operated at around 900°C, and the product gas from the gasifier contains H_2, CO, CO_2, H_2O, CH_4, C_2H_4, benzene, and tars. Gasification uses oxygen (or air) and steam to help the process conversion, just as the advanced gasification of coal [5]. The effluent gas from the fluidized bed gasifier contains a decent amount of syngas compositions, however, the hydrocarbon contents are also quite substantial. Therefore, the gasifier effluent gas cannot be directly used as syngas for further processing for other liquid fuels or chemicals without major purification steps. This is why the gasifier is coupled with a catalytic reformer, where hydrocarbons are further reformed to synthesis gas. In this stage, the hydrocarbon content including methane is reduced by 95% or better. A very successful example is Chrisgas, an E.U.-funded project, which operates an 18 MW_{th} circulating fluidized gasification reactor at Värnamo, Sweden. They use a pressurized circulating fluidized bed gasifier operating on oxygen/steam, a catalytic reformer, and a water gas shift conversion reactor that enriches the hydrogen content of the product gas. The process also uses a high-temperature filter. The project has been carried out by the VVBGC (Växjö Värnamo Biomass

Gasification Centre). The use of oxygen instead of air is to avoid a nitrogen dilution effect that, if not avoided, adds an additional burden of nitrogen removal as a downstream processing [83].

Another CFB gasification process by Termiska Processor AB (TPS) in Nyköping, Sweden developed for small- to medium-scale electric power generation is using biomass and refuse-derived fuel (RDF) as their feedstock [84]. The process is based on an air-blown low-pressure CFB gasifier which operates at 850–900°C and at 1.8 bar. The raw product gas has a tar content of 0.5–2.0% of dry gas with a heating value of 107–188 BTU/scf. As such, the raw product gas is a low-BTU gas, if the coal syngas classification scheme is followed. The process has merits of good fuel flexibility, good process controllability, and low-load operation characteristics, uniform gasifier temperature due to highly turbulent movement of biomass solids, high gasification yield, additional features of catalytic tar cracking, and fines recycling from a secondary solids separation. The tar in the syngas is catalytically cracked by dolomite [$CaCO_3 \cdot MgCO_3$] in a separate reactor vessel at 900°C immediately following the gasifier. The full calcination of dolomite is active at this temperature, as the chemical equilibrium constant (K_p) for calcite ($CaCO_3$) decomposition becomes unity at approximately 885°C [5]. A schematic of a pilot-scale TPS process with a tar cracker is shown in Figure 5.13. A waste-fueled gasification plant was

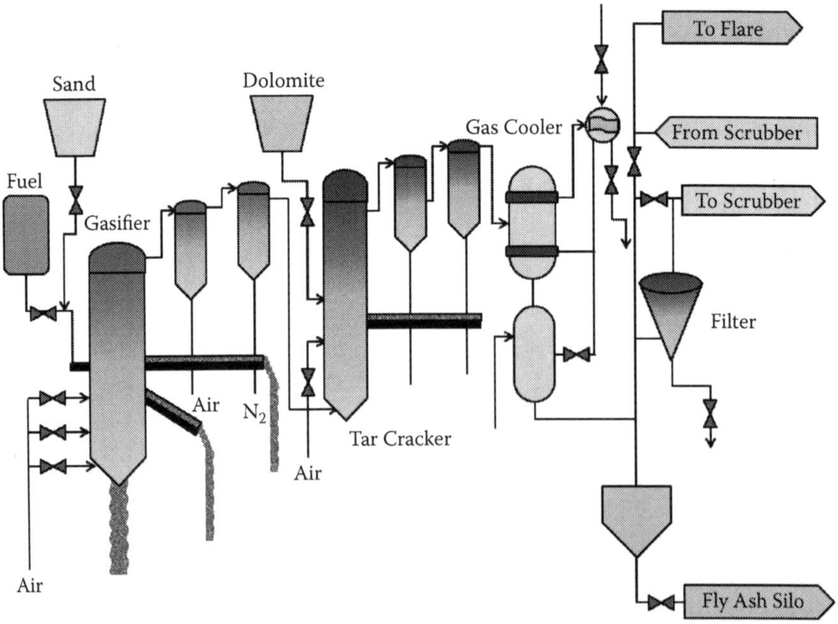

FIGURE 5.13
A schematic of TPS biomass gasification system with a tar cracking unit.

constructed based on the TPS CFB process by Ansaldo Aerimpianti SpA in Grève-in-Chianti, Italy [85].

Indirect gasification is another gasification process technology that takes advantage of the unique properties associated with biomass feedstock. As such, indirect gasification of biomass is substantially different from most coal-based gasification process technologies. For example, biomass is low in sulfur, low in ash, highly reactive, and highly volatile. In an indirect gasification process, biomass is heated indirectly using an external means such as heated sands as in the Battelle process [86]. A typical gaseous product from an indirect gasifier is close to medium-BTU gas. Battelle began this process R&D in 1980 and has continued till now, accumulating a substantial amount of valuable data regarding biomass gasification and utilization through demonstration plant operation. Battelle's process is known as the FERCO SilvaGas process, which has been commercialized by FERCO Enterprise. The principal gasifying medium for the process is steam. A commercial-scale demonstration plant of the SilvaGas process was constructed in 1997 at Burlington, Vermont, at a Burlington Electric Department (BED) McNeil Station. The design capacity of this plant is 200 tons/day of biomass feed (dry basis). McNeil Station uses conventional biomass combustion technology, a stoker gate, conventional steam power cycle, and an ESP (electrostatic precipitator) based particulate matter (PM) removal system. The gas produced by the SilvaGas gasifier is used as a cofired fuel in the existing McNeil power boilers [86]. The product gas has a heating value of about 450–500 BTU/scf, which is in a medium-BTU gas range. A schematic of the FERCO SilvaGas process is shown in Figure 5.14.

CUTEC-Institut Gmb of Germany recently constructed and operated an oxygen-blown circulating fluidized bed (CFB) gasifier of 0.4 MW_{th} capacity coupled with a catalytic reformer [87]. Part of their product gas is, after compression, directly sent to a Fischer–Tropsch synthesis (FTS) reactor for liquid hydrocarbon synthesis. The CFB gasifier of the CUTEC process was operated at 870°C, whereas the fixed bed Fischer–Tropsch synthesis reactor was operated at 150–350°C and 0.5–4.0 MPa using a fused iron catalyst (Fe/Al/Ca/K/Mg = 100/1.7/2.5/0.7). Their pilot-scale process system was successfully demonstrated with a variety of biomass feedstock with wide ranges of particle sizes and moisture levels, including sawdust (~3 mm), wood pellets (6–18 mm), wood chips (~10 mm), and chipboard residues (~30 mm). This process, once fully developed for large-scale operation, has a good potential for a single-train biomass-to-liquid (BtL) fuel conversion process [87]. The CUTEC's idea of direct linking between the biomass gasification and Fischer–Tropsch synthesis is very similar to that adopted by the indirect coal liquefaction based on Fischer–Tropsch synthesis.

Another process option for biomass gasification for syngas production involves the use of an entrained flow reactor. This type of process is operated at a very high temperature, around 1,300°C, and without use of catalyst. The high temperature is necessary due to the fast reaction rate required for an

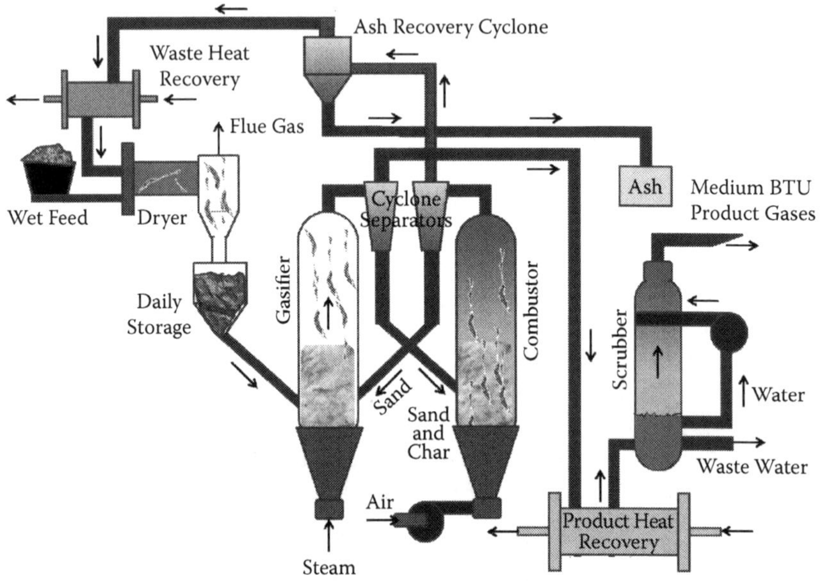

FIGURE 5.14
FERCO SilvaGas biomass gasification process. (Modified from Paisley, Irving, and Overend, 2001. A promising power option - The FERCO silvagas biomass gasification process - Operating experience at the burlington gasifier, *Proceedings of ASME Turbo Expo 2001 June 4–7, 2001*, New Orleans, pp. 1–7.)

entrained flow reactor whose reactor residence time is inherently very short. If a specific biomass feed has a high ash content, which is not very typical for biomass, slag can be formed at such a high temperature. Learning from the research developments in coal gasification [14], a slagging entrained flow gasifier may be adopted for high-ash biomass conversion. Another important process requirement in addition to high temperature and short residence time is the particle size of the solid feed; it must be very fine for efficient entrainment as well as for better conversion without mass transfer limitations. However, pulverization or milling of biomass to very fine particle sizes is energy-intensive and costly, in general. To facilitate an efficient size reduction of biomass feed, two options are most commonly adopted, viz., torrefaction and pyrolysis. *Torrefaction* is a mild thermal treatment at a temperature of 250–300°C, which converts solid biomass into a more brittle and easily pulverizable material that can be treated and handled just like coal [88]. This torrefied product is often called *biocoal*. Thus, pulverized torrefied biomass can be treated just as coal and most entrained flow gasifiers designed for coal can be smoothly converted for torrefied biocoal without much adaptation. Torrefaction as a process has long been utilized in many applications including the coffee industry. Torrefaction of biomass can alleviate some of the logistical problems involved with biomass feedstock collection and transportation. However, more study is needed for the biomass industry to make it more tuned for biomass and optimized as an efficient pretreatment technique. Gases produced during the torrefaction process may be used as an energy source for torrefaction, thus accomplishing a self-energy supply cycle.

An example of entrained flow biomass gasification can be found from the Buggenum IGCC plant whose capacity is 250 MW_e [89]. NUON has operated this process and their demonstration test program using, from 2001 through 2004, 6,000 M/T of sewage sludge, 1,200 M/T of chicken litter, 1,200 M/T of wood, 3,200 M/T of paper pulp, 50 M/T of coffee, and 40 M/T of carbon black as cofeeds with coal. A typical particle size of biomass feed was smaller than 1 mm and pulverization of wood was more difficult than that of chicken litter and sewage [89]. In their test program, they also mentioned torrefaction as a pretreatment option. Their experience with a variety of biomass feedstock provides valuable operation data for future development in this area.

As explained in the fast pyrolysis section of this chapter, biomass pyrolysis takes place actively at around 500°C and produces a liquid product via fast cooling (shorter than two seconds) of volatile pyrolytic products. As also mentioned earlier, the liquid product produced is called bio-oil. The produced bio-oil can be mixed with char (biomass char or biochar) to produce a bioslurry. Bioslurry can be more easily fed, as a pumpable slurry, to the gasifier for efficient conversion. Bioslurry is somewhat analogous to coal-oil slurry (COM) [5]. A successful example of using bioslurry is found from the FZK process [90, 91]. FZK (Forschungszentrum Karlsruhe) developed a process that produces syngas from agricultural waste feeds such as straw. They developed a flash pyrolysis process that is based on twin screws for pyrolysis, as explained earlier as an auger pyrolyzer. The process concept is based on the Lurgi–Ruhrgas coal gasification process [5, 15]. A 5–10 kg/hr PDU (process development unit) is available at the FZK company site. In this process, straw is flash-pyrolyzed into a liquid that is subsequently mixed with char to form a bio-oil/biochar slurry. The slurry is pumpable and alleviates technical difficulties involved in solid biomass feeding and handling. This slurry is transported and added to a pressurized oxygen-blown entrained gasifier. The operating conditions of the gasifier at Freiberg, Germany involve a slurry throughput of 0.35–0.6 tons/day, 26 bars, and 1,200–1,600°C. The current FZK process concepts involve gasification of flash-pyrolyzed wood products, slow-pyrolyzed straw char slurry (with water condensate), and slow-pyrolyzed straw char slurry (with pyrolysis bio-oil) [90]. Slurries from straw have been efficiently converted into syngas with high conversion and near-zero methane content [83]. Their ultimate objective is development of an efficient biomass-to-liquid plant. A simplified block diagram of the FZK process concept leading to BtL is shown in Figure 5.15.

Canadian developments in biomass gasification for the production of medium- and high-BTU gases have also received worldwide technological

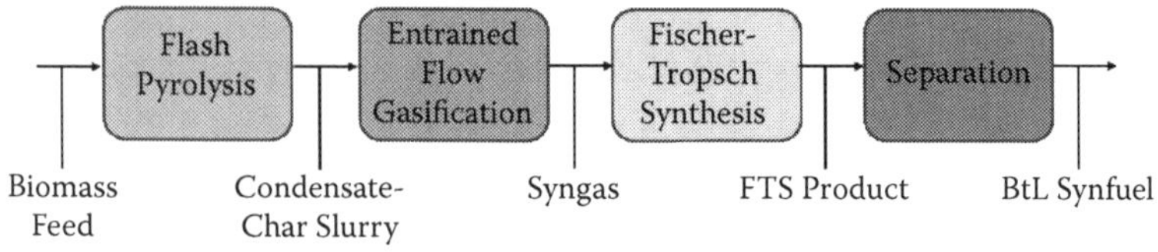

FIGURE 5.15
FZK process concept for BtL synfuel.

attention. The BIOSYN gasification process was developed by Biosyn Inc., a subsidiary of Nouveler Inc., a division of Hydro-Quebec. The process is based on a bubbling fluidized bed gasifier containing a bed of silica (or alumina) and can be operated at a pressure as high as 1.6 MPa. They tested the process extensively during 1984 till 1988 on a 10 ton/hr demonstration plant that was comprised of a pressurized air- or oxygen-fed fluidized-bed gasifier [92]. The system has the ability to utilize a diversified array of feedstock including: whole biomass, fractionated biomass, peat, and municipal solid waste. The primary end-use for the biogas is replacing the oil currently used in industrial boilers. It also has the added capability for producing synthesis gas for methanol or low-energy gas production. Later, they used a 50 kg/hr BIOSYN gasification process development unit and the test program also proved the feasibility of gasifying a variety of other feedstock, such as primary sludges, refuse-derived fuels, rubber residues containing 5–15% Kevlar, granulated polyethylene, and polypropylene [93].

Another emerging process option for biomass gasification involves supercritical water gasification of biomass. Supercritical water (SCW) is water existing under a condition where both temperature and pressure are above critical temperature and pressure; that is, $T > 374°C$ and $P > 218$ atm. At supercritical conditions, water exhibits extraordinary properties that are quite different and distinct from those of ambient water, as compared qualitatively in Table 5.12 [94].

Biomass feedstock can be gasified in a supercritical water medium at a temperature higher than about 650°C and pressure higher than 22.1 MPa [95, 96]. Although common gasification technology requires wet biomass to be sufficiently dried before gasification treatment, the technology based on gasification in supercritical water can handle wet biomass as is without energy- and cost-intensive drying of the feed material. Boukis et al. [96] studied biomass gasification in near-critical and supercritical conditions using a pilot-scale process system called VERNA (a German acronym for "experimental facility for the energetic exploitation of agricultural matter") that had a throughput capacity of 100 kg/h and a maximum reaction temperature of 660°C at 28 MPa. The process system was capable of preparing large biomass particles into about <1 mm particle size using a cutting mill followed by a colloidal mill. The reactor was a downflow type and the reactor system could handle the separation of brines and solids from the bottom of the reactor [96].

TABLE 5.12

Physicochemical and Transport Properties of Water at Supercritical and Ambient Conditions

Ambient Water	Properties Compared	Supercritical Water
Negligible to Low	Organic Solubility	Very High
Very High to High	Inorganic Solubility	Negligible to Very Low
Higher	Density	Medium to High
Higher	Viscosity	Lower
Lower	Diffusivity	Higher
~80	Dielectric Constant	5.7 at Critical Point
High	Polarity	Low
Not	Corrosivity	Somewhat
Lower	Energetics	Highly Energized
Fire Extinguishing	Oxidation	Ideal Combustion Medium
9.2 mg/L	Oxygen Solubility	In any Proportion
Low	H_2 Solubility	Very Low

The principal gaseous products of supercritical water gasification are hydrogen, carbon monoxide, carbon dioxide, methane, and ethane. A generalized reaction scheme involves *reformation* (steam gasification) reaction of hydrocarbons and oxygenates as well as pyrolytic decomposition reactions involving the cleavages of both C–C and C–H bonds. Carbon dioxide concentration usually increases with the temperature of reaction, whereas carbon monoxide decreases. The increase of carbon dioxide in the product stream is due to the results of the forward water gas shift reaction that converts carbon monoxide and water into carbon dioxide and hydrogen [97]. The methane formation for all ranges of gasification temperature is believed to have originated from pyrolytic decomposition of hydrocarbons and their intermediates, not by methanation reaction of syngas; that is, $CO + 3H_2 = CH_4 + H_2O$ [98].

From the kinetics standpoint, the reformation reaction is more active than pyrolytic decomposition at higher temperatures, whereas the pyrolysis reaction is faster and more active than the reformation reaction at lower temperatures [99]. As such, the two pseudo first-order reaction rates meet and cross each other at some point in the temperature domain, when the rates are plotted against the temperature or a reciprocal of temperature. This is a crossover point in the rates between the two representative chemical reactions, where the two reactions have an identical pseudo first-order reaction rate. The higher the crossover temperature is, the more difficult the gasification (or reformation). The location of this kinetic cross-over point is different from chemical to chemical and from one biomass type to another.

The extent of the gasification or gasification efficiency, which is defined as the total carbon appearing in gas phase products divided by the total carbon in biomass feedstock entering the reactor, depends very strongly upon imposed reaction conditions such as the reaction temperature, space time, feed biomass/water ratio, molecular structures of biomass pertaining to the H/C ratio, O and –OH content, isothermality/nonisothermality of the reactor, monolithic catalytic effect from the reactor wall materials, and so on. As the gasification temperature increases, the gasification efficiency generally increases to a certain point. However, gasification efficiency does not change much with the pressure, as long as the water is in its supercritical fluid region. Picou et al. [100] demonstrated that the hydrocarbon reformation in supercritical water can be operated in an autothermal mode with numerous advantages including (i) 100% gasification efficiency, (ii) energywise self-sustainable operation, and (iii) comparable or enhanced product gas composition and yield. They demonstrated their technology using jet fuel in supercritical water at temperatures up to 775°C on a high-nickel Haynes Alloy 282 reactor system. Autothermal reformation (ATR) has been successfully practiced in the field of steam reformation of methane (SMR) [101], wherein a substoichiometric amount of air (or oxygen) is fed to the reformer for sacrificial oxidation reaction of hydrocarbons, thus generating the exothermic heat and promoting reformation reaction. The conventional ATR has been operated catalytically on an industrial-scale subcritical reformation process at much higher temperatures, unlike the process Picou et al. demonstrated using noncatalytic supercritical reformation technology [100]. As shown in Table 5.12, oxygen fully dissolves in supercritical water, thereby easily facilitating an autothermal mode of operation in a supercritical water reactor.

Bouquet et al. [98] showed direct experimental evidence that hydroxyl functional groups present in the molecular structure of a feed chemical play an important role by making the reformation (or gasification) reaction proceed more efficiently via a mechanistically simpler pathway leading to syngas: hydrogen and carbon oxides. They experimentally compared the kinetic results of the supercritical water gasification reactions of three different kinds of C_3 alcohols: iso-propyl alcohol, propylene glycol, and glycerin. The three alcohols used in the experiment represent monhydric, dihydric, and trihydric alcohols of C_3 hydrocarbons, respectively. The results bear significance in biomass gasification in supercritical water, because biomass is rich in hydroxyl groups due to its abundant cellulosic ingredients. In particular, biomass feedstock pretreatment should be carried out in such a way that hydroxyl groups in the molecular structure should not be prematurely destroyed or extracted if the biomass is to be gasified in supercritical water.

5.6 Utilization of Biomass Synthesis Gas

Synthesis gas obtained by biomass gasification can be utilized in a variety of ways and end-uses. Although all conventional syngas utilization methods

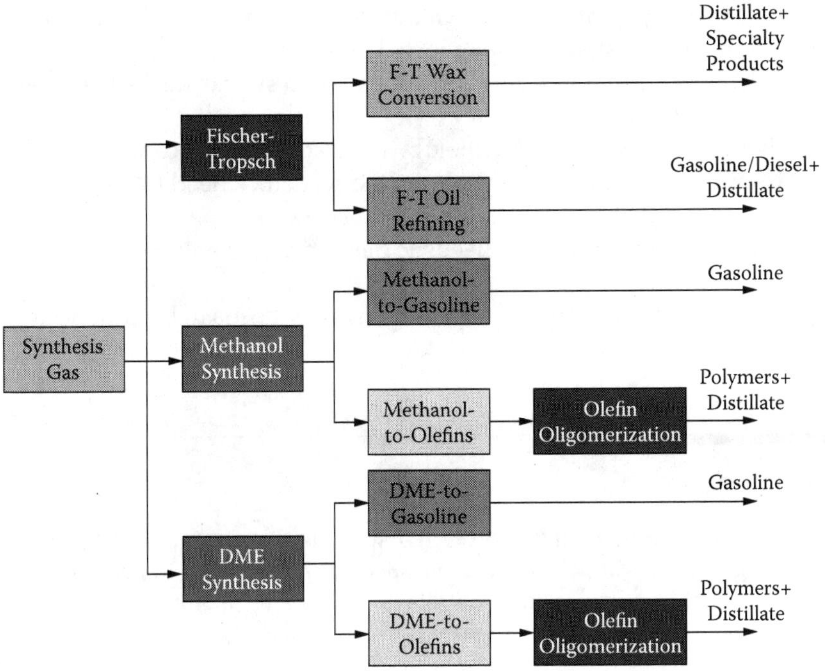

FIGURE 5.16
Transformation of syngas into liquid transportation fuels.

are conceivable and applicable, the process economics, based on the current fuel market and the current level of available technology, may not be favorable for the manufacture of bulk petrochemicals that have traditionally relied on the syngas derived from either natural gas (NG) or coal. However, in certain niche markets, biomass synthesis gas can favorably compete against the NG-based syngas or coal-syngas. Furthermore, biomass syngas is CO_2 neutral and renewable as well as possibly being better suited for small- to medium-scale operations.

Syngas obtained from biomass gasification can be used for indirect liquefaction processes by which syngas is converted to liquid transportation fuels such as methanol, dimethylether (DME), ethanol, higher alcohols, gasoline, diesel, and jet fuels, as shown in Figure 5.16. The following will have to be accomplished in order to convert the biomass syngas into clean liquid fuels that can be used in the conventional energy/fuel infrastructure:

(a) Innovative process integration schemes need to be devised.
(b) Highly efficient energy integration between the intraprocess and interprocess steps needs to be devised and achieved.
(c) Effects of trace minerals in the biomass syngas on the process catalysts need to be fully understood and managed.

(d) Robust and highly effective catalyst systems need to be developed and demonstrated on a long-term basis.
(e) Efficient conversion technology of CO_2-rich syngas needs to be developed and refined.
(f) Biomass pretreatment technology needs to be enhanced.
(g) Conversion technologies using mixed feedstock need to be developed and refined.
(h) Gas clean-up technology needs to be sufficiently enhanced in terms of the efficacy and cost.
(i) Gas separation and purification technology needs to be enhanced.

References

1. Riedy, M.J. and Stone, T.C. 2010. Defining biomass - A comparison of definitions in legislation," Mintz, Levin, Cohn, Ferris, Glovsky, and Popeo, PC.http://www.mintz.com/newsletter/2009/Special/Defining_Biomass.pdf.
2. Sampson, R.N. 1993. Biomass management and energy, *Water Air Soil Pollution*, 70: 139–159.
3. MacCleery, D.W. 1993. *American Forests: A History of Resiliency and Recovery*, U.S. Dept. of Agriculture Forest Service, Tech. Rep. p. 59.
4. Trebbi, G. 1993. Power-production options from biomass: The vision of a Southern European utility, *Bioresource Technol.*, 46: 23–29.
5. Lee, S., Speight, J.G., and Loyalka, S.K. 2007. *Handbook of Alternative Fuel Technology*. Boca Raton, FL: CRC Press.
6. Merriam–Webster On-line Dictionary, 2011.
7. Sampson, R.N. 1993. Forest management & biomass in the USA, *Water, Air, Soil Pollution*, 70: 519–532.
8. Reese, R.A. 1993. Herbaceous biomass feedstock production, *Energy Policy*, 21: 726–734.
9. Nurmi, J. 1993. Heating values of the above ground biomass of small-sized trees, *ACTA Forestalia Fennica*, 236: 2–30.
10. World Energy Council (WEC). 1993. *Energy for Tomorrow's World*, New York: St. Martin's Press.
11. REN21 (Renewable Energy Policy Network for the 21st Century). 2011. Renewables 2011: Global status report.
12. U.S. Forest Service. 2011. Woody biomass utilization. http://www.fs.fed.us/woodybiomass/whatis.shtml.
13. United States Department of Agriculture, United States Department of Energy and United States Department of the Interior. 2003. Memorandum of understanding on policy principles for woody biomass utilization for restoration and fuel treatments on forests, woodlands, and rangelands. http://www.fs.fed.us/woodybiomass/documents/BiomassMOU_060303_final_web.pdf.
14. Lee, S. 1996. *Alternative Fuels*. Philadelphia: Taylor & Francis.

15. Speight, J.G. 1994. *The Chemistry and Technology of Coal.* New York: Marcel Dekker.
16. U.S. Department of Energy, Energy Efficiency & Renewable Energy, Biomass Program, 2011. Biomass program: Trace metal scavenging from biomass syngas using novel sorbents, http://www1.eere.energy.gov/biomass/pdfs/trace_metal.pdf.
17. Channiwala, S.A. 1992. On biomass gasification process and technology development - Some analytical and experimental investigation, PhD Thesis, Mechanical Engineering Dept., IIT, Mumbay.
18. Parikh, J., Channiwala, S.A., and Ghosal, G.K. 2005. A correlation for calculating HHV from proximate analysis of solid fuels, *FUEL,* 84: 487–494.
19. Demirbas, A. 1997. Calculation of higher heating values of biomass fuels, *FUEL,* 76: 431–434.
20. Jenkins, B.M. 1980. Downdraft gasification characteristics of major California residue derived fuels, Ph D dissertation, University of California-Davis.
21. Jenkins, B.M. and Ebeling, J.M. 1985. Correlation of physical and chemical properties of terrestrial biomass with conversion. *Symposium Energy from Biomass and Waste IX,* Institute of Gas Technology, pp. 371.
22. Tillman, D.A. 1978. *Wood as an Energy Resource.*
23. Bridgwater, A.V. 2004. Biomass fast pyrolysis, *Thermal Sci.,* 8: 21–49.
24. Lee, S. 1990. *Methanol Synthesis Technology.* Boca Raton, FL: CRC Press.
25. Lee, S. 1997. *Methane and its Derivatives.* New York: Marcel Dekker.
26. Wright, M.W., Satrio, J.A., Brown, R.C., Daugaard, D.E., and Hsu, D.D. 2010. *Technoeconomic Analysis of Biomass Fast Pyrolysis to Transportation Fuels,* Tech. Rep. NREL/TP-6A20-146586, November, Golden, CO: National Renewable Energy Laboratory.
27. Mullen, C.A., Boateng, A.A., Goldberg, N., Lima, I.M., and Hicks, K.B. 2010. Bio-oil and bio-char production from corn cobs and stover by fast pyrolysis, *Biomass Bioenergy,* 34: 67–74.
28. KBR Homepage. 2011. Coal gasification. http://www.kbr.com/Technologies/Coal-Gasification/ August.
29. U.S. Department of Energy, National Energy Technology Laboratory (NETL). 2011. Gasification systems: Key area - gasifier optimization, http://www.netl.doe.gov/technologies/coalpower/gasification/adv-gas/index.html.
30. Walker, P.L., Rusinko, F., and Austin, L.G. 1959, Gas reactions in carbon. In D.D. Eley, P.W. Selwood, and P.B. Weisz (Eds.), *Advances in Catalysis,* New York: Academic Press, pp. 133.
31. Asadullah, M., Miyazawa, T., Ito, S., Kunimori, K., and Tomishige, K. 2003. Catalyst performance of $Rh/CeO_2/SiO_2$ in the pyrogasification of biomass, *Energy Fuels,* 17: 842–849.
32. Arauzo, J., Radlein, D., Piskorz, J., and Scott, D.S. 1997. Catalytic pyrogasification of biomass. Evaluation of modified nickel catalyst, *Ind. Eng. Chem. Res.,* 36: 67–75.
33. Wen, C.Y. and Huebler, J. 1965. Kinetic study of coal char hydrogasification. Rapid initial reaction, *Ind. Eng. Chem. Process Des. Dev.,* 4: 142–147.
34. Lee, S., Angus, J.C., Edwards, R.V., and Gardner, N.C. 1984. Non-catalytic coal char gasification, *AIChE ,* 30: 583–593.

35. von Frederdorff, C.G. and Elliott, M.A. 1963. Coal gasification. In H.H. Lowry (Ed.), *Chemistry of Coal Utilization, Supplementary Volume*, New York: Wiley, pp. 892.
36. Hottel, H.C., Williams, G.C., and Wu, P.C. 1977. *Am. Chem. Soc. Div. Fuel Chem.*, 22: 1.
37. Strange, J.F. and Walker, P.L. 1976. Carbon-carbon dioxide reaction - Langmuir-Hinshelwood kinetics at intermediate pressures, *Carbon*, 14: 345–350.
38. National Energy Technology Laboratory (NETL), U.S. Department of Energy. 2011. Gasifipedia. http://www.netl.doe.gov/technologies/coalpower/gasification/gasifipedia/4-gasifiers/4-1-4-3_hydro.html September.
39. Scott, D.S. 1986. *Catalytic Hydrogasification of Wood*. Canada: Renewable Energy in Canada.
40. Walter, K., Friedman, S., Frank, L.V., and Hiteshu, R.W. 1968. Coal hydrogasification catalyzed by aluminum chloride, *ACS Fuel Preprint*, 12: 43-47.
41. Asami, K. and Ohtsuka, Y. 1992. Hydrogasification of brown coal with active iron catalysis, *ACS Fuel Preprint*, 37: 1951–1956.
42. Nishyama, Y. and Haga, T. 1980. Low temperature hydrogasification of carbons using nickel based catalyst. In T. Seiyama and K. Tanabe (Eds.), *New Horizons in Catalysis: Part 7B. Proceedings of the 7th International Congress on Catalysis*, Tokyo: Elsevier, pp. 1434–1438.
43. Haga, T. and Nishyama, Y. 1987. Promotion of iron-group catalysts by a calcium salt in hydrogasification of carbons at elevated pressures, *Ind. Eng. Chem. Res.*, 26: 1202–1206.
44. Porada, S. 2009. A comparison of basket willow and coal hydrogasification and pyrolysis, *Fuel Process. Technol.*, 90: 717–721.
45. Milne, T.A. and Evans, R.J. 1998. Biomass gasifier "tars"; Their nature, formation, and conversion. U.S. Department of Energy, National Energy Technology Laboratory.
46. Elliott, D.C. (Ed.) 1988. *Relation of Reaction Time and Temperature to Chemical Composition of Pyrolysis Oils*. Washington, DC: American Chemical Society.
47. Wu, C. and Williams, P.T. 2011. Nickel-based catalysts for tar reduction in biomass gasification, *Biofuels*, 2: 451–464.
48. Brage, C., Yu, Q., Chen, G., and Sjöström, K. 1997. Use of amino phase adsorbent for biomass tar sampling and separation, *Fuel*, 76: 137–142.
49. Technical committee CEN/BT/TF 143. 2004. Measurement of organic contaminants (tar) in biomass producer gas. Biomass gasification - Tar and particles in product gases - sampling and analysis. http://www.eeci.net/results/pdf/CEN-Tar-Standard-draft-version-2_1-new-template-version-05-11-04.pdf.
50. Dynamotive Energy Systems. 2011. BioOil. http://www.dynamotive.com/industrialfuels/biooil/.
51. Lee, S. 1991. *Oil Shale Technology*. Boca Raton, FL: CRC Press.
52. Czernik, S. 2002. Reviews of fast pyrolysis of biomass. National Renewable Energy Laboratory. http://www.nh.gov/oep/programs/energy/documents/biooil-nrel.pdf.
53. Czernik, S. and Bridgwater, A.V. 2004. Overview of applications of biomass fast pyrolysis, *Energy Fuels*, 18: 590–598.
54. Bain, R.L. 2004. An introduction to biomass thermochemical conversion. Presented at *DOE/NASLUGC Biomass and Solar Energy Workshops*. http://www.nrel.gov/docs/gen/fy04/36831e.pdf.

55. Ringer, M., Putsche, V., and Scahill, J. 2006. *Large-Scale Pyrolysis Oil Production: A Technology Assessment and Economic Analysis*, Tech. Rep. NREL/TP-510-37779, November. Golden, CO: U.S. Department of Energy, National Energy Research Laboratory.
56. Conversion and Resource Evolution (CARE), Ltd. 1998–2002. 250 kg/h biomass fast pyrolysis plant for power generation, Wellman Process Engineering Ltd., Oldbury, England. http://www.care.demon.co.uk/projectprofile07.pdf.
57. Scott, D.S., Majerskib, P., Piskor J., and Radlein, D. 1999. A second look at fast pyrolysis of biomass - The RTI process, *J. Anal. Appl. Pyrolysis*, 51: 23–37.
58. VTT Technical Research Centre of Finland, 2002. *Review of Finnish Biomass Gasification Technologies*, Tech. Rep. Report No. 4.
59. Gradassi, A.T. 2002. Fast pyrolysis of biomass: Work in progress at ENEL produzione, *IEA Clean Coal Sciences Agreement 23rd Meeting*, Pisa, Italy.
60. Biomass Technology Group (BTG). 2011. Pyrolysis oil. http://www.btgworld.com/index.php?id=22&rid=8&r=rd, November.
61. Ray, C. 2000. Vacuum pyrolysis breakthrough, *Pyrolysis Network*.
62. Bridgwater, A.V. 1999. Principles and practice of biomass fast pyrolysis processes for liquids, *J. Anal. Appl. Pyrolysis*, 51: 3–22.
63. U.S. Department of Energy–Energy Efficiency and Renewable Energy–Biomass Program. 2011. Pyrolysis and other thermal processing. http://www1.eere.energy.gov/biomass/printable_versions/pyrolysis.html]. November.
64. Bain, R.L. 1993. Electricity from biomass in the United States; Status and future direction, *Bioresource Technol.*, 46: 86–93.
65. Solantausta, Y. 1993. Wood-pyrolysis oil as fuel in diesel-power plant, *Bioresource Technol.*, 46: 177–188.
66. Rick, F. and Vix, U. 1991. Product standards for pyrolysis products for use as a fuel. In A.V. Bridgwater and G. Grassi (Eds.), *Biomass Pyrolysis Liquids Upgrading & Utilization*, London: Elsevier, pp. 177–218.
67. Skelley, W.W., Chrostowski, J.W., and Davis, R.S. 1982. The energy resources fluidized bed process for converting biomass to electricity, *Symposium on Energy from Biomass & Wastes VI*, pp. 665–705.
68. Brown, R.C. and Holmgren, J. 2011. Fast pyrolysis and bio-oil upgrading. http://www.ascension-publishing.com/BIZ/HD50.pdf.
69. Levenspiel, O. 1999. *Chemical Reaction Engineering*. Hoboken, NJ: Wiley.
70. Agri-Therm. 2011. Agri-therm pyrolysis systems. http://agri-therm.com/.
71. Piskorz, J., Majerski, P., Radlein, D., Scott, D.S., and Bridgwater, A.V. 1998. Fast pyrolysis of sweet sorghum and sweet sorghum bagasse, *J. Anal. Appl. Pyrolysis*, 46: 15–29.
72. Piskorz, J. and Scott, D.S. 1987. The composition of oils obtained by the fast pyrolysis of different woods, *ACS Fuel Preprint*, 32: 215–222.
73. Kaushal, P., Mirhidi S.A. and Abedi, J. 2011. Fast pyrolysis of biomass in bubbling fluidized bed: A model study, *Chem. Product Process Model.*, 6: Article 24.
74. Bridgwater, A.V., Czernik, S., and Piskorz, J. 2001. An overview of fast pyrolysis. In A. V. Bridgewater (Ed.), *Progress in Thermochemical Biomass Conversion*, Oxford: Blackwell Science, pp. 977–997.
75. van der Velden, M., Fan, X., Ingramz, A., and Baeyens, J. 2007. Fast pyrolysis of biomass in a circulating fluidized bed, *2007 ECI Conference on the 12th International Conference on Fluidization - New Horizons in Fluidization Engineering*.

76. Anderson, P.G. 2007. Twin screw extrusion. In S. Lee (Ed.), *Encyclopedia of Chemical Processing*, New York: Taylor & Francis.
77. Benson, W R. 1977. Biomass potential from agricultural production. *Proceedings: Biomass-a Cash Crop for the Future? Conference on the Production of Biomass from Grains, Crop Residues, Forages and Grasses for Conversion to Fuels and Chemicals,* Kansas City, MO.
78. The Guardian Data Blog. 2010. U.S. corn production and use for fuel ethanol. http://www.guardian.co.uk/environment/datablog/2010/jan/22/us-corn-production-biofuel-ethanol January.
79. Kadam, K.L. and McMillan, J.D. 2003. Availability of corn stover as a sustainable feedstocks for bioethanol production, *Biores. Technol.,* 88: 17-25.
80. Raman, K.P., Walawender, W.P., Shimizu, Y., and Fan, L.T. 1980. Gasification of corn stover in a fluidized bed, *Bio-Energy 80,* Atlanta, GA.
81. Bodland, B. and Bergman, J. 1993. Bioenergy in Sweden: Potential, technology, and application, *Biores. Technol.* 46: 31–36.
82. Finnish Ministry of the Environment, Communications Unit. 2007. Gasified biomass for efficient power and heat generation - FACTS on environmental protection. http://www.ymparisto.fi/download.asp?contentid=68164&lan=fi.
83. van der Drift, A. and Boerrigter, H. 2006. Synthesis gas from biomass for fuels and chemicals. *Report of Workshop on Hydrogen and Synthesis Gas for Fuels and Chemicals, Organized IEA Bioenergy Task 33, SYNBIOS Conference,* Stockholm, Sweden.
84. Craig, K.R. and Mann, M.K. 1996. Cost and performance analysis of biomass-based integrated gasification combined-cycle (BIGCC) power systems. National Renewable Energy Laboratory, U.S. Department of Energy. http://www.nrel.gov/docs/legosti/fy97/21657.pdf.
85. Yan, J., Alvfors, P., Eidensten, L., and Svedberg, G. 1997. A future for biomass, *MEMagazine,* October.
86. Paisley, M.A., Irving, J.M., and Overend, R.P. 2001. A promising power option - The FERCO silvagas biomass gasification process - Operating experience at the burlington gasifier, *Proceedings of ASME Turbo Expo 2001 June 4–7, 2001,* New Orleans, pp. 1–7.
87. Claussen, M. and Vodegel, S. 2005. The CUTEC concept to produce BtL-fuels for advanced power trains. *International Freiberg Conference on IGCC and XtL Technologies,* Freiberg, Germany.
88. Shah, Y.T., Ontko, J., Gardner, T., Barry, D., and Summers, W. 2012. Torrefaction. In S. Lee (Ed.), *Encyclopedia of Chemical Processing*.
89. Wolters, C., Canaar, M., and Kiel, J. 2004. Co-generation of biomass in 250 MWe IGCC plant 'willem-alexander-centrale.' *Biomass Gasification Workshop* at Rome, Italy. http://www.gastechnology.org/webroot/downloads/en/IEA/IEARomeWSKiel.pdf.
90. Dinjus, E. 2005. German developments in biomass gasification. *IEA Renewable Working Party Bioenergy Task 33: Thermal Gasification*, Innsbruck, Austria, September 26, http://www.gastechnology.org/webroot/downloads/en/IEA/Fall05AustriaTaskMeeting/GermanyGasificationActivities.pdf.
91. Henrich, E. 2007. The status of the FZK concept of biomass gasification. *Second European Summer School on Renewable Motor Fuels.* http://www.baumgroup.de/Renew/download/5%20-%20Henrich%20-%20slides.pdf.

92. Hayes, R.D. 1991. Overview of thermochemical conversion of biomass in Canada. In A.V. Bridgwater and G. Grassi (Eds.), *Biomass Pyrolysis Liquids: Upgrading and Utilization*, Amsterdam, The Netherlands: Elsevier Science.
93. Babu, S. 2011. Biomass gasification for hydrogen production, December.
94. Lee, S., Lanterman, H.B., Wenzel J.E., and Picou, J. 2009. Noncatalytic reformation of JP-8 fuel in supercritical water for production of hydrogen, *Energy Sources: Part A: Recovery, Utilization, Environ. Effects*, 31: 1750–1758.
95. Antal, M.J., Jr., Allen, S.G., Schulman, D., and Xu, X. 2000. Biomass gasification in supercritical water, *Ind. Eng. Chem. Res.*, 39: 4040–4053.
96. Boukis, N., Galla, U., Muller, H., and Dinjus, E. 2007. Biomass gasification in supercritical water experimental process achieved with the VERENA pilot plant, *15th European Biomass Conference & Exhibition*, Berlin, pp. 1013–1016.
97. Picou, J., Stever, M.S., Bouquet, J.S., Wenzel, J.E., and Lee, S. 2012. Kinetics of the noncatalytic water gas shift reaction in supercritical water production, *Energy Sources, Part A: Recovery, Utilization, Environ. Effects*, in press.
98. Bouquet, J.S., Tschannen, R.E., Gonzales, A.C., and Lee, S. 2011. The effects of carbon-to-oxygen ratio upon supercritical water reformation for hydrogen production. *2011 AIChE Annual Meeting, Symposium on Alternative Fuels I*, Minneapolis.
99. Lee, S., Lanterman, H.B., Picou, J., and Wenzel, J.E. 2009. Kinetic modeling of JP-8 reforming by supercritical water, *Energy Sources, Part A: Recovery, Utilization, Environ. Effects*, 31: 1813–1821.
100. Picou, J., Wenzel, J.E., Lanterman, H.B., and Lee, S. 2009. Hydrogen production by non-catalytic autothermal reformation of aviation fuel using supercritical water, *Energy Fuels*, 23: 6089–6094.
101. Aasberg-Petersen, K. 2005, Synthesis gas. In S. Lee (Ed.), *Encyclopedia of Chemical Processing*, London: Taylor & Francis, pp. 2933–2946.

6

Conversion of Waste to Biofuels, Bioproducts, and Bioenergy

6.1 Introduction

As the world population grows and its natural resources diminish, the concept of waste is changing. Generally *waste* is considered as those chemicals and materials that are either used and discarded or those that are perceived to have little direct use potential for human or animal needs. Currently, the preferred mode is to recycle and reuse waste or discard it in landfills. Not all materials are recyclable or can be reused without further treatment or conversion. The new concepts of Enhanced Waste Management (EWM) and Enhanced Landfill Mining (ELFM) put landfilling of waste in a sustainable context [1, 2]. In these new concepts, a landfill is no longer considered as a final solution but a temporary storage place before the stored waste is reused through an appropriate conversion process [3]. Thus, ELFM offers an opportunity to select an appropriate path for the conversion of waste into either materials (waste to product, WtP) or energy (waste to energy, WtE) and thereby reuse all waste to the extent new technologies and environmental regulations allow. The success of these new concepts depends not only on the technological improvements and breakthroughs but also on a multitude of socioeconomic barriers such as regulations, social acceptance, economic uncertainty, and feasibility of a particular technology in the given environment which prevents the emissions of CO_2 and pollutants [2].

An integrated solid waste management is typically governed by the process (often known as the "Ladder of Lansink") which specifies a generally accepted hierarchy of preferred methods of dealing with different types of waste. Direct recycling and reuse of waste is preferred, however, this is not always possible. Numerous technologies are now available to convert each type of waste either to energy or to a reusable product. This chapter examines these technologies and associated processes to obtain the desirable outcomes. The chapter also briefly examines the environmental and economic issues associated with various conversion processes.

6.2 Types of Waste and Their Distributions

The waste originates from numerous sources: residential community (i.e., municipal solid waste or MSW), commercial and light-industrial communities, manufacturing activities such as heavy-industrial and chemical industries (generally classified as hazardous waste), agricultural and forestry waste, human and animal waste, paper and pulp industry waste, automobile and other transportation waste, hospital waste (generally considered as infectious waste), nuclear waste, and so on. In addition to man-made waste, there are numerous naturally occurring waste materials generally classified under the category of "lignocellulosic materials (LCM)" and certain forms of crop oils (e.g., algae, waterweed, water hyacinth). These are biomasses that do not have a useful purpose for food or a direct use for human or animal purposes.

Table 6.1 lists the heating values of various waste-derived fuels. This table shows the valuable energy content of various types of waste materials. This chapter does not consider special types of waste such as nuclear and infectious waste as well as several types of hazardous and nonhazardous industrial waste such as glass, metals, and other noncombustible waste. The chapter does, however, examine the appropriate conversion processes for all cellulosic-based waste as well as some polymeric waste such as plastic and rubber tires. A typical material distribution of MSW collected in the United States is illustrated in Figures 6.1a and b. In 1988, approximately 80% of 180 million tons of waste generated in the United States was cellulose-based [10]. This percentage has not changed in the 1990s and 2000s [4]. In Europe, MSW is expected to increase up to 300 million tons by 2015 [5].

TABLE 6.1

Comparison of Heating Values of Various Waste-Derived Fuels

Fuel Source	BTU/Lb
Yard wastes	3,000
Municipal solid waste	6,000
Combustible paper products	8,500
Textiles and plastics	8,000
Bituminous coal (average)	11,300
Anthracite coal (average)	12,000
Spent tires	13,000–15,000
Crude oil (average)	17,000
Natural gas (425 ft3)	13,500

Source: Lee, Speight, and Loyalka, 2007. Energy generation from waste sources, *Handbook of Alternate Fuel Technologies*, Ch. 13. New York: CRC Press, pp. 395–419.

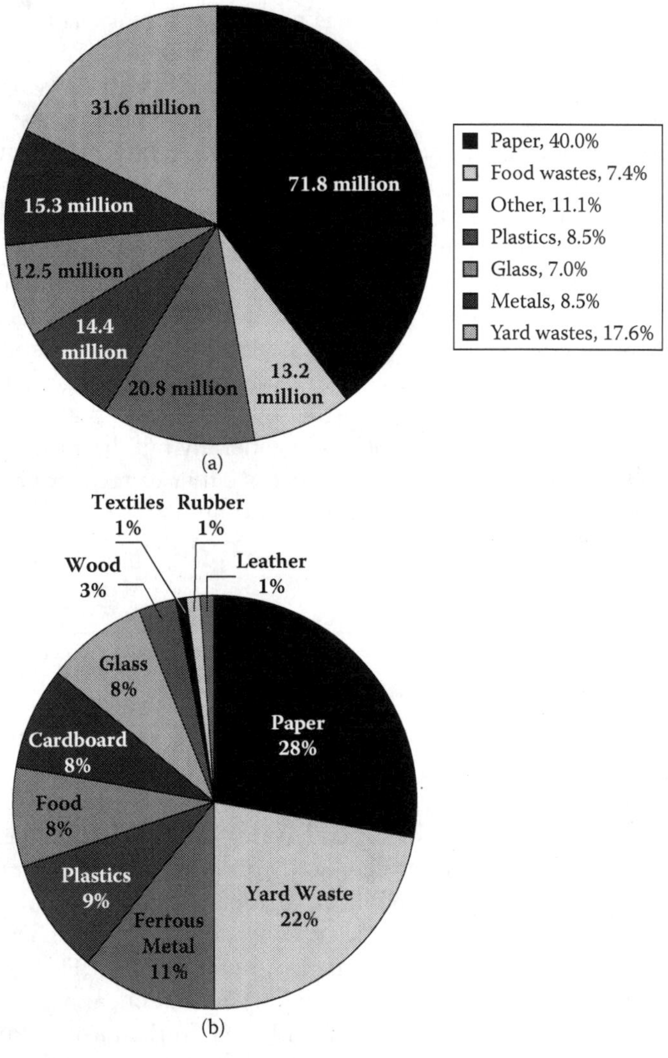

FIGURE 6.1
(a) Material distribution of MSW collected in the United States by weight (tons). (b) Breakdown of material distribution of MSW collected in the United States in 2002. (After Lee, Speight, and Loyalka, 2007. Energy generation from waste sources, *Handbook of Alternate Fuel Technologies*, New York: CRC Press, Ch. 13., pp. 395–419. With permission.)

In processing waste, noncombustible materials such as glass and dirt are removed. The glass is either recycled or sent to glass-melting furnaces. The heavy metals such as ceramics, heavy metals, and aluminum are routed to the landfill for disposal. A significantly important component of MSW is polymeric materials. Although polymer waste only accounts for 8.5% by mass of the total MSW disposed of in the United States, plastic represents over 28% by volume [6]. Plastic waste is not biodegradable. Thus it is mostly recycled either for reuse or to recover basic monomers. Polymeric waste ranges from packaging materials used in the food industry to various parts in automobiles to high-density polyethylene (HDPE) containers such as pop bottles, laundry detergent bottles, milk jugs, and so on. It is estimated that over 65% of the food packaging in the United States is from plastics. As of 1978, the Ford Motor Company estimated that the average junked car contained 80 kg of plastic and nontire rubber. This number is increasing every year because of the increased use of polymers in various automobile parts.

Some types of plastic waste are recycled directly. For example, milk bottles, juice containers, laundry detergent bottles, motor oil cans, spring water bottles, and other similar containers are subjected to a thorough cleaning process and are reused. Some rubber tires are retread and put back for reuse. Polymeric materials including rubber are generally not discarded in a landfill and are subjected to a conversion process either to recover basic chemicals and materials or to generate energy.

6.3 Strategies for Waste Management

As indicated earlier, in a new approach to strategic resource management, the concepts of WtP and WtE are implemented for every different type of waste. As shown in Figure 6.2, there are numerous technologies now available to convert waste to heat, electricity, transportation fuels, chemicals, or materials. These technologies are generally broken down into three categories: thermochemical, physicochemical, and biochemical. In each of these categories, a process can be catalytic or noncatalytic. Thermochemical conversion of waste to energy is illustrated in more detail in Figure 6.3 [7, 8]. All advanced technologies such as pyrolysis, gasification, and plasma-based technologies have been developed since 1970 [9]. In the past, thermochemical techniques were predominantly used to generate energy. In recent years, these techniques are also used to generate chemicals and materials [9] via various methods of product upgrading. For example, gasification of biomass can produce syngas which can be further converted to a host of liquid products such as methanol, diesel fuel, gasoline, and jet fuel via Fischer-Tropsch and related syntheses such as methanol and iso- or oxy-synthesis. Liquefaction can produce liquids that can be upgraded by hydro-deoxygenation, hydrogenation, hydrocracking, or catalytic cracking to produce a host of chemicals and fuels. Pyrolysis can produce gases, liquids, or solids depending upon the reaction conditions. The gases can be used as fuel or raw material for polymers such as polyethylene and polypropylene, and others. Pyrolysis oil can also be upgraded to use as fuels or refined into a number of chemicals.

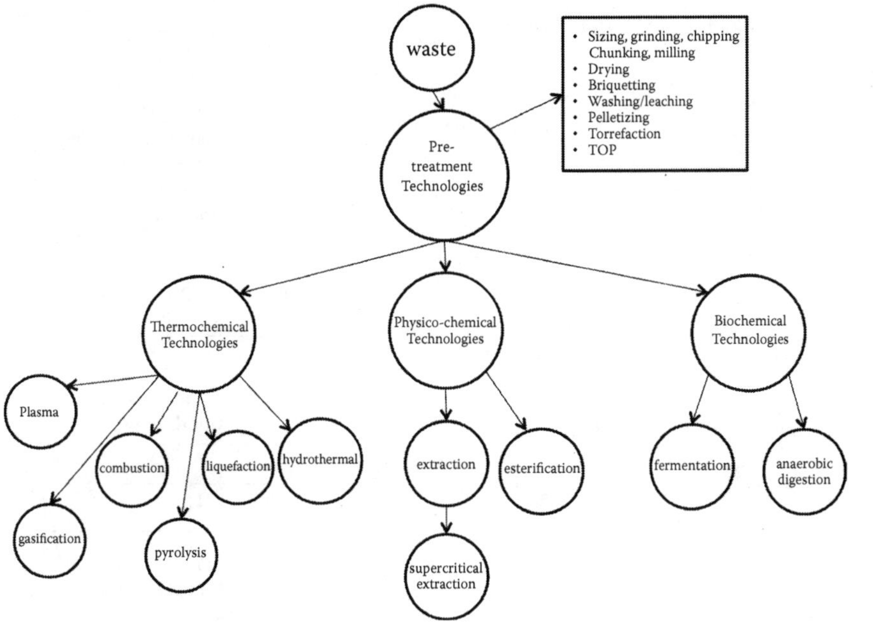

FIGURE 6.2
Waste-to-energy conversion technologies. (Adapted from M. Kaltschmitt, and G. Reinhardt, eds. (1997), *Nachwachsende Energietrager-Grundlagen, Verfahren, okologische Bilanzierung*, Braunschweig/Wiesbaden, Vieweg Verlagsgesellschaft. With kind permission of Springer Science+Business Media.)

The solid residues from pyrolysis can be important raw materials for the construction and fertilizer industries [10].

Biochemical conversion techniques used to convert lignocellulosic waste can also generate chemicals and materials. The LCM can be fractionated into hemicellulose, cellulose, and lignin by selective solubilization of hemicelluloses via hydrothermal processing with water or prehydrolysis with externally added mineral acids. The liquor produced by this process can contain oligosaccharides, fermentable sugar, furfural, low molecular weight phenolics, and levulinic acid depending on the reaction operating conditions. Various products can be extracted from the liquor. The sugar can be further fermented to produce a host of alcohols and acids such as ethanol, butanol, xylitol, butanediol, and lactic acid, among others. The solids coming from the solubilization step mainly contain cellulose and lignin and can be further hydrolyzed by acids or enzymes to give a fermentable glucose solution and a solid phase that largely contains lignin. This solid material can either be used as fuel or a raw material for gasification or pyrolysis. The fermentable glucose solution can be converted to a host of products such as lactic acid, citric acid, succinic acid, itaconic acid, or bioplastics by the suitable fermentation process [10].

The strategy for each type of waste is to apply the appropriate technology for the desired end product. The choice of the best technology will

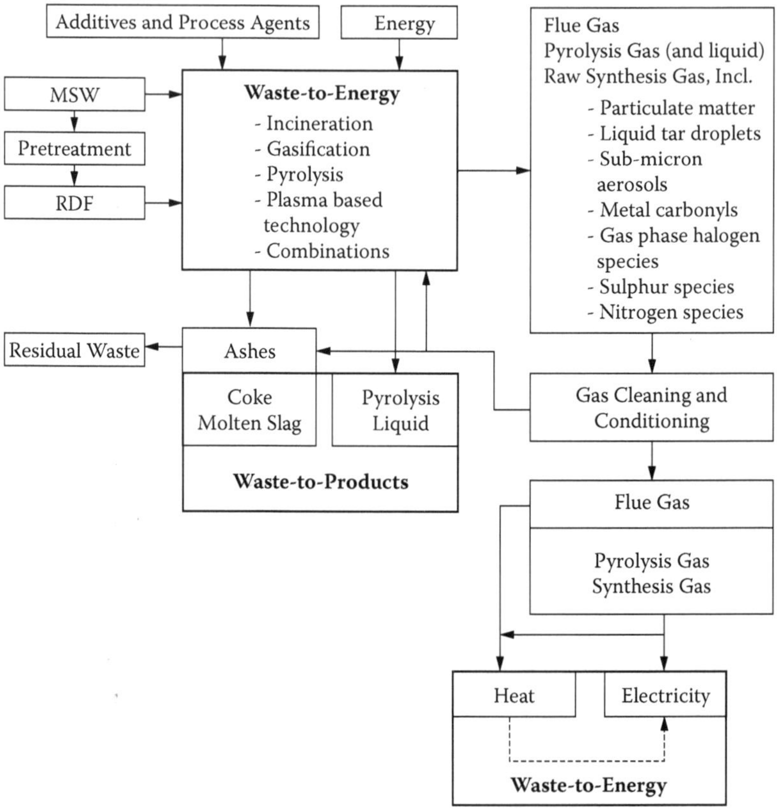

FIGURE 6.3
Schematic overview WtE concept. (After Helsen and Bosmans, 2010. Waste to energy through thermochemical processes: Matching waste with process, *Conference Proceedings on Enhanced Landfill Mining and Transition to Sustainable Materials Management,* Molenheide, Houthalen-Heichteren, Belgium, October 4–6. With permission.)

depend on various factors such as environmental regulations, local economics, resources available to use the technology, and the market for the end-product. Some of the major technologies are further described later in this chapter.

6.4 Waste Preparation and Pretreatment for Conversion

If the MSW is used in an as-received condition as input to WtE or WtP processes, it can lead to variable (and even unstable) operating conditions due to variable properties of the feedstock. This, at least, will result in fluctuating product quality. Depending upon the technology used, convenient and stable feeding of waste to the conversion process is important. Refuse-derived fuel (RDF), which is a processed form of MSW, is often used to prepare waste for various WtE and WtP processes. This preparation or pretreatment usually consists of size reduction, screening, sorting, and in some cases, drying or pelletization to improve the handling characteristics and homogeneity of the waste materials. Therefore, for a given technology, a trade-off between the increased cost of producing RDF from MSW and potential cost reductions from system design and operations needs to be found. The main advantages of RDF are higher heating value; homogeneous physical and chemical structure of the feed; easy storage, handling, and transportation; lower pollutant emissions; and reduced excess air requirement during combustion. In addition to pelletization, biomass, in general, can undergo a variety of pretreatments. The advantages and disadvantages of these pretreatment methods are described in Table 6.2.

Because waste can contain numerous impurities, which during the conversion process mostly end up either in the gas phase or the solid phase, these impurities can also be removed as a part of waste preparation methods. Once again it is a choice of economics between removing the impurities up front or at the end of the conversion process. Waste can, in general, contain both inorganic and organic impurities. Depending upon the technology used, some CO, HCl, HF, HBr, HI, NO_x, SO_2, VOCs, PCDD/F, PCBs, and heavy metal compounds (among others) can be formed or remain [7]. Solids from various conversion processes can also contain numerous impurities including ash.

6.5 Technologies for Conversion of Waste to Energy and Products

As mentioned before, the technologies for the conversion of waste to energy and products can be broken down into three categories: thermochemical, biochemical, and physicochemical technologies. Combustion/incineration is a special type of thermochemical technology which is mainly used for the generation of heat and electricity. As indicated earlier, various types of upgrading technologies are used to generate transportation fuels, chemicals, and materials [10]. The end results for the applications of these technologies are heat, electricity, transportation fuel, chemicals, or materials. In the following sections, we briefly evaluate each of these technologies. Some of the detailed descriptions of the combustion, pyrolysis, gasification, and plasma technology outlined below closely follow the excellent review of Helsen and Bosmans [7].

TABLE 6.2

Summary of Advantages and Disadvantages of Various Biomass Pretreatments

Biomass Pretreatment	Advantages	Disadvantages
Sizing (grinding, chipping, chunking, milling)	Adjusts the feedstock to the size requirement of the downstream use.	Nonbrittle character of biomass creates problems for sizing. Should be done before transportation but storage of sized materials increases dry matter losses and microbiological activities leading to GHG (CH_4, N_2O) emissions.
Drying	Reduces dry matter losses, decomposition, self-ignition, and fungi developments during storage. Increases potential energy input for steam generation.	Natural drying is weather-dependent; drying in dryers requires sizing.
Bailing	Better for storage and transportation; higher density and lower moisture content.	Cannot be used without sizing for gasification.
Briquetting	Higher energy density, possibility for more efficient transport and storage. Possibility for utilization of coal infrastructure for storage, milling, and feeding; rate of combustion comparable with coal. Reduces spontaneous combustion.	Easy moisture uptake leading to biological degradation and losses of structure. Require special storage conditions. Hydrophobic agents can be added to briquetting process, but these increase their costs significantly.
Washing/leaching	Reduction of corrosion, slagging, fouling, sintering, and agglomeration of the bed-washing is especially important in case of herbaceous feedstock. Reduced wearing-out of equipment, and system shutdown risks.	Increased moisture content of biomass. Addition of dolomite or kaolin, which increase ash melting point, can also reduce negative effects of alkali compounds.
Pelletizing	Higher energy density leads to better transportation, storage, and grinding and reduced health risks. Possible utilization of coal infrastructure for feeding and milling (permits automatic handling and feeding).	Sensitive to mechanical damaging and can absorb moisture and swell, lose shape and consistency. Demanding with regard to storage conditions.

(Continued)

TABLE 6.2 (CONTINUED)

Summary of Advantages and Disadvantages of Various Biomass Pretreatments

Biomass Pretreatment	Advantages	Disadvantages
Torrefaction	Possibility for utilization of coal infrastructure for feeding and milling. Improved hydrophobic nature – easy and safe storage, biological degradation almost impossible. Improved grinding properties resulting in reduction of power consumption during sizing. Increased uniformity and durability.	No commercial process. Torrefied biomass has low volumetric energy density.
TOP process	Combines the advantages of torrefaction and pelletizing. Better volumetric energy density leading to better storage and cheaper transportation. Desired production capacity can be established with smaller equipment. Easy utilization of coal infrastructure for feeding and milling.	No commercial process. Does not address the problems related to biomass chemical propertied, that is, corrosion, slagging, fouling, sintering, or agglomeration.

Source: Modified from Shah and Gardner, in press. Biomass Torrefaction: Applications in Renewable Energy and Fuels. In *Encylopedia of Chemical Processes*, Boca Raton, FL: CRC Press.

Note: This information is repeated in Table 7.6.

6.5.1 Combustion/Incineration

Combustion or incineration technology is used for a very wide range of waste to reduce its volume and hazardous characteristics as well as to generate heat and electricity. It is most widely applied and can be implemented on a large as well as small scale. Waste generally contains organic matter, minerals, metals, and water. Metals are separated before incineration. During incineration, flue gases are generated that contain energy which can be transformed into heat and electricity. The organic waste burns when it has reached the ignition temperature and come into contact with oxygen. The overall oxidative process of combustion is highly exothermic, although it occurs in the stages of drying, degassing, pyrolysis, gasification, and combustion; not all of which are exothermic in nature. Initially, heat may be needed to start the process. However, once the chain reaction of the combustion process is started [11], no external heat or additional fuel is required. Although the stages of the combustion process are inseparable, furnace design, air distribution, and control system can affect these stages and reduce pollutant emissions.

TABLE 6.3

Typical Reaction Conditions and Products from Pyrolysis, Gasification, Incineration, and Plasma-Based Processes

	Pyrolysis	Gasification	Combustion	Plasma Treatment
Temperature [°C]	250–900	500–1,800	800–1,450	1,200–2,000
Pressure [bar]	1	1–45	1	1
Atmosphere	Inert/nitrogen	Gasification agent: O_2, H_2O	Air	Gasification agent: O_2, H_2O Plasma gas: O_2, N_2, Ar
Stoichiometric ratio	0	<1	>1	<1
Products from the process:				
Gas phase	H_2, CO, H_2O, N_2, Hydrocarbons	H_2, CO, CO_2, CH_4, H_2O, N_2	CO_2, H_2O, O_2, N_2	H_2, CO, CO_2, CH_4,H_2O, N_2
Solid phase	Ash, coke	Slag, ash Some tar	Ash, slag	Slag, ash
Liquid phase	Pyrolysis oil and water		—	—

Source: From Helsen and Bosmans. 2010. Waste to energy through thermochemical processes: Matching waste with process, *Conference Proceedings on Enhanced Landfill Mining and Transition to Sustainable Materials Management*, Molenheide (Houthalen-Heichteren, Belgium), October 4–6; Kolb and Seifert. 2002, Thermal waste treatment: State of the art – A summary. *Waste Management 2002: The Future of Waste Management in Europe*, October 7–8, Strasbourg (France), Edited by VDI GVC (Düsseldorf, Germany), and ETC/RWM. 2007. *Environmental Outlooks: Municipal Waste, Working Paper* no. 1/2007, European Topic Centre on Resource and Waste Management, Retrieved July 27, 2010, from http://waste.eionet.europa.eu/publications.

A typical set of reaction conditions for the combustion process is outlined in Table 6.3 [7, 9]. The table also compares the typical operating conditions for combustion with those of pyrolysis, gasification, and plasma treatment (i.e., other important thermochemical processes). Air in the combustion process can be replaced by oxygen, and it is the only thermochemical process where air (or oxygen) is in stoichiometric excess. In fully oxidative combustion, the flue gas contains water vapor, nitrogen, carbon dioxide, and oxygen. However, depending upon the nature of the waste and the operating conditions, smaller amounts of CO, HCl, HF, HBr, HI, NO_x, SO_2, VOCs, PCDD/F, PCBs, and heavy metal compounds can be a part of flue gas [7]. The presence of these compounds can cause environmental issues, and they should be removed before the flue gas is emitted to the environment. Depending upon the combustion temperature, heavy metals and inorganic matters (salts) can end up in the flue gas or fly ash. The amount and composition of solid residue during combustion depends on the nature of waste, combustion temperature,

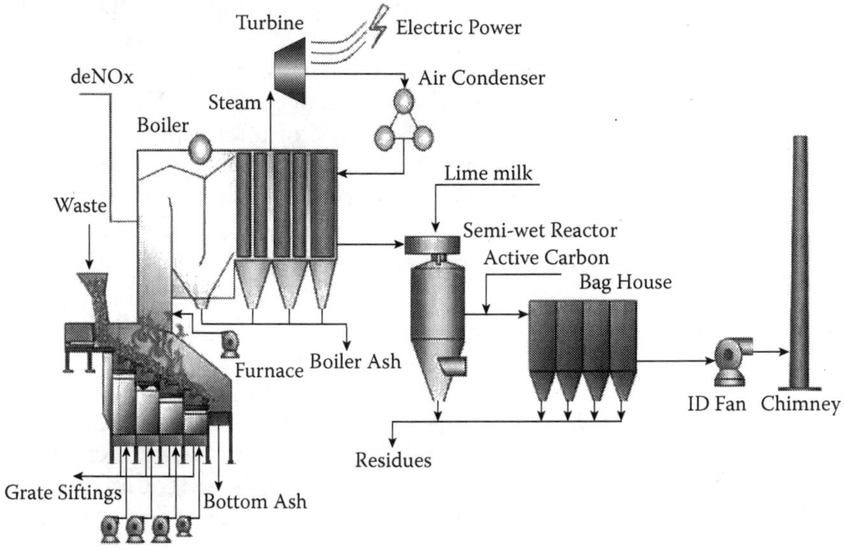

FIGURE 6.4
Example layout of a MSW incineration plant. (From BIC. (2010), BIC group-moving bed incinerator, retrieved 19 Aug., 2010 from http://www.bicgroup.com.sg/.)

and the detailed process design. In MSW incineration, the bottom ash is about 25 to 30% by weight of solid waste input. Fly ash can be 1 to 5% by weight of the waste input [12]. Additional treatment of the bottom ash can allow its use in the concrete and other construction industries. Fly ash needs to be immobilized or vitrified to make it environmentally safe for landfill disposal.

For an effective incineration, an excess of air or oxygen (generally 1.2 to 2.5 times stoichiometric requirement) is necessary. This number depends upon the nature of fuel (gas, liquid, or solid) and the incinerator design. In principle, the incinerator generates excess energy from MSW combustion to produce heat and electricity. The incinerator size may vary depending on the availability of waste. In Europe, the average capacity is about 193 k tons per year [7, 11]. A typical schematic of the incineration plant is shown in Figure 6.4. Such a plant includes (a) waste reception, storage, and preparation unit; (b) a combustor; (c) boiler or other type of energy recovery unit; (d) flue gas cleaning and emission monitoring and control system; (f) waste water control and management system (e.g., from site drainage, flue gas treatment, and waste storage); and (g) residue management and discharge system which includes bottom ash management and treatment system as well as solid residue discharge and disposal system [13]. The plant design also depends on the nature of the waste being treated (its chemical composition and physical and thermal characteristics), types and quantities of residues which in turn depend on the installation design, its operation, and the waste input.

Process stability and optimization depend significantly upon the variability in the waste input. Although a narrow variability of input will give more stability and better environmental performance, this may require expensive waste pretreatment and the selective collection of waste. On the other hand, gas cleaning equipment is generally 15 to 35% of total capital investment, so its optimization is also essential [7]. Thus, a prudent optimization of the overall waste management system is often needed.

There are basically three types of incinerators used in commercial operations: grate incinerators, rotary kilns, and fluidized beds [7].

6.5.1.1 Grate Incinerators

About 90% of incinerators treating MSW use grate incinerators because of their simplicity and ability to handle a wide range of waste particle sizes. Grate incinerators are also used for commercial and industrial nonhazardous waste. This type of incinerator contains (a) a waste feeder unit, (b) incinerator grate on which waste materials are placed, (c) incinerator chamber, (d) incinerator air duct system, (e) auxiliary burner, and (f) bottom ash discharger [13].

6.5.1.2 Rotary Kilns

The rotary kiln consists of a cylindrical vessel slightly inclined on its horizontal axis. The vessel is usually located on rollers, allowing the kiln to rotate or oscillate around its axis. Solid, liquid, gaseous, or sludge waste is conveyed through the kiln as it rotates. For most fluids, direct injection is preferred. Rotary kilns are used more for hazardous and clinical wastes and less for MSW. Although the rotary kiln is normally operated at 850°C, the temperature can vary from 500°C (as a gasifier) to 1,450°C for an ash-melting kiln [7]. For hazardous waste, the kiln is operated between 900–1,200°C. The residence time in the kiln normally varies between 30 to 90 minutes. For complete destruction of toxic compounds, a postcombustion chamber may be required [7, 11].

6.5.1.3 Fluidized Beds

For RDF, sewage sludge, and finely divided waste, fluidized bed incinerators are widely used. These types of incinerators are normally used for large-scale operations. Various types of fluidized bed—stationary, bubbling (atmospheric or pressurized), rotating, or circulating—have been used in industrial practice. These fluidized beds differ in gas velocities, the design of the nozzle plate, and their internal design. The waste is injected from the top in an inert sand or ash fluidized bed (by air) which is preheated at the desired level [15]. Sometimes the waste is fluidized by simple injection of air (no inert solids) through holes in the bed plate. The inert solids provide a better heat transfer to the waste. For most waste, the space above the bed is kept at 850–950°C and the temperature of the bed is kept around 650°C [7]. A fluidized bed combustor generally gives uniform temperature and oxygen concentration distributions as well as more stable operation due to a high level of mixing within the bed.

A fluidized bed generally operates with waste particle size of 50 mm. For heterogeneous waste, this may require a pretreatment and sizing step that may add to the overall cost. For a rotating fluidized bed, a particle size of 200–300 mm [16] is possible due to additional mixing caused by the bed rotation. The unburned waste and ash in a fluidized bed incinerator are removed at the bottom of the reactor [11, 13]. The combustion heat can be either captured by an internal device, at the exit, or both together.

6.5.2 Gasification

Unlike combustion, gasification is carried out with less than a stoichiometric amount of required oxygen such that only partial oxidation of waste materials occurs. Gasification generates a producer gas of low, medium, or high BTU content depending on the operating conditions and amount of oxygen. The gasification temperature varies from 500°C to about 1,800°C depending upon the desired gas composition and the nature of the slag. The producer gas generally contains CO, CO_2, H_2, H_2O, CH_4, trace amounts of higher hydrocarbons such as ethane and ethylene, inert gases originating from the gasification agent, and various contaminants such as small particles and others depending upon the impurities present in the waste [12, 17, 18]. The partial oxidation can be carried out using either air, oxygen, carbon dioxide, steam, or a mixture of these substances. Generally oxygen produces medium BTU producer gas whereas air produces low BTU producer gas. The producer gas can either be used for heating and the generation of electricity or it can be reformed to produce syngas that largely contains CO and H_2 which can then be used as a raw material for second-generation biofuels via Fischer–Tropsch syntheses. Syngas can also be produced by operating a gasification reactor at a very high temperature (greater than 1,400°C). Several types of gasification reactors have been developed to process MSW, hazardous waste, and dried sewage sludge. A stable and optimum operation of the gasification reactor with minimum generation of tar or slag formation requires feedstock of uniform size with some consistencies in its composition. This may require pretreatment of the feedstock which can be expensive.

The gasification process involves smaller gas volume (by a factor of 10 if pure oxygen is used) and smaller waste water from the producer gas cleaning process compared to incineration. High operating pressures applied in some gasification processes can also lead to smaller and more compact aggregates. Unlike incinerators, a gasification process mainly produces carbon monoxide instead of carbon dioxide. High-temperature

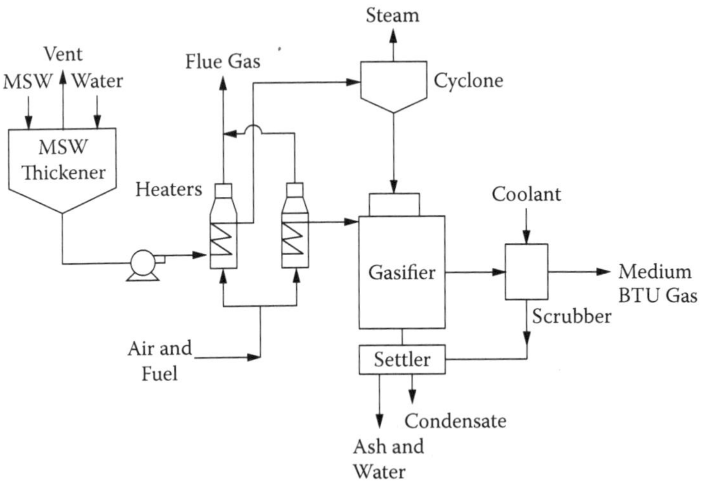

FIGURE 6.5
Simplified Texaco gasification process for the conversion of MSW to a medium-BTU gas. (After Lee, Speight, and Loyalka, 2007. Energy generation from waste sources, *Handbook of Alternate Fuel Technologies*. New York: CRC Press, Ch. 13, pp. 395–419.)

gasifiers capture inorganic impurities within the slag. Incinerators only produce heat (and thereby electricity), but syngas produced from a gasifier can be used to produce materials and transportation fuel along with heat and electricity.

Various types of gasification reactors (packed bed with upflow or downflow mode of operation, bubbling or circulating fluidized bed, entrained bed, and cyclone) are used in commercial operations. The gasification technology is very versatile and well developed, and it can process all types of waste. In recent years, new gasifiers tend to be entrained bed with both low- and high-pressure operational flexibility. The most preferred feedstock for the gasifier is high-energy density solids and for efficient operation of all gasifiers, waste materials must be finely ground before feeding into the gasifier. Hazardous waste may be gasified directly if it is liquid or finely granulated.

A typical simplified Texaco gasification process for conversion of MSW to a medium BTU gas is illustrated in Figure 6.5. Numerous other commercial processes for waste gasification are available, and they are described by Lee, Speight, and Loyalka [4]. SVZ Schwarze Pumpe GMbH operates both a packed bed gasifier for coal–waste mixtures (with waste up to 85%) and an entrained flow gasifier for hazardous waste. The entrained flow gasifier is operated at temperatures between 1,600°C and 1,800°C. The packed bed gasifier has a capacity of 8–14 tons per hour, and it operates between 800 and 1,300°C and 25 atm pressure and produces syngas using steam and oxygen as the gasification agents. A slag bed gasifier operates up to 1,600°C with a

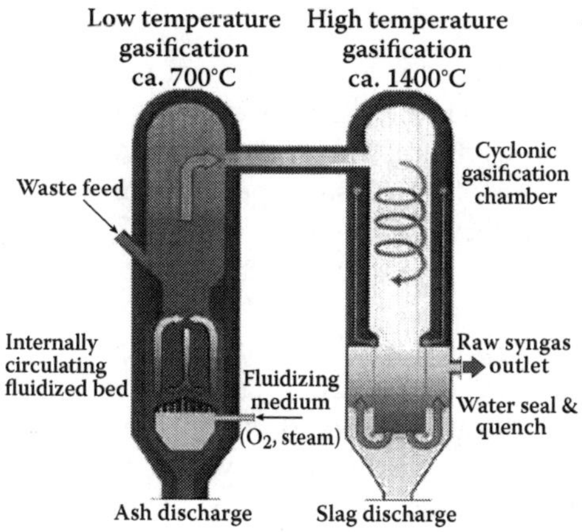

FIGURE 6.6
Fluidized bed gasifier with high-temperature slagging furnace. (From EBARA. (2003). EUP-EBARAUBE. *Process for gasification of waste plastics*. Retrieved May, 2010 from http://www.ebarra.ch/)

throughput rate of 30 tons per hour and slag is discharged as liquid [19, 20]. Recently, gasification technology has been used for numerous types of waste in addition to MSW. Among others, hazelnut shell, rice husk, salmon waste, and several other types of solids and liquid organic waste have been successfully gasified to generate producer gas or syngas [21–25]. In all cases, gasification technology produced good quality producer gas. In some instances, producer gas was subsequently transformed into biomethanol.

A gasification process can also use two stages. An example of a two-stage waste gasification process using a fluidized bed and an entrained flow reactor (see Figure 6.6) is used in Japan for waste conversion to syngas. The fluidized bed gasifier operates at a lower temperature, and it converts heterogeneous waste into syngas. The ash produced in this reactor is then passed onto a high-temperature cyclone gasifier where slag is collected. The syngas produced from this process is used for ammonia production and other applications. Other modifications of this process for different types of wastes are described by Bridgewater [26].

A two-stage gasification system sometimes also uses a gasification reactor in combination with a combustion reactor. For example, a combination of fluidized bed gasifier and a high-temperature combustor is used to process shredded MSW, plastics, and residues. In this process, the gasifier is generally operated at 580°C to produce gas and the combustor is operated at 1,350–1,450°C for melting ash and other solid materials to further recover energy [7]. Generally particle size of 300 mm is preferred in such a process [20].

6.5.3 Pyrolysis

Pyrolysis is a major process for waste disposal in which waste is thermally degraded in an inert environment. Pyrolysis is usually carried out at a lower temperature (250–900°C) and at low pressure. As shown in Table 6.4 [27–29], there are various modes of pyrolysis depending on the method of pyrolysis and the operating conditions. The heating value of pyrolysis gas typically lies between 5 and 15 MJ/m^3 based on MSW and between 15 and 30 MJ/m^3 based on RDF [27].

Conventional pyrolysis reactors are: fixed bed, fluidized bed, entrained flow, moving bed, rotary kiln, ablative reactor, and so on. Generally waste pyrolysis generates a large number of compounds, and these need to be tracked for their effective uses. Generally, the pyrolysis process involves three stages: (a) A feed preparation and pretreatment step that includes grinding and drying. The grinding improves the feed quality and subsequent heat transfer. The drying improves gas–solid contact, heat and mass transfer, and reactions in the reactor. (b) A pyrolysis reactor that generates

TABLE 6.4

Waste Pyrolysis Technologies, Operating Conditions, and Major Products

Technology	Residence Time	Temp. (°C)	Heating Rate	Major Products
Conventional carbonization	hrs–days	300–500	Very Low	Charcoal
Pressurized carbonization	15 min–2 hrs	450	Medium	Charcoal
Slow pyrolysis	5–30 min	About 600	Low	Charcoal, oil, gas
Conventional pyrolysis	hr	400–600	Low	Charcoal, oil, gas
Conventional pyrolysis	5–30 min	700–900	Medium	Charcoal, gases
Fast pyrolysis	0.5–5 sec	About 650	Fairly high	Oil
Fast pyrolysis	0.1–2 sec	400–650	High	Oil
Flash pyrolysis	Less than 1 sec	650–900	High	Oil, gases
Ultra pyrolysis	Less than 0.5 sec	1000–3000	Very high	Gases
Vacuum pyrolysis	2–30 sec	350–450	Medium	Oil
Pressurized hydropyrolysis	Less than 10 sec	Less than 500	High	Oil
Methanopyrolysis		Less than 10 sec	Greater than 700	High Oil, chemicals

Source: Modified from Bridgwater and Bridge, 1991. A review of biomas pyrolysis and pyrolysis technologies. In A.V. Bridgwater and G. Grassi, (Eds.) *Biomass Pyrolysis Liquids Upgrading and Utilisation*, London: Elsevier Applied Science, pp.11–92; Bridgwater, 1995/ Thermal biomass conversion technologies, *Biomass and Renewable Energy Seminar*, Loughborough University, IK, March; and Huber, Iborra, and Corma, 2006. Synthesis of transportation fuels from biomass: chemistry, catalysis, and engineering, *Chem. Rev., ACS.*

gas, solids containing mineral and metallic compounds, and liquids. (c) An upgrading of pyrolysis gas, liquids, and solids to generate more useful fuel, chemicals, and materials. The impurities in the products such as the presence of arsenic and other materials can make the product very difficult to use and the process useless. These impurities are removed using appropriate purification technologies [30].

The pyrolysis process can also (a) recover organic fractions such as methanol as a material/fuel, (b) recover char for external use, (c) generate more efficient electricity using gas engines or turbines, and (d) reduce flue gas volume after combustion which may reduce flue gas treatment capital costs. The pyrolysis process is used for MSW and sewage sludge, decontamination of soil, synthetic waste, and used tires, cable tails, and metal and plastic materials for substance recovery. In general, pyrolysis is a very versatile process and has been extensively used for waste conversion. When the pyrolysis process is used to generate fuels or chemicals, a subsequent upgrading of pyrolysis oil is often carried out. Also, as shown in Table 6.4 [27–29], pyrolysis is often carried out in the presence of hydrogen (i.e., hydropyrolysis) to improve quality and quantity of the oil production.

A novel Chartherm process is a pyrolytic distillation process designed to maximize the useful recovery of both materials and energy from waste [7]. The process was developed by Thermya Co. in France with an industrial capacity of 1,500 kg/hr wood waste. The advantages of this process are (a) low operating temperature (300–400°C) compared to conventional pyrolysis and gasification temperature of 500–1500°C, (b) avoidance of tar and dioxin emissions and thereby reducing gas cleaning equipment, and (c) possibilities of recovering both energy and materials from the waste feed [31, 32].

The process of pyrolysis in its different formats has been extensively used to treat different types of waste. Some of the typical recently reported literature studies on this subject are depicted in Table 6.5. This table by no means gives a complete account of all reported studies.

6.5.4 Plasma Technology

The use of plasma to treat waste is a relatively new concept. Plasma is generated when gaseous molecules are forced into high-energy collisions with charged electrons resulting in the generation of charged particles. There are fundamentally two types of plasmas: high-temperature or fusion plasmas or low-temperature or gas discharge plasmas [60]. The low-temperature plasmas can be further divided into thermal plasmas in which a quasi-equilibrium state occurs (characterized by high electron density and temperature between 2,000 and 30,000°C) and cold plasmas characterized by a nonequilibrium state [61]. Thermal and gas plasmas are most widely used for waste treatment. As shown in Table 6.3, the plasma process operates at a higher temperature than other thermochemical processes. Plasma technology also

TABLE 6.5
Some Typical Literature Studies on Pyrolysis of Waste Materials

Type of Waste	Authors	Comments
Poultry litter and pine woody biomass	Das et al., 2008 [33]	Slow pyrolysis-effects of proteins and ash on the products
Corn stover	Wu et al., 2009 [34] Yang et al., 2010 [35]	Microwave assisted pyrolysis
Local crop waste	Nardin and Catanzaro, 2007 [36]	Bio-oil from local crop waste
Agricultural residue	Demirbas, 2008 [37]	Bio-oil, biogas, and biochar
Mosses and algae	Demirbas, 2006 [38]	Maximum yield at 775 K
Black alderwood	Balat and Demirbas, 2009 [39]	Yield plateaued after 25 mins.
Rice straw in China	Xiao et al., 2009 [40]	Production of biomethanol
Wood waste	Demirbas, 2010 [41]	Biofuels and biochemical
Waste materials	Irvine et al., 2010 [42]	Compost heat
Corn straw	Liu et al., 2010 [43]	Fast pyrolysis
Bagasse	Anto and Thomas, 2009 [44]	Pyrolysis in a fixed bed reactor
Soft shell of pistachio	Demiral et al., 2009 [45]	Fixed bed reactor 350-500°C
Hazelnut bagasse	Demiral and Sensoz, 2006 [46, 47]	Detailed structural analysis
Corn residues (cobs and stalks)	Ioannidou et al., 2009 [48]	Catalytic and no-catalytic pyrolysis
Biomass residues in Turkey	Kar and Tekeli, 2008 [49]	Energy source for Turkey
Pine chips	Sensoz and Can, 2002 [50]	Fixed bed pyrolysis
Miscanthus x giganteus	Yorgan and Simsek, 2003 [51]	Yields and bio-oil characterization
Rapeseed cake	Culcuoglu et al., 2005 [52]	Pyrolysis at 65°C
Olive cake	Demirbas, 2008, 2009 [53,54]	Fast pyrolysis
Cottonseed cake	Ozbay et al., 2006 [55]	Product yields and compositions
Oilseed by-product	Gercel and Gercel, 2007 [56]	Fixed bed pyrolysis of olive cake
Chicken manure	Schnitzer et al., 2008 [57]	Fast pyrolysis
Euphorbia rigida and sunflower	Ozcun et al., 2000 [58]	Structural analysis
Pressed bagasse		
Cotton straw and stalk	Putun, 2002 [59]	Fast pyrolysis at 550°C

uses less than required stoichiometric oxygen and generally operates at low residence time [62–64].

Haberlein and Murphy [61] indicated that the most important advantages of plasma technology are (a) high energy density and high temperatures, and (b) use of electricity as the energy source. The first advantage allows

(i) rapid heating and reactor startup, (ii) high heat and kinetic rates, (iii) smaller installation size (due to smaller residence time), and (iv) processing of materials with high melting or boiling point. The second advantage allows increased process controllability and flexibility due to decoupling of heat generation from the oxygen potential and lower off-gas flow rates resulting in lower gas cleaning costs. Because electricity can be expensive, plasma technology is most desirable for waste streams that contain most organic materials with high heating value and for the waste that generates valuable coproducts such as synthesis gas, hydrogen, or electricity. Plasma technology is also valuable for treating waste materials containing inorganic solids, because these materials can either be recovered or reduced in volume or can be oxidized and immobilized in a vitrified nonleaching slag. In general, plasma technology is capable of processing a wide variety of waste materials.

Waste treatment by plasma technology can be divided into three categories: plasma pyrolysis, plasma gasification, and plasma compaction and vitrification of solid wastes. For solid wastes with high organic content, a combination of these three categories is often used [7].

6.5.4.1 Plasma Pyrolysis

Of the three categories mentioned above, the most extensive scientific studies are performed on plasma pyrolysis [24]. Different types of organic waste such as plastic, used tires, agricultural residue, and medical waste have been studied both at the laboratory and pilot-scale level [60]. Plasma pyrolysis generally produces two products, a combustible gas and a carbonaceous residue (char), both of which can recover useful materials. It can recover valuable chemicals (e.g., ethylene and propylene) and carbon black from tires. Although plasma pyrolysis of solid waste still needs some technical development, plasma pyrolysis of hazardous gases and liquids is a proven commercial technology such as the PLASCON process (developed by CSIRO and SRL Plasma Ltd. in Australia which is now owned by Dolomatrix International Ltd.). For MSW and RDF, plasma pyrolysis is combined with plasma gasification to produce useful synthesis gas. Also, for these types of solid waste, plasma gasification and vitrification are preferred over plasma pyrolysis. Small-scale plasma pyrolysis is practiced to treat polymers, medical waste, and low-level radioactive wastes [7].

6.5.4.2 Plasma Gasification and Vitrification

The energy contained in a plasma allows gasification of low-energy fuels such as household and industrial waste without the need for an additional fuel. Thermal plasma is an excellent technology for the conversion of waste into valuable synthesis gas and a vitrified slag. The high-temperature condition that exists in the plasma gasification produces a cleaner synthesis gas compared to what is achieved in the conventional gasification process. The inorganic matter (such as glass, metals, silicate, heavy metals, etc.) contained in MSW is converted into a dense, inert, no-leaching vitrified slag. The synthesis gas can be used for heat and electricity or it can be converted to biofuels via Fischer–Tropsch synthesis. The vitrified slag can be used as a building material additive [65].

The synthesis gas in the plasma gasification process contains the plasma gas components. Air is often used as gas for economical reasons and for providing oxygen, but other gases such as nitrogen, carbon dioxide, steam, and argon have also been tested. The literature [66, 67] has shown that the plasma torches operating with steam offer definite advantages for waste processing applications. Gasplasma™ technologies for waste treatment use electricity as the energy source and that makes the system more flexible and controllable and variable waste input does not pose problems. Thermal arc plasmas dominate in waste treatment because they are relatively insensitive to changes in process conditions. Inasmuch as solid waste treatment requires decontamination in combination with volume reduction and immobilization of inorganic contaminants, most plasma-based waste treatment systems make use of transferred arc reactors offering high-heat fluxes which facilitates solid melting [61].

Fundamentally, plasma technology application to waste treatment is divided into two categories, single-stage and two-stage. Here we briefly examine some specific examples of each of these categories [7].

6.5.4.2.1 Single-Stage Process

The single-stage process was developed and commercialized by Westinghouse Corporation (Madison, Wisconsin, United States) and Europlasma Corporation (Mocenx and Cenon, France and Shimonoseki, Japan). The scale of the process varied from about 10 tons/day to about 42 tons/day in different locations [68, 69]. The largest commercial plant was established at Utashinai in Japan (180 tons/day) and a smaller (22 tons/day) at Mihama-Mikata in Japan both using Westinghouse plasma gasification technology [7]. Both of these plants are called Alter NRG/Westinghouse plasma gasification process [68–70]. In 2007, NRG acquired the Westinghouse Plasma Corporation and combined the Westinghouse updraft gasification reactor concept that uses plasma torches to provide part of the energy input, with synthesis gas cleaning in order to convert the synthesis gas to heat and electricity and other value-added products. The process can use a variety of feedstock such as MSW, MSW plus tires, RDF, ASR, coal plus wood, petcoke, and other hazardous wastes. The ability to process heterogeneous, unsorted, or differently sized feedstock reduces the cost required for feed handling prior to gasification.

The plasma gasification reactor (PGR) used by Alter NGR is graphically illustrated in Figure 6.7. The reactor has a refractory lining to withstand high temperatures and corrosive conditions. The reactor is first packed

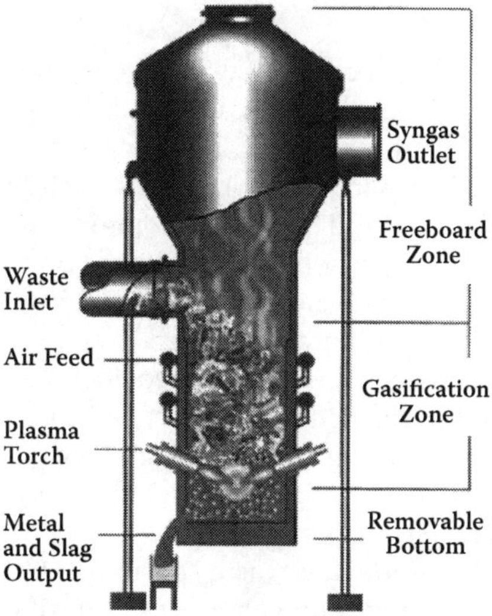

FIGURE 6.7
Alter NRG plasma gasification reactor. (Modified from Helsen and Bosmans, 2010. Waste to energy through thermochemical processes: Matching waste with process, *Conference Proceedings on Enhanced Landfill Mining and Transition to Sustainable Materials Management*, Molenheide, Houthalen-Heichteren, Belgium, October 4–6.)

with metallurgic coke which absorbs and retains the thermal energies from plasma torches and creates an environment to melt inorganic materials. The coke is consumed as the reaction proceeds. The temperature of the plasma plume varies from 5,000 to 7,000°C and the temperature at the bottom is about 2,000°C [7].

The reactor uses a mixture of oxygen and steam to improve hydrogen yield in the synthesis gas. As shown in Figure 6.7, whereas the synthesis gas leaves the reactor at the top at about 890–1,100°C, the molten slag at the bottom containing nocombustible inorganics and recoverable metals leaves the reactor at about 1,650°C. The molten slag then goes through a slag handling system for further processing. High residence time in the reactor assures the conversion of tar and it minimizes the particulate carryover. The electrical energy supplied by plasma torches counterbalances the heating value of the waste feed and thereby controls the temperature and the quality of the synthesis gas. The vitrified slag can be used for the construction industries, although the quality of this slag can be deteriorated by the presence of ASR in the waste feed. The major problem with the single-stage process is the different types of harmful contaminants present in the synthesis gas which requires a series of downstream processes to remove them. These downstream processes add significantly to the overall cost of the gasification process [7].

6.5.4.2.2 *Two-Stage Process*

A two-stage Gasplasma™ process [70] was developed by Tetronics Corp. as well as Plasco Energy Group (Ottawa, Canada). Integrated Environmental Technologies of the United States and Pyrogenesis Corporation of Canada also use this process. The pilot-scale process is used by Advanced Plasma Power of Swindon, UK [72–74]. The process contains two stages, gasification followed by plasma conversion, which are followed by a number of processes to clean and cool the synthesis gas prior to delivering it to gas engines for conversion to mechanical/electrical energy.

The Gasplasma process converts waste feedstock into a clean, hydrogen-rich synthesis gas and a vitrified recyclate called Plasmarok™ that can be used for building material or replacement aggregate. The process is capable of producing synthesis gas which, after passing through further treatment, is suitable for use as fuel in a gas engine [75].

In the first stage, the pretreated waste stream (RDF, pretreated commercial and municipal waste and refined biomass) is gasified in a fluidized bed gasifier in the presence of oxygen and steam at a temperature around 800–900°C. The reactor uses a portion of heat contained in the waste material. The synthesis gas generated in the reactor contains tar and soot and solid char and ash contained in the feed material are removed at the bottom of the reactor and processed in the plasma converter [7].

In the second stage, the plasma converter cracks tars and soot to synthesis gas to form a gas comprised primarily of hydrogen, carbon monoxide, carbon dioxide, and nitrogen. The ash and inorganic fraction from the gasifier are vitrified to form Plasmarok™. The plasma converter is designed in such a way as to allow the maximum amount of residence time for the synthesis gas under most energy-intensive conditions. The synthesis gas leaves the reactor around 1,200°C, and it is then cooled to about 200°C [7]. The heat release during cooling is recovered for further use in the reactor via steam. The synthesis gas is further cooled and the acidic components in the gas are absorbed by the alkaline solutions. The final synthesis gas is introduced in the engines at constant pressure to generate electricity. The two-stage process shows great promise for the conversion of waste into heat, electricity, and other valuable products. It gives higher throughput rate, higher conversion efficiency to a clean high calorific syngas, and better control over VOCs/tars compared to single-stage operation. Process control and engineering are critical in both single- and two-stage processes [7, 70].

6.5.5 Liquefaction

The process of liquefaction has been widely used for fossil energy such as direct liquefaction of coal, shale, bitumen, heavy oil, and the like. The same concept can be easily applied to biomass which produces a water-insoluble bio-oil at high pressure (50–200 atm) and low temperature (250–450°C). These process conditions are less severe than the ones normally used for coal. The liquefaction involves all kinds of processes such as solvolysis, depolymerization, decarboxylation, hydrogenolysis, and hydrogenation (which can be accompanied by mild hydrocracking). The overall objective of biomass (and waste) liquefaction is to control the reaction rate and reaction mechanisms using pressure, gases, and catalysts to produce a high-quality liquid oil. The reactor feed generally consists of solid biomass feed (or a suitable waste), solvent, reducing gas such as H_2 or CO, or a catalyst. The bio-oil produced via liquefaction has a lower oxygen content, lower viscosity, and higher energy density than pyrolysis-derived oil. Fundamentally, the nature of the liquefaction process can be broken down into several categories depending on the nature of the solvent and gas. For example, hydrothermal liquefaction uses water as the solvent, hydropyrolysis uses H_2 or a reducing gas but no solvent, and solvolysis uses a reacting or hydrogen donor solvent. A review of biomass liquefaction research done between 1920 and 1980 is presented by Moffatt and Overend [77].

The nature of solvent, gas, and catalyst dictate the operating conditions and the quality of the liquefaction product. A number of solvents such as water, creosote oil, ethylene glycol, methanol, and recycled oil have been tested. Water is the most attractive because it is cheaper, and it does not require drying of waste or biomass. Hydrothermal liquefaction is separately described in this section. The recycled oil increases the selectivity, and it also makes the process self-sufficient with no need to add new solvent in the process. Creosote oil, ethylene glycol, tetralin, phenathrene, alcohols, and phenols among other hydrogen donor solvents are used in the solvolysis process. They react with the solid materials during the liquefaction process. Hydrogen, carbon monoxide, and even methane act as reducing agents during hydropyrolysis in the presence of a catalyst and give a higher quality bio-oil. A number of catalysts have been used for liquefaction including alkali (from the alkaline ash components in the wood, alkaline oxides, carbonates, and bicarbonates) and metals such as zinc, copper, nickel, formate, iodine, cobalt sulfide, zinc chloride, and ferric hydroxide, as well as Ni, Mo, Ru, Co (which aid in hydrogenation/hydrocracking).

Akhtar and Amin [80] studied the effects of various reaction variables on the hydroliquefaction process. They concluded that the major parameters influencing the yield and composition of bio-oil are temperature, properties of solvent, solvent density, and type of biomass or waste. Temperature is the most important parameter, and they recommended a temperature range of 300–374°C depending upon the biomass type and specifications for the composition of bio-oil. At temperatures higher than 350°C, too much gas is formed and for temperature less than 280°C, conversion to oil is low. They recommended the use of water, methanol, ethanol, acetone, tetralin, and benzene, among others as suitable solvents. Residence time, heating rates, pressure, biomass particle size, presence of reducing gas, or hydrogen donor species are found to be of secondary importance in the hydroliquefaction process. The nature of the biomass waste also influences the yield of liquid production.

6.5.5.1 Hydrothermal Liquefaction

Waste feedstock containing lignocellulose, fatty acids, and protein derivatives can be hydrothermally transformed to produce a range of products such as biocrude, methane, hydrogen, biodiesel, and biogasoline. Many waste products such as agricultural residues, food processing wastes, and municipal and agricultural sludge contain large amounts of water. The removal of this water in gasification, pyrolysis, and other thermochemical processes consumes a significant amount of energy. The hydrothermal process avoids the need for this water removal. Also, the considerable variations in the physical properties of water that occur with changes in temperature and pressure can facilitate efficient separations of product and by-product streams with low energy requirements. For feedstock that contains inorganics such as sulphates, nitrates, and phosphates, the hydrothermal method can facilitate recovery and recycling of these chemicals in their ionic form for eventual use as fertilizer. The product streams from hydrothermal processing are also completely sterilized against any possible pathogens including biotoxins, bacteria, and viruses.

In general, hydrothermal processing can be divided into three regions: liquefaction, catalytic gasification, and high-temperature gasification [76]. These three regions are graphically illustrated in Figure 6.8. Both catalytic and high-temperature gasifications occur under supercritical conditions producing gases with high hydrogen content. Also, depending upon the operating conditions, a significant amount of reforming reactions can take place under gasification conditions, particularly when a suitable reforming catalyst is used. The gasification process under supercritical water is covered in the later section on supercritical technology. Hydrothermal liquefaction generally occurs at temperatures between 200 and 370°C and pressure between 4 and 20 MPa, and it can be applied to a stream of mixed waste. A significant literature on hydrothermal processing of biomass and waste has been reported over the last two decades [78–83].

Two other variations within hydrothermal liquefaction (HTL) have also been examined in the literature [83, 84]. At low temperatures (between about 180–220°C) and at saturated pressure, hydrothermal carbonization of biomass occurs. Although carbonized biomass is inferior to liquid or gaseous fuels, process requirements for hydrothermal carbonization are comparably low while producing a fuel that is easy to handle and store because it is stable and nontoxic. Thus, HTC may provide some advantages when considering small-scale, decentralized applications.

HTC is produced for residence time between 1 and 72 hours and with water pH below 7. Alkaline conditions produce substantially different products

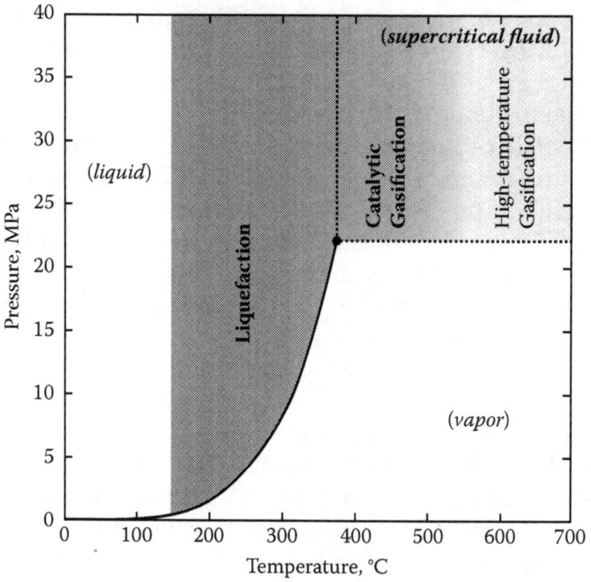

FIGURE 6.8
Hydrothermal processing regions referenced to the pressure-temperature phase diagram of water. (After Peterson et al. 2008. *Energy Environ. Sci.*, 1: 32–65.)

[81–83]. The process goes through numerous reactions such as hydrolysis, dehydration, decarboxylation, polymerization, and aromatization. For a variety of feedstock such as cellulose, lignin, wood, peat bog, and the like, as reaction severity increases, more carbonization occurs reducing H/C and O/C ratios of the product. In a typical HTC operation, 48–50% of HTC coal (which contains lignitelike material in a dispersed powder form), about 35–37% of water and total organic carbon (which contains sugars and derivatives, organic acids, furanoid, and phenolic compounds), and about 15–16% of gas (which contains mainly CO_2, with some CH_4 and CO and traces of H_2 and C_nH_m) are generated.

Unlike HTC, HTL (hydrothermal liquefaction) is carried out in a temperature range of about 200 to 400°C, and it produces products often called bio-oil or biocrude. Although HTL is a promising technology to treat waste streams from various sources, its commercial growth is inhibited due to the high transportation costs of cellulosic waste, poor conversion efficiency, and lack of understanding of complex reaction mechanisms. Kranich [85] was one of the first to use an HTL process to convert MSW. He used three different types of materials from MSW; primary sewage sludge, settled digester sludge, and digester effluent. In a laboratory autoclave, he performed experiments at temperatures ranging from 570–720 K, pressure of 14 MPa, and residence time ranging from 20 to 90 minutes. The results showed the organic conversion rates from 45 to 99% and oil production rates from 35 to 63.3%.

Subsequent works were reported by Suzuki et al. [86], Itoh et al. [87], and Inoue et al. [88]. Most recently, Changing World Technology Co. in Carthage, Missouri, United States [89] reported the effectiveness of hydrothermal liquefaction for the treatment of turkey waste.

The process of hydrothermal upgrading (HTU) combines the liquefaction and upgrading steps. A schematic of this HTU process is described by Demirbas [84]. In this process the feed is pretreated, preheated, and pumped into the reactor. The products are separated in gas, liquid, and solid streams. In normal operating conditions the product composition is: gaseous product (>90% CO_2) about 25%, process water about 20%, water-soluble organics about 10%, and biocrude about 45%. The gaseous products, which have some heating value, are catalytically combusted with air to generate flue gas. The process water stream and soluble organics go through an anaerobic digestion to produce biogas (largely containing methane) which can go through a combustion process to generate heat and electricity. The solid biocrude is separated into light and heavy biocrude. The heavy biocrude is co-combusted with coal to generate electricity. The lighter biocrude is hydrodeoxygenated to produce upgraded products such as premium diesel fuel, kerosene, and other feedstock for biorefineries. The process has an overall thermal effciciency of about 70–90%.

The hydrothermal upgrading process is generally carried out at 575 K; however, the process can be operated in the temperature range of 575–625 K, pressure range of 12–18 MPa, and residence time range of 5–20 minutes. The process carries out a series of depolymerization, decarboxylation, dehydration, hydrodeoxygenation, and hydrogenation reactions. The oxygen is removed as water and carbon dioxide. Typically the feed slurry contains 25% solids in water which may include wood and forest wastes, agricultural and domestic residues, municipal solid waste, or organic industrial residues. Kumar and Gupta [90] examined the effect of temperature on molecular structure and enzymatic activity of cellulose in subcritical hydrothermal technology. They indicated that the percentage of crystallinity of microcrystalline cellulose increased with treatment with water.

A thorough and excellent review of biofuel production in hydrothermal media was given recently by Peterson et al. [76].

6.5.6 Supercritical Technology

In recent years the application of supercritical technology for waste treatment has gained significant momentum. Two major areas have been the application of supercritical water for waste gasification and supercritical extraction by carbon dioxide, water, and other solvents. The application of supercritical technology to transesterification is briefly examined in the subsequent section on transesterification. Lastly, the interest in reforming (and in tri-reforming) under supercritical conditions has also gained significant momentum over last decade.

6.5.6.1 Supercritical Water Gasification

Hydrothermal treatment of waste materials in supercritical conditions has gained a great deal of momentum ever since the pioneering work of Modell and coworkers from M.I.T. in the late 1970s. As indicated earlier, Figure 6.8 illustrates the region of supercritical water gasification for waste treatment in the pressure-temperature diagram. The three regions shown in the diagram take advantage of substantial changes in the properties of water that occur in the vicinity of its critical point at 374°C (T_c) and 22 MPa (P_c). The behavior of the important properties of water such as density, ion dissociation constant, and dielectric constant with respect to temperature is illustrated in Figure 6.9 [76]. In supercritical conditions, more chemically and energetically favorable pathways to gaseous and liquid fuels can be achieved by better control of the rate of hydrolysis and phase partitioning and solubility of components in supercritical water.

Water at ambient conditions (25°C and 0.1 MPa) is a good solvent for electrolytes because of its high dielectric constant [76], whereas most organic matter is sparingly soluble [76]. As water is heated, the H-bonding starts weakening, allowing dissociation of water into acidic hydronium ions (H_3O^+) and basic hydroxide ions (OH^-). The structure of water changes significantly near the critical point because of the breakage of infinite networks of hydrogen bonds, and water exists as separate clusters with a chain structure. In fact, the dielectric constant of water decreases considerably near the critical

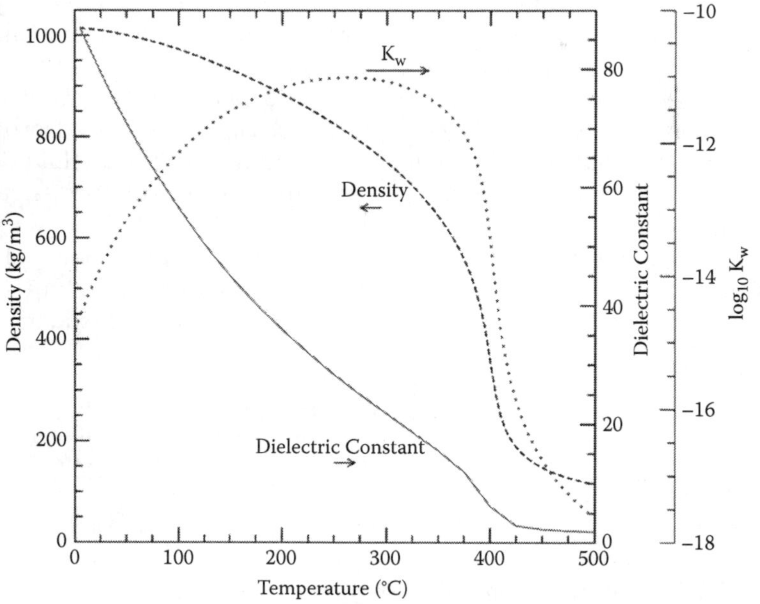

FIGURE 6.9
Variations of density, dielectric constant, and ion dissociation constant as a function of temperature for water. (From Peterson et al. 2008. *Energy Environ. Sci.*, 1: 32–65.)

point, which causes a change in the dynamic viscosity and also increases the self-diffusion coefficient of water [76].

Supercritical water has liquid-like density and gaslike transport properties, and behaves very differently than water at room temperature. For example, it is highly nonpolar, permitting complete solubilization of most organic compounds. The resulting single-phase mixture does not have many of the conventional transport limitations that are encountered in multiphase reactors. However, the polar species present, such as inorganic salts, are no longer soluble and start precipitating. The physicochemical properties of water, such as viscosity, ion product, density, and heat capacity, also change dramatically in the supercritical region with only a small change in the temperature or pressure, resulting in a substantial increase in the rates of chemical reactions.

It is important to mention that the dielectric behavior of 200°C water is similar to that of ambient methanol, 300°C water is similar to ambient acetone, 370°C water is similar to methylene chloride, and 500°C water is similar to ambient hexane [76]. In addition to the unusual dielectric behavior, the transport properties of water are significantly different than the ambient water as shown in Table 6.6.

Supercritical processes are often not considered to be economical because of high capital costs associated with high-pressure equipment and high operating costs associated with the compression or pumping of supercritical media, however, the above-described unique properties of supercritical fluids offer some very interesting possibilities. Furthermore, in recent years the prices of high-pressure equipment has come down. In conjunction with catalysis, a supercritical fluid can dissolve unwarranted hydrocarbons from the catalyst surface into the supercritical fluid phase. Supercritical fluids have a better capacity to handle heat due to high capacities. The adsorption/desorption phenomena can be better handled in a supercritical fluid due to higher solubility. The oligomeric coke precursors or sulfur species can be easily dissolved by the supercritical fluids.

An excellent review of supercritical water (SCW) gasification of biomass and organic wastes was recently published by Guo, Cao, and Liu [91]. Numerous studies have examined supercritical water partial oxidation [92,

TABLE 6.6

Comparison of Ambient and Supercritical Water

	Ambient Water	Supercritical Water
Dielectric constant	78	<5
Solubility of organic compounds	Very low	Fully miscible
Solubility of oxygen	6 ppm	Fully miscible
Solubility of inorganic compounds	Very high	~0
Diffusivity ($cm^2\ s^{-1}$)	10^{-5}	10^{-3}
Viscosity ($g\ cm^{-1}\ s^{-1}$)	10^{-2}	10^{-4}
Density ($g\ cm^{-3}$)	1	0.2–0.9

93]. These and other studies found that the yields of H_2O and CO increased with increasing water density. Yields of H_2 were 4 times better with NaOH and 1.5 times better with ZrO_2 compared to reaction without a catalyst. Supercritical fluids gave increased pore accessibility, enhanced catalyst ability to coking, and increased desired product selectivity.

An extensive amount of work on supercritical water gasification of organic wastes has been reported in the literature [78, 79, 94–96]. The studies have shown that gasification generally produces a hydrogen and carbon dioxide mixture with simultaneous decontamination of waste. The homogeneous solution of waste and water makes it easy to pump to the high-pressure reactor without pretreatment. Xu and Antal [97] studied gasification of 7.69 wt% digested sewage sludge in supercritical water and obtained gas that largely contained H_2, CO_2, a smaller amount of CH_4, and a trace of CO. Other waste materials show similar behavior. The equilibrium yields as functions of temperature and pressure for SCWG of 5% sawdust reported by Guo et al. [91] indicate the main products to be hydrogen, carbon dioxide, and methane at low temperatures and hydrogen and carbon dioxide at high temperatures.

An increase in pressure significantly decreases the product concentration of carbon monoxide and slightly decreases the product concentration of the hydrogen. The pressure change has very little effect on the product concentrations of carbon dioxide and methane. In addition to temperature and pressure, other parameters that affect the gas yield are feedstock concentration, oxidant, reaction time, feedstock composition, inorganic impurities in the feedstock, and biomass particle size. Several catalysts such as alkali (NaOH, KOH, Na_2CO_3, K_2CO_3, $Ca(OH)_2$), an activated carbon, metal oxide, and metals also affect the conversion and gas yields. The last two are important for reforming under supercritical conditions. Although high-temperature supercritical water gasification produces hydrogen and carbon dioxide, Sinag, Kruse, and Schwarzkopf [98] showed that a combination of two technologies, supercritical water and hydropyrolysis on glucose in the presence of K_2CO_3, produces phenols, furfurals, organic acids, aldehydes, and gases.

The generation of hydrogen from waste has long-term and strategic implications inasmuch as hydrogen is the purest form of energy and is very useful for product upgrading, fuel cells, and many other applications. Hydrogen can be produced from waste via numerous high-temperature technologies such as conventional or fast pyrolysis (e.g., olive husk, tea waste, crop straw, etc.), high-temperature or steam gasification (e.g., bionutshell, black liquor, wood waste, etc.), supercritical fluid extraction (e.g., swine manure, orange peel waste, crop grain residue, petroleum basis plastic waste, etc.), and supercritical water gasification (e.g., all types of organic waste, agricultural and forestry waste, etc.), as well as low-temperature technologies such as anaerobic digestion and fermentation (e.g., manure slurry, agricultural residue, MSW, tofu wastewater, starch from food waste, etc.).

For high-temperature technologies, supercritical water gasification generates more hydrogen at a lower temperature than pyrolysis or gasification [99, 100]. Supercritical water gasification also does not require drying, sizing, and other methods of feed preparation thereby requiring less expense for the overall process. The temperature of the pyrolysis and gasification processes can be reduced if the gases coming from them are further steam reformed. This, however, adds to the overall cost. The rates for the low-temperature processes such as anaerobic digestion and fermentation can be enhanced with the use of suitable microbes and enzymes. The development of a future hydrogen economy will require further research in the improvement of these technologies.

6.5.6.2 Supercritical Extraction

In addition to water, carbon dioxide and alcohols have been used for supercritical extraction. The use of supercritical carbon dioxide to extract oil from coal, tar, and various crops or crop waste has been extensively studied [101–105]. These studies have shown that supercritical carbon dioxide can be very effective in extracting certain types of chemicals from various carbohydrates, lignin, and organic materials. More work is, however, needed.

Xiu et al [106] reported an interesting study of supercritical extraction of swine manure by ethanol in the temperature range of 240°C to 360°C and at a pressure of 6.37 MPa under noncatalytic conditions. The maximum yield of oil at 26.7% was obtained at 340°C. At the same temperature, the highest liquefaction yield of 62.77% was obtained. The study concluded that the supercritical ethanol liquefaction was an effective way to remove oxygen and utilize carbon and hydrogen in swine manure to produce energy-condensed biofuel.

6.5.7 Transesterification

Transesterification is the reaction of triglycerides (or other esters) with alcohols to produce alkyl esters (biodiesel) and glycerol, typically in the presence of acid and base catalysts (see Figure 6.10). Triglycerides are one of the three types of biomass obtained in 350 types of crops such as soybean, cotton seed, rapeseed, and algae, among others, as well as in fatty acids present in a variety of fresh and waste cooking oils. The basic mechanism of transesterification of triglycerides is described in Figure 6.10. Methanol is most commonly used because of its low cost, although 2-propanol gives better biodiesel and ethanol is preferred in Brazil because of its easy and inexpensive availability. Alkyl esters or biodiesel are also called fatty acid methyl esters (FAME) and they can be directly used in diesel engines.

For the purpose of this chapter, it is the transesterification of waste cooking or frying oil that is of interest. Worldwide there is significant production of waste cooking and frying vegetable oils. The process of transestrification allows the conversion of these waste oils into useful biodiesel. The literature

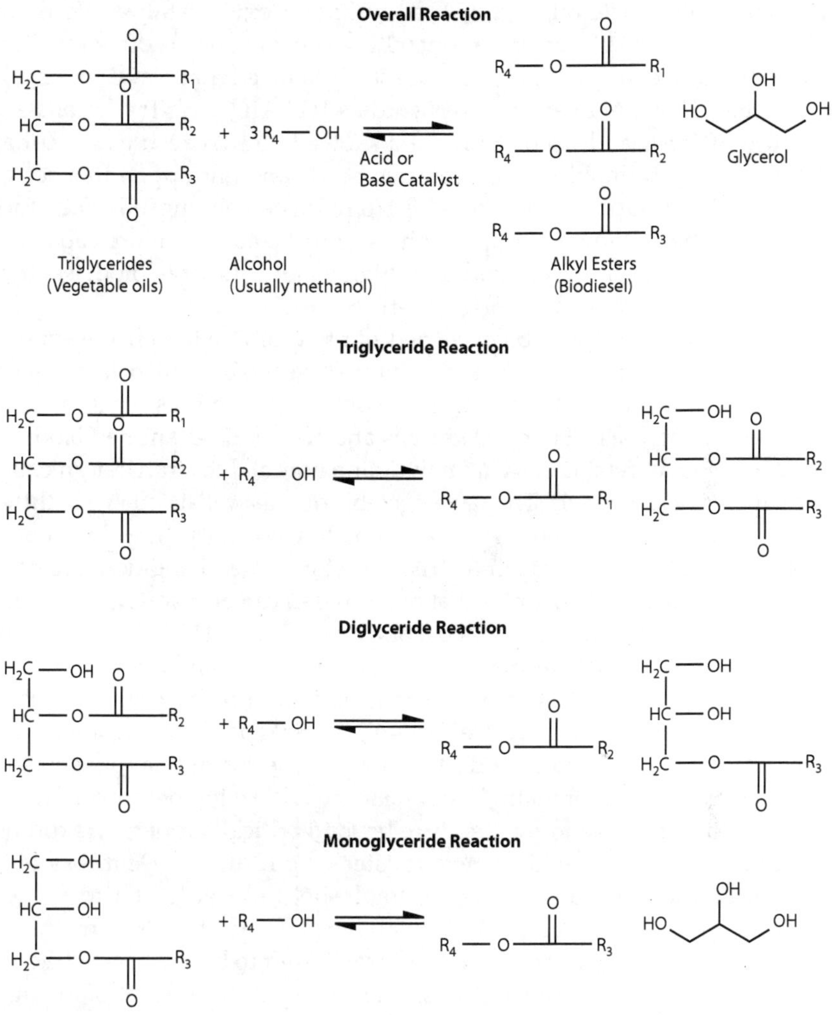

FIGURE 6.10
Overall and intermediate reactions for transesterification of triglyceride and alcohol to produce alkylester (biodiesel) and glycerol. (After Huber et al., 2006. Synthesis of transportation fuels from biomass: Chemistry, catalysis and engineering, *Chem. Rev.*, ACS.

reported [99–117] for the conversion of waste oils into biodiesel indicates that the efficiency of this conversion process is lower than the one for fresh vegetable oil. This is due to the presence of water and free fatty acid in the waste oils that create a soap film and reduce the effectiveness of the transesterification process. Hossain and Mekhled [111] showed that the transesterification of waste canola (often called oilseed rape), which is the second largest oilseed crop in the world, results in the biodiesel yield of about 49.5% when transesterification is carried out for two hours with 1:1 molar ratio of methanol and waste oil, and 0.5% sodium hydroxide catalyst. Normally, methanol/waste oil ratio used is about 6:1 or even 9:1 and this often results in nearly 100% conversion to biodiesel. The study showed that NaOH is a better catalyst than KOH. Numerous other studies [112–116] with frying oils and a mixture of fresh and used frying oils showed similar results. Al-Zuhair [114] used lipase immobilized on ceramic beads and entrapped in a sol-gel matrix for the production of biodiesel from waste cooking oil. The study showed that the immobilized lipase on ceramic beads was more capable of transesterifying waste cooking oil with high water content to biodiesel than lipase in free or entrapped in sol-gel matrix forms.

Studies [99, 117] have also been carried out to examine the effectiveness of supercritical alcohols for transesterification of vegetable oil in the absence of catalyst. These studies have produced some promising results. Demirbas, Ozturk, and Demirbas [99] and Demirbas and Kara [117] examined biodiesel production from vegetable oils via catalytic and noncatalytic transesterification using supercritical alcohols. The raw vegetable oils (as well as other 350 different types of crop oils) have high viscosity and low volatility; they do not burn completely and form deposits in the fuel injector of diesel engines. Vegetable oil viscosity (which is 11–17 times that of biodiesel) can be reduced by (a) dilution, (b) microemulsion, (c) thermal decomposition, (d) catalytic cracking, and (e) transesterification with alcohols, preferably methanol and ethanol.

Demirbas et al. [99, 117] showed the transesterification of vegetable oils by supercritical alcohols to be a very effective process. Their results for hazelnut kernel oil under sub- and supercritical conditions for methanol showed a sharp increase in yield of methyl ester near the critical temperature of methanol. A further increase in temperature beyond critical temperature did not significantly affect the yield of methyl ester. The results of Demirbas et al. also indicate the molar ratio of waste to alcohol to be at least 1 to 9 to get significant yield of methyl ester. Within all the lower alcohols (e.g., methanol, ethanol, propanol, etc.) tested, methanol was found to be the best extracting agent. Afify et al. [108], among others, examined transesterification of algae (crop waste) to biodiesel and they found two-solvent systems to work more effectively. More work for the conversion of algae to diesel oil via improved transesterification process is needed.

6.5.8 Anaerobic Digestion

Anaerobic digestion (in the absence of oxygen) with anaerobic bacteria or methane fermentation is used worldwide for disposal of domestic, municipal, agricultural, and industrial biomass waste. This reaction generally produces methane and carbon dioxide and it also occurs in the ecosystem and in the digestive tract. As shown in the following reactions (6.1) and (6.2), hydrogen along with acetic and butyric acids can also be produced by dark fermentation processes using anaerobic and facultative anaerobic chemohetrotrophs [118].

$$C_6O_6H_{12} + 2H_2O \rightarrow 2CH_3COOH + 4H_2 \quad (6.1)$$

$$C_6O_6H_{12} \rightarrow CH_3CH_2CH_2COOH + 2CO_2 + 2H_2 \quad (6.2)$$

Different types of waste materials can also be used for hydrogen fermentation. Hydrogen production is highly dependent on the pH, retention time, and gas partial pressure. Generally, hydrogen production increases with retention time. Biogas produced from landfills generally contains methane (about 55%) and carbon dioxide with traces of hydrogen, ethane, and other impurities. The following description of the sequence of biochemical reactions that occur to convert complex molecules to methane closely follow the excellent review of Weiland [118].

In general, methane fermentation can be divided into four phases: hydrolysis, acidogenesis, acetogenesis/dehydrogenation, and methanation. As shown by Weiland [118], the degradation of complex polymers such as polysaccharides, proteins, and lipids results in the formation of monomers and oligomers such as sugars, amino acids, and long-chain fatty acids. The individual degradation steps are carried out by different consortia of micro-organisms, which place different requirements on the environment [119]. Hydrolyzing and fermenting micro-organisms are responsible for the initial attack on polymers and monomers and produce mainly acetate, hydrogen, and varying amounts of fatty acids such as propionate and butyrate [118]. Hydrolytic micro-organisms excrete hydrolytic enzymes such as cellulose, amylase, lipase, and the like. Thus, a complex consortium of micro-organisms most of which are strict anaerobes such as Bacteriocides, Clostridia, and Bifidobacteria [118] participate in the hydrolysis and fermentation of organic material [118].

The higher volatile fatty acids are converted into acetate and hydrogen by obligate hydrogen-producing acetogenic bacteria. The maintenance of an extremely low partial pressure of hydrogen is very important for the acetogenic and hydrogen-producing bacteria. The present state of knowledge indicates that hydrogen may be a limiting substrate for methanogens [120]. This is because an addition of hydrogen-producing bacteria to the natural biogas-producing consortium increases daily biogas production. Studies [118] have shown that only two groups of methanogenic bacteria produce methane from acetate, hydrogen, and carbon dioxide. These bacteria are strictly anaerobes and require a lower radox potential for growth than most other anaerobic bacteria. Only few species are able to degrade acetate into CH_4 and CO_2, for example, *Methanosarcina barkeri, Methanonococcus mazei,* and *Methanotrix soehngenii,* whereas all methanogenic bacteria are able to use hydrogen to form methane [118].

The overall process of methane fermentation can be accomplished in two stages and a balanced anaerobic digestion process demands that in both stages the rates of degradation must be equal in size. If the first degradation step runs too fast, the acid concentration rises and pH drops below 7.0 which inhibits methanogenic bacteria. If the second phase runs too fast, methane production is limited by the hydrolytic stage. Thus the rate-limiting step

TABLE 6.7

Some Typical Literature Studies on Anaerobic Digestion of Waste Materials

Types of Waste	Authors
Swine waste	Chen et al., 2008 [121]
Coir pith	Kunchikannan et al., 2007 [122]
Wastewater and organic kitchen waste	Weichgrebe et al., 2008 [123]
Distillary spent wash	Pathe et al., 2002 [124]
Biodiesel by-products	Kolesarova et al., 2011 [125]
Whey (a component of dairy product or an additive for food product)	Beszedes et al., 2010 [126]
Palm oil effluent	Yusoff et al., 2010 [127]
Tofu wastewater	Zheng et al., 2008 [128]
Starch food waste	Ding et al. 2008 [129]
Municipal solid waste	Ismail and Abderrezaq, 2007 [130]
Solid organic waste and energy crops	Angelidaki et al., 2009 [131]
Food residuals	Shin et al., 2000 [132] Haug et al., 2000 [133]
Dairy effluent	Desai et al., 2009 [134]
Organic solid waste	Zhang, 2002 [135]
Household organic waste	Narra et al., 2009 [136]
Distillery spent waste	Nandy et al., 1992 [137]
Long fatty acids	Alves et al., 2009 [138]

depends on the compounds of the substrate used for biogas production. Undissolved compounds such as cellulose, proteins, and fats take several days to crack whereas soluble carbohydrates crack in a few hours. Therefore the process design must be well adapted to the substrate properties for achieving complete degradation without process failure.

Numerous studies have been reported for anaerobic digestion to produce either methane or hydrogen from a variety of waste streams. Some of these studies are summarized in Table 6.7. Maximum gas yields and theoretical methane contents that can be generated from carbohydrates, raw protein, raw fat, and lignin are summarized in Table 6.8 [118]. The biogas generated from landfills generally contains about 50 to 55% methane and the remaining composition consists largely of CO_2 and traces of water, hydrogen, and other impurities. It is clear from these studies that anaerobic digestion is a very widely used process to generate methane or hydrogen from a variety of organic wastes, both of which are important gaseous biofuels.

6.5.9 Fermentation

Fermentation, one of the oldest and most widely used chemical processes, is used to produce a variety of products and chemicals. Currently, the fermentation industry competes with the thermocatalytic synthesis from petroleum

TABLE 6.8

Maximal Gas Yields and Theoretical Methane Contents

Substrate	Biogas (Nm^3/t TS)	CH_4/CO_2
Carbohydrates (not including inulins and single hexoses)	790–800	1/1
Raw proteins	700	Approx. 70/30
Raw fat	1200–1250	Approx. 67/33
Lignin	0	Both 0

Source: Referred by Weiland, 2010. *Appl. Microbiol Technol.*, 85: 849–860.

products. The future of the fermentation process thus depends on its ability to produce complex products with high efficiency through the use of various types of hydrolysis pretreatments followed by enzyme-catalyzed processes and its ability to overcome variations in the quality and availability of the raw materials.

The fermentation process is most commonly used to produce ethanol. Any materials that contain sugars or sugar precursors can be converted to ethanol. Most commonly used raw materials for the fermentation process are sugars, starches, and cellulosic materials. Sugars can be directly converted to ethanol, starches must first be hydrolyzed to fermentable sugars by the action of enzymes, and cellulose must likewise be converted to sugars generally by the action of mineral (i.e., inorganic) acids. Once the simple sugars are formed, enzymes from yeasts can readily ferment them to ethanol.

The most widely used form of sugar for ethanol fermentation is blackstrap molasses, which contains 30–40% sucrose, 15–20 wt% invert sugars such as glucose and fructose, and 28–35% nonsugar solids. In industrial-based production of ethanol, sucrose-based substances such as sugarcane and sugarbeet juices offer many advantages including their abundance and renewable nature. Ethanol and gasohol produced in Brazil are largely produced from sugarcane. Although bacteria such as *Zymomonos mobilis* have been extensively tested [1], yeasts of the saccharomyces genus are mainly used in industrial processes. A schematic of the synthesis of ethanol from grains and sugar crops is illustrated in Figure 6.11. In both Brazil and Europe, ethanol is used as a substitute fuel for the automobile industry.

Cellulosic waste can also be converted to ethanol via fermentation. Cellulose from wood, agricultural residue, and waste sulfite liquor from pulp and paper mills must first be converted to sugar before it can be fermented. Although the technology for converting waste materials to ethanol is available, a significant amount of cellulosic mass disappears during the conversion to ethanol in the form of carbon dioxide which leads to its disposal problem. Another problem is that the aqueous acid used to hydrolyze the cellulose in wood to glucose and other simple sugars destroys much of sugars in the process. New methods use less corrosive acids and have a lower hydrolysis time.

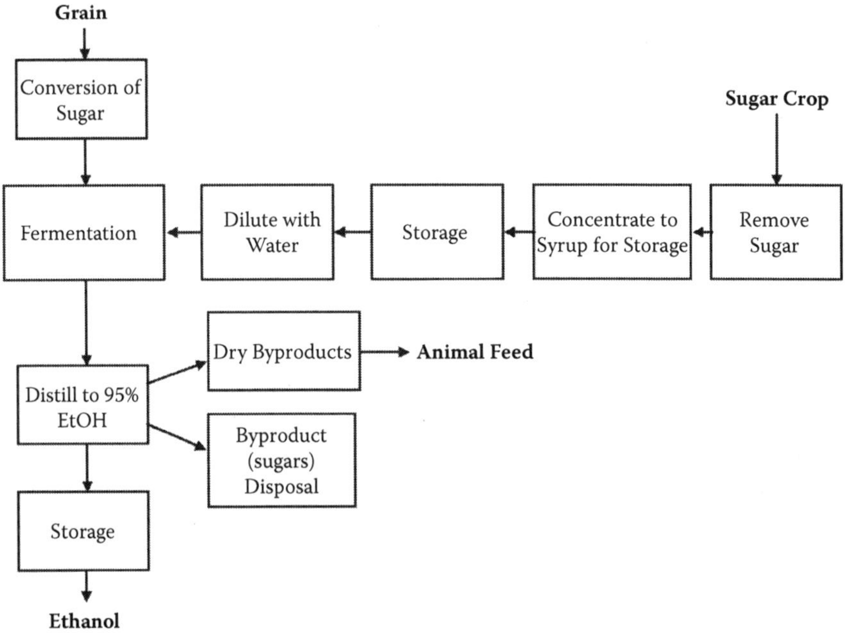

FIGURE 6.11
Synthesis of ethanol from grains and sugar crops. (After Lee, Speight, and Loyalka, 2007. Ethanol from lignocelluloses, *Handbook of Alternate Fuel Technologies*, New York: CRC Press, Ch. 11, p. 343.)

The conversion of cellulose waste to ethanol requires two steps, hydrolysis and fermentation. Within these two sequential steps, hydrolysis which converts cellulose into glucose is more critical and rate determining. Cellulose hydrolysis (saccharification) is often carried out in the presence of an enzyme (i.e., enzymatic hydrolysis), which is not as corrosive as acid. However, glucose product inhibits the effectiveness of the enzymes and therefore it must be removed during the reaction. Gulf researchers [2] commercialized the two-step process using mutant bacteria. In general, the production of ethanol from the mixture of sugars present in lignocellulosic biomass remains challenging. More robust micro-organisms are needed with a higher rate of conversion and yield to allow process simplification by consolidating process steps. This development would reduce both capital and operating costs, which remain high by comparison with those of corn.

Most methods for biomass pretreatment to produce hydrolyzates also produce side products such as acetate, furfural, and lignin that are inhibitory to micro-organisms. These inhibitory side products often significantly reduce the growth of biocatalysts, rate of sugar metabolism, and final ethanol production. In all cases, the impact of hydrolyzates on xylose metabolism is much greater than that of glucose. Present studies have focused on increasing the robustness of ethanologenic biocatalysts that utilize all sugars (hexoses and pentoses) produced from biomass saccharification at rates and production that match or exceed glucose fermentation with yeast. The critical parameters needed for a cost-competitive process are (a) high yield with complete sugar utilization, minimal by-product formation, and minimal loss of carbon into cell mass; (b) higher final ethanol production; (c) higher overall volumetric productivity, especially under high-solid conditions; and (d) tolerance to inhibitors present in hydrolyzates. Waste from grain crops, agricultural residues, and energy biomass crops as well as animal wastes are most suitable for the fermentation process. Recently sugarbeet [139], molasses, and cassava [140], municipal solid waste [141], sugarcane bagasse [142], and energy-rich carbohydrates [143] have also been used as raw materials for the production of ethanol by fermentation.

6.5.10 Products Upgrading Technologies

As indicated earlier, liquids and solids produced from waste conversion can be used either as fuel for power generation or they can be upgraded by numerous conventional and novel technologies such as hydroprocessing, catalytic cracking, product blending, aqueous phase reforming, tri-reforming, and gas to liquid conversion (Fischer–Tropsch and related syntheses) among others, to make useful transportation fuels, chemicals, and materials. These technologies are extensively discussed in the literature [29, 84, 145], and they can be applied to waste conversion products.

The effectiveness of a particular type of upgrading technology to waste conversion products depends on the quality of the products which in turn depends on the nature of the waste. For example, pyrolysis of rubber tires can generate gases that contain significant amounts of ethylene and propylene which are the basic building blocks of polymers. Landfill gas that predominantly contains methane and carbon dioxide can be upgraded by dry reforming to make syngas, which is the basic building block for Fischer–Tropsch synthesis. Aqueous phase reforming of lignocellulosic waste can produce a variety of useful chemicals. The upgrading technologies are now increasingly being applied as more efforts are being made to convert waste into useful products.

6.6 Economic and Environmental Issues Related to Waste Conversion

Waste is a relatively inexpensive material, but there are some economic and environmental issues associated with its treatment that can make it unattractive for commercial use. For example, landfills are useful sources for methane, however, a continuous collection of methane can be an issue. Depending on the size of the landfill, single or multiple sources of collection points of methane need to be equipped and this will require building a carefully designed methane collection infrastructure. Also, if a power plant is to be built on or near landfill, its size will depend on the amount of the continuous methane supply from the landfill. Generally the size of the power plant will vary from 1 to 100 MW, a size that is not always most economical. Getting a local permit for such an operation can also be an issue. Landfill gas generally contains about 50 to 55% methane, remaining carbon dioxide, and some other trace materials including hydrogen, nitrogen, and the like. This gas cannot be transported by a natural gas pipeline, which generally contains about 95% methane, without a further purification and upgrading process. Such a process can be expensive.

Numerous localities (particularly on the East Coast) have used MSW to support local power needs. Like landfill gas, the size of such a power plant will depend on the amount and continuous supply of MSW. For these reasons, these plants are now largely located in densely populated areas where a significant amount of MSW can be easily transported to a common collection point. An installation of a power plant using MSW as raw material in remote areas can be economically unattractive. The same principle applies to other kinds of waste such as rubber tires, agricultural and forestry residues, and so on, where transportation to a central location can be very expensive. This problem is further compounded because of low mass and energy densities of cellulosic waste. Sizing and separation of waste materials for a particular treatment technology can also become a significant economic issue.

The treatment of waste can also lead to significant environmental issues. Depending on the composition of waste, thermochemical treatment can lead to significant amounts of impurities in the exit gas or remaining solids. Both of these impurities need to be eliminated either by pretreating waste or purifying the gaseous product and solid streams. These processes can be expensive depending upon the nature and amount of impurities. The problem can be further compounded for a mixed waste.

6.7 Future of the Waste Industry

As the world population grows, so does the amount of waste produced by society. Scarcity of natural resources, particularly in some parts of the world, will force us to develop more and more innovative technologies to convert waste into energy and products. The new paradigm of sustainable resource management will put the use of landfill, recycling, and resource recovery in a different light and waste will be another source of raw material that will have to be used to produce energy and products. Local communities and public policies will play very important roles in making the waste industry profitable and environmentally acceptable. Overall, the future of the waste industry is very promising and ready for the development of new and innovative technologies.

References

1. Geysen, D., Jones, P.T., and Van Acker, K. 2009. Enhanced landfill mining - A future perspective for landfilling. In R. Cossu, L.F. Diaz, and R. Stegmann (Eds.), *Proceedings of Twelfth International Waste Management and Landfill Symposium*, Sardinia, Italy: Cagliari, pp. 255–256.
2. Jones, P.T., Geysen, D., Rossy, A., and Bienge, K. 2010. Enhanced landfill mining (ELFM) and enhanced waste management (EWM); essential components for the transition to sustainable materials management (SMM). In *Proceedings of the First International Symposium on Enhanced Landfill Mining*, Houthalen-Helchteren, Belgium.
3. Van Oost, G., Hrabovsky, M., Kopecky, V., Konrad, M., Hlina, M., Kavka, T., Chumak, A., Beeckman, E., and Verstraeten, J. 2006. Pyrolysis of waste using a hybrid argon-water stabilized torch, *Vacuum*, 80(11–12): 1132–1137.
4. Lee, S., Speight, J.G., and Loyalka, S.K. 2007. Energy generation from waste sources, *Handbook of Alternate Fuel Technologies*, New York: CRC Press, Ch. 13, pp. 395–419.
5. ETC/RWM. 2007. *Environmental Outlooks: Municipal Waste, Working Paper* no. 1/2007, European Topic Centre on Resource and Waste Management, Retrieved 27 July 2010, from http://waste.eionet.europa.eu/publications.
6. Scott, D.S., Czernik, S.R., Piskorz, J., and Radlein, A.G. 1990. Fast pyrolysis of plastic wastes, *Energy Fuels*, 4: 407–411.
7. Helsen, L. and Bosmans, A. 2010. Waste to energy through thermochemical processes: Matching waste with process, *Conference Proceedings on Enhanced Landfill Mining and Transition to Sustainable Materials Management*, Molenheide (Houthalen-Heichteren, Belgium), October 4–6.
8. Kaltschmitt, M. and Reinhardt, G. (Eds.). 1997. *Nachwachsende Energieträger - Grundlagen, Verfahren, ökologische Bilanzierung*, Braunschweig/Wiesbaden: Vieweg Verlagsgesellschaft.
9. Kolb, T. and Seifert, H. 2002. Thermal waste treatment: State of the art – A summary. *Waste Management 2002: The Future of Waste Management in Europe*, October 7–8, 2002, Strasbourg (France), Düsseldorf, Germany: VDI GVC.
10. Gullon, P., Conde, E., Moure, A., Dominguez, H., and Purajo, C. 2010. The open, *Agric. J.*, 4: 135–144.
11. BREF (2006). Reference document on the best available techniques for waste incineration. European Commission, Retrieved 27 July 2010, from http://eippcb.jrc.es/reference/wi.html.

12. EMIS. 2010. Vlaamse Instelling voor Technologisch Onderzoek (VITO). Energie- & milieu-informatiesysteem voor het Vlaams Gewest. Retrieved July 27, 2010, from http://www.emis.vito.be/.
13. UBA. 2001. Draft of a German report with basic information for a BREF-Document *Waste Incineration*. Umweltbundesamt Berlin, Retrieved July 27, 2010, from http://193.219.133.6/aaa/Tipk/tipk/4_kiti%20GPGB/63.pdf.
14. BIC. 2010. *BIC Group - Moving Grate Incinerator*. Retrieved August 19, 2010, from http://www.bicgroup.com.sg/.
15. Ullmanns Encyclopedia. 2001. *Ullmann's Encyclopedia of Industrial Chemistry*, Kellersohn T. Weinheim: Wiley-VCH.
16. Limerick. 2005. *Feasibility Study of Thermal Treatment Options for Waste in the Limerick/Clare/Kerry Region*, Retrieved July 27, 2010, from http://www.managewaste.ie/docs/WMPNov2005/FeasabilityStudy/LCK%20Thermal%20Feasibility%20Report-Ful%20%28web%29.pdf.
17. Bridgwater, A.V. 1994. Catalysis in thermal biomass conversion, *Appl. Catalysis A: Gen.*, 116(1–2): 5–47.
18. Helsen, L. 2000. Low Temperature Pyrolysis of Chromate Copper Arsenate (CCA) Treated Wood Waste, PhD, University of Leuven, Belgium.
19. GL. (2009). Germanischer Lloyd (GL). *BGL Gasifier*. Retrieved August 19, 2010, from http://www.gl-group.com/.
20. EBARA. 2003. EUP - EBARA UBE *Process for Gasification of Waste Plastics*. Retrieved May 2010, from http://www.ebara.ch/.
21. Rowland, S., Bower, C.K., Patil, K.N., and DeWitt, C.A.M. 2009. *J. Food Sci.*, 74(8): E426–E431.
22. Lopez, J.A.S., Li, Q., and Thompson, I.P. 2010. *Chem. Rev. Biotechnol.*, 30(1): 63–69.
23. Lugina, E.S., Dmitruk, A.F., Lyubchik, S.B., and Tret'yakov, V.F. 2009. *Solid Fuel Chem.*, 43(4): 247–266.
24. Demirbas, A. 2008. *Energy Sources, A*, 30: 565–572.
25. Jain, A.K. 2006. Design parameters for a rice husk throatless gasifier, *Agric. Engrg. Int. CIGR EJournal*, manuscript EE 05 012, VIII (May).
26. Bridgwater, A.V. 2003. Pyrolysis and gasification of biomass and waste. In A.V. Bridgwater (Ed.), *Proceedings of an Expert Meeting*, Strasbourg, France, Ashton University, Bio-Energy Research Group and CPL Press, UK.
27. Bridgwater, A.V. and Bridge, S.A. 1991. A review of biomass pyrolysis and pyrolysis technologies. In A.V. Bridgwater and G. Grassi (Eds.), *Biomass Pyrolysis Liquids Upgrading and Utilisation*, London: Elsevier Applied Science, pp. 11–92.
28. Bridgwater, A.V. 1995. Thermal biomass conversion technologies, *Biomass and Renewable Energy Seminar*, Loughborough University, IK, March.
29. Huber, G.H., Iborra, S., and Corma, A., 2006. Synthesis of transporation fuels from biomass: Chemistry, catalysis, and engineering, *Chem. Rev., ACS* 106(9): 4044–4098.
30. Helsen, L. and Van den Bulck, E. 2005. Review of disposal technologies for chromated copper arsenate (CCA) treated wood waste, with detailed analyses of thermochemical conversion processes, *Environ. Pollution*, 134: 301–314.
31. Themelis, N. 2007. Thermal treatment review, *Waste Manage. World*, July–August.
32. Helsen, L., Van den Bulck, E., and Hery, J.S. 1998. Total recycling of CCA treated wood waste by low-temperature pyrolysis, *Waste Manage.*, 18(6): 571–578.
33. Das, K.C., Perez, M.G., Bibens, B., and Melear, N. 2008. *J. Environ Sci. Health, A*, 43: 714–724.

34. Wu, J., Wang, Y., Wan, Y., Lei, H., Yu, F., Liu, Y., Chen, P., Yang, L., and Ruan, R. 2009. *Int. J. Agric. Biol. Engrg.*, 2(1, March): 40–50.
35. Yang, C., Zhang, B., Moen, J., Hennessy, K., Liu, Y., Lin, X., Wan, Y., Lei, H., Chen, P., and Ruan, R. 2010. *Int. J. Agric. Bio. Engrg.*, 3(3, Sept.): 54–61.
36. Nardin, G. and Catanzaro, G. 2007. *Helia*, 30(46): 143–156.
37. Demirbas, A. 2008. "Biofuels from agricultural residue: Recovery, utilization, and environmental effects," *Energy Sources A*, 30: 101–109.
38. Demirbas, A. 2006. *Energy Sources A*, 28: 933–940.
39. Balat, M. and Demirbas, M.F. 2009. *Energy Resources. A: Recovery, Utilization Environ. Effects*, 31(19): 1719–1727.
40. Xiao, J., Shen, L., Zhang, Y., and Gu, J. 2009. *Ind. Eng. Chem. Res.*, 48(22), 9999–10007.
41. Demirbas, A. 2010. *Energy Sources A*, 32: 1–9.
42. Irvine, G., Lamont, E.R., and Antizar-Ladislao, B. 2010. *Int. J. Chem. Engrg.*, Article ID 627930: 1–10.
43. Liu, R., Deng, C., and Wang, J. 2010. *Energy Sources A*, 32: 10–19.
44. Anto, L.P. and Thomas, S. 2009. *Proceedings of the International Conference on Energy and Environment*, March 19–21, pp. 558–559.
45. Demiral, I., Atilgon, N.G., and Sensoz, S. 2009. *Chem. Eng. Comm.*, 196: 104–115.
46. Demiral, I. and Sensoz, S. 2006. *Energy Sources A*, 28: 1159–1168.
47. Demiral, I. and Sensoz, S. 2006. *Energy Sources A*, 28: 1149–1158.
48. Ioannidou, O., Zabaniotou, A., Antonakou, E.V., Papazisi, K.M., Lappas, A., and Athanassiou, C. 2009. *Renew. Sustain. Energy Rev.*, 13: 750–762.
49. Kar, Y. and Tekeli, Y. 2008. *Energy Sources A*, 30: 483–493.
50. Sensoz, S. and Can, M. 2002. *Energy Sources*, 24: 357–364.
51. Yorgun, S. and Simsek, Y.E. 2003. *Energy Sources*, 25: 779–790.
52. Culcuoglu, E., Unay, E., Karaosmanoglu, F., Angin, D., and Sensoz, S. 2005. *Energy Sources*, 27: 1217–1223.
53. Demirbas, A. 2008. "Producing bio-oil from olive cake by fast pyrolysis," *Energy Sources A*, 30: 38–44.
54. Demirbas, M.F. 2009. "Evaluation of olive cake for bio-oil," *Energy Sources A*, 31: 1236–1241.
55. Ozbay, N., Putun, A.E., and Putun, E. 2006. *Int. J. Energy Res.*, 30: 501–510.
56. Gercel, H.F. and Gercel, O. 2007. *Energy Sources A*, 29: 695–704.
57. Schnitzer, M., Montreal, C.M., and Jandl, G. 2008. *J. Environ. Sci. Health B*, 43: 81–95.
58. Ozcan, A., Bartle, K.D., and Putun, E. 2000. *Energy Sources*, 22: 809–824.
59. Putun, A.E. 2002. *Energy Sources*, 24: 275–285.
60. Huang, H. and Tang, L. 2007. Treatment of organic waste using thermal plasma pyrolysis technology, *Energy Conversion Manage.*, 48(4): 1331–1337.
61. Heberlein, J. and Murphy, A.B. 2008. Thermal plasma waste treatment, *J. Phys. D: Appl. Phys.*, 41(5): 053001.
62. Murphy, A.B. and McAllister, T. 2001. Modeling of the physics and chemistry of thermal plasma waste destruction, *Phys. Plasmas*, 8(5): 2565–2571.
63. Murphy, A.B. and McAllister, T. 1998. Destruction of ozone-depleting substances in a thermal plasma reactor, *Appl. Phys. Lett.*, 73(4): 459–461.
64. Murphy, A.B. and Kovitya, P. 1993. Mathematical model and laser-scattering temperature measurements of a direct-current plasma torch discharging into air, *J. Appl. Phys.*, 73(10): 4759–4769.

65. Lapa, N., Oliveira, S., Camacho, S.L., and Circeo, L.J., 2002. An ecotoxic risk assessment of residue materials produced by plasma pyrolysis/vitrification (PP/V) process, *Waste Manage.*, 22(3): 335–342.
66. Hrabovsky, M. 2002. Generation of thermal plasmas in liquid stabilized and hybrid dc-arc torches, *Pure Appl. Chem.*, 74(3): 429–433.
67. Watanabe, T. and Shimbara, S. 2003. Halogenated hydrocarbon decomposition by steam thermal plasmas, *High Temp. Mater. Process.*, 7(4): 455–474.
68. Lemmens, B., Elslander, H., Vanderreydt, I., Peys, K., Diels, L., Oosterlinck, M., and Joos, M. 2007. Assessment of plasma gasification of high caloric waste streams, *Waste Manage.*, 27(11): 1562–1569.
69. EUROPLASMA. 2008. Production d'électricité par gazéification - Energies renouvelables et développement durable, Projet CHO-Power – Morcenx. Retrieved December 21, 2009, from http://www.paris-dechets.com/medias/interventions/a10/a10_marc_lefour_production_electricite_gazeification.pdf.
70. Advanced Plasma Power (APP). 2010. *What Is Gasplasma? – The Process*, Retrieved March 29, 2010, from http://www.advancedplasmapower.com/.
71. Westinghouse Plasma Corporation WPC. 2010. Technology and solutions – Plasma gasification vitrification reactor, Retrieved July 22, 2010, from http://www.westinghouse-plasma.com/.
72. Taylor, R. 2009. Advanced plasma power (Swindon, UK), personal communication.
73. Cleiren, D. and GOM-Antwerpen. 2000. Objectieve keuze inzake verwerkingstechnieken. Het objectiveren van de techniekkeuze voor de eindverwijdering van huishoudelijk afval: Het praktijkvoorbeeld binnen de provincie Antwerpen". *20ste Internationaal Seminarie – Het Beheer van Afvalstoffen*, Vrije Universiteit Brussel (Belgium) in samenwerking met RDC.
74. Vanderreydt, I. 2010. Mol, Belgium, Personal communication.
75. Fichtner, K. 2008. *Advanced Plasma Power-Gasplasma TM- Stage 5*, assessment report, Cheshire (UK), Fichtner Eng. and Consulting Services, Retrieved January 11, 2009 from http://www.advancedplasmapower.com.
76. Peterson, A.A., Vogel, F., Lachance, R.P., Froling, M., Antal, M.J., and Tester, J.W. 2008. "Thermochemical biofuel production in hydrothermal media: A review of sub- and supercritical water technologies," *Energy Environ. Sci.*, 1: 32–65.
77. Moffatt, J.M. and Overend, R.P. 1985. *Biomass*, 7: 99.
78. Kumar, S. and Gupta, R.B. 2009. "Biocrude production from switchgrass using sub-critical water," *Energy Fuels*, 23: 5151–5159.
79. Byrd, A.J., Kumar, S., Kong, L., Ramsurn, H., and Gupta, R.B. 2011. *Int. J. Hydrogen Energy*, 36: 3426–3433.
80. Akhtar, J. and Amin, N.A.S. 2011. *Renew. Sustain. Energy Rev.*, 15: 1615–1824.
81. Sevilla, M. and Fuertes, A.B. 2009. *Carbon*, 47: 2281–2289.
82. Jena, U., Das, K.C., and Kastner, J.R. "Effect of operating conditions of thermochemical liquefaction on biocrude production from spirulina platensis," *Biores. Technol.*, 102(10), 6221–6229.
83. Funke, A. and Ziegler, F. 2010. Hydrothermal carbonization of biomass: A summary and discussion of chemical mechanisms for process engineering, *Biofuels, Bioproducts Biorefining*, 4: 160–177.
84. Demirbas, A. 2010. Green energy and technology. *Biorefineries for Biomass Upgrading Facilities*, London: Springer-Verlag.

85. Kranich, W.L. 1984. *Conversion of Sewage Sludge to Oil by Hydroliquefaction*, EPA-600/2 84-010. Report for the U.S. EPA, Cincinnati, Ohio.
86. Suzuki, A., Yokoyama, S., Murakami, M., Ogi, T., and Koguchi, K. 1986. New treatment of sewage sludge by direct thermochemical liquefaction, *Chem. Lett.*, 9: 1425–1428.
87. Itoh, S., Suzuki, A., Nakamura, T., and Yokoyama, S. 1994. Production of heavy oil from sewage sludge by direct thermochemical liquefaction. *Proceedings of the IDA and WRPC World Congress on Desalination and Water Treatment.*
88. Inoue, S., Sawayama, S., Dote, Y., and Ogi, T. 1997. Behavior of nitrogen during liquefaction of dewatered sewage sludge, *Biomass Bioenergy*, 12: 473–475.
89. *Changing World Technology*. 2008. Company Report, Carthage, MO.
90. Kumar, S. and Gupta, R. 2008. Hydrolysis of microcrystalline cellulose in subcritical and supercritical water in a continuous flow reactor, *Ind. Eng. Chem. Res.*, 47: 9321–9329.
91. Guo, L., Cao, C., and Liu, Y. 2010. "Supercritical water gasification of biomass and organic wastes," *Biomass*. M. Momba and F. Bux (Eds.), Ch 9, p. 165.
92. Hong, G.T. and Spritzer, M.H. 2002. Super critical water partial oxidation. *Proceedings of the 2002 U.S. DOE Hydrogen Program Review.*
93. Johanson, N.W., Spritzer, M.H., Hong, G.T., and Rickman, W.S. 2001. Supercritical water partial oxidation. *Proceedings of 2001 DOE Hydrogen Program Review.*
94. Lu, Y., Guo, L., Zhang, X., and Yan, Q. 2007. *Chem. Eng. J.*, 131: 233–244.
95. Nath, K. and Das, D. 2003. Hydrogen from biomass, *Curr. Sci.*, 85(3, Aug.): 265–271.
96. Picou, J.W., Wenzel, J.E., Lanterman, H.B., and Lee, S. 2009. Hydrogen production by non-catalytic autothermal reformation of aviation fuel using supercritical water, *ACS Journal* (electronic publication).
97. Xu, X.D. and Antal, M.J. 1998. Gasification of sewage sludge and other biomass for hydrogen production in supercritical water, *Environ. Prog.*, 17(4): 215–220.
98. Sinag, A., Kruse, A., and Schwarzkopf, V. 2003. *Ind. Eng. Chem. Res.*, 42: 3516-3522.
99. Demirbas, A., Ozturk, T., and Demirbas, M.F. 2006. *Energy Sources, A*, 28: 1473–1482.
100. Demirbas, A. 2000. Recent advances in biomass conversion technologies, *Energy Edu. Sci. Technol.*, 6: 19:40.
101. Valenzuela, M.B., Jones, C.W., and Agarwal, P.K. 2006. *Energy Fuels*, 20: 1744–1752.
102. Markovic, Z., Markovic, S., Engelbrecht, J.P., and Visser, F.D. 2000. *S. African J. Chem.*, 53(3):2 (2000).
103. Bimakr, M., Rahman, R., Taip, F., Chuan, L., Ganjloo, A., Selamat, J., and Hamid, A. 2009. *European J. Sci. Res.*, 33(4): 679–690.
104. Sajfrtova, M., Lickova, I., Wimmmerova, M., Sevova, H., and Wimmer, Z. 2010. *Int. J. Mol. Sci.*, 11: 1842–1850.
105. Stahl, E., Schultz, E., and Mangold, H.K. 1980. *J. Agric. Food Chem.*, 28: 1153–1157.
106. Xiu, S., Shabhazi, A., Wang, L., and Wallace, C.W. 2010. *Amer. J. Engrg. Appl. Sci.*, 3(2): 494–500.
107. Demirbas, A. 2005. *Progr. Energy Combustion Sci.*, 31: 466–487.
108. Afify, A.M.R., Shalaby, E.A., Shanab, S.M.H., and Grasas, Y. 2010. *Aceities*, 61(4): 416–422.

109. Nakpong, P., Wiotthikanokkhan, S., and Suranaree, S. 2010. *J. Sci. Technol.*, 17(2): 193–202.
110. Balat, M. 2007. *Energy Sources A*, 29: 895–913.
111. Hossain, A.B.M.S. and Mekhled, M.A. 2010. *Australian J. Crop Sci.*, 4(7): 543–549.
112. Li, L., Cappola, E., Rine, J., Miller, J.L., and Walker, D. 2010. *Energy Fuels*, 24: 1305–1315.
113. Kraemer, V., Araujo, W.S., Hamacher, S., and Scavarda, L.F. 2010. *Bioresource Technol.*, 101: 4415–4422.
114. Al-Zuhair, 2008. The open. *Chem. Engrg. J.*, 2: 84–88.
115. Ozata, I., Ciliz, N., Mammadov, A., Buyukbay, B., and Ekinci, E. 2008. Comparative life cycle assessment approach for sustainable transport fuel production from waste cooking oil and rapseed (personal communication).
116. Hossain, A.B.M.S. and Mazen, M.A. 2010. *Australian J. Crop Sci.*, 4(7): 550–555.
117. Demirbas, A. and Kara, H. 2006. "New options for conversion of vegetable oils to alternative fuels," *Energy Sources Recovery, Utilization and Environmental Effects, A.*, 28(7): 619–626.
118. Weiland, P. 2010. *Appl. Microbiol Technol.*, 85: 849–860.
119. Abdoun, E. and Weiland, P. 2009. *Bermimer Agrartecnische Berichte*, 68: 69–78.
120. Ahrens, T. and Weiland, P. 2007. *Land-bauforschung Volkenrode*, 57: 71–79.
121. Chen, Y.S.A., Zuckerman, G.J., and Zering, K. 2008. *Engrg. Econ.*, 53: 156–170.
122. Kunchikannan, L.K.N., Mande, S.P., Kishore, V.V.N., and Jain, K.L. 2007. *Energy Sources A*, 29: 293–301.
123. Weichgrebe, D., Urban, I., and Friedrich, K. 2008. *Water Sci. Technol.*, 379–384.
124. Pathe, P.P., Rao, M.N., Kharwade, M.R., Lakhe, S.B., and Kaul, S.N. 2002. *Int. J. Env. Stud.*, 59(4): 415–437.
125. Kolesarova, N., Hutnan, M., Bodik, I., and Spalkova, V. 2011. *J. Biomed. Biotechnol.*, 2011: 1–15.
126. Beszedes, S., Laszlo, Z., Szabo, G., and Hodur, C. 2010. *J. Agric. Sci. Technol.*, 4(1, Feb.), serial no. 26: 62–68.
127. Yusoff, M., Rehman, N.A., Aziz, S.A., Ling, C.M., Hassan, M.A., and Shirai, Y. 2010. *Australian J. Basic Appl. Sci.*, 4(4): 577–587.
128. Zeng, G.H., Kang, Z.H., Qian, Y.F., Wang, L., Zhou, Q., and Zhu, H.G. 2008. In K. Tohji, N. Tsuchiya, and B. Jeyadevan (Eds.), *Int. Workshop on Waterdynamics*, AIP, 143–148.
129. Ding, H.B., Liu, X.Y., Stabnikova, O., and Wang, J.Y. 2008. *Water Sci. Technol.*, 1031–1036.
130. Ismail, Z.K. and Abderrezaq, S.K. 2007. *Energy Sources A*, 29: 657–668.
131. Angelidaki, I., Alves, M., Bolzonella, D., Borzacconi, L., Campos, J.L., Guwy, A.J., Kalyozhnyi, S., Jenicek, P., and Van Lier, J.B. 2009. *Water Sci. Technol.*, 927–934.
132. Shin, H.K., Han, S.K., Song, Y.C., and Hwan, E.J. 2000. *Biocycle*, Aug.: 82–86.
133. Haug, R.T., Hernandez, G., Scrullo, T., and Gerringer, F. 2000. *Biocycle*, Sept.: 74–77.
134. Desai, H., Nagori, G.P., and Vahora, S. 2009. *Proceedings of the International Conference on Energy and Environment*, March 19–21, pp. 457–459.
135. Zhang, R. 2002. *Biocycle* (Jan.): 56–59.
136. Narra, M., Nagon, G.P., and Pushalkar, M. 2009. *Proceedings of the International Conference on Energy and Environment*, March 19–21, pp. 477–480.

137. Nandy, T., Kaul, S.N., Pathe, P.P., Deshpande, C.V., and Daryapurkar, R.A. 1992. *Int. J. Environ. Stud.*, 41: 87–107.
138. Alves, M., Pereira, M., Sousa, D., Cavaleiro, A., Picavet, M., Smidt, H., and Stams, A.J.M. 2009. *Microbial Biotechnol.*, 2(5): 538–550.
139. Kranjc, D. and Glavic, P. 2009. *Chem. Engrg. Res. Des.*, 87: 1217–1231.
140. Maryana, R. and Wahono, S.K. 2009. In L.T. Handoko and M.R.T. Siregar (Eds.), *International Workshop on Advanced Material for New and Renewable Energy*, Advanced Institute of Physics, pp. 174–178.
141. Li, A., Antizar-Ladislao, B., and Khraisheh, M. 2007. *Bioprocess Biosyst. Eng.*, 30: 189–196.
142. Dias, M., Ensinas, A., Nebra, S., Filho, R., Russell, C., and Maciel, M. 2009. *Chem. Eng. Res. Des.*, 87: 1206–1216.
143. Antoni, D., Zverlov, V., and Schwarz, W. 2007. *Appl. Microbial Biotechnol.*, 77: 23–35.
144. Lee, S., Speight, J.G., and Loyalka, S.K. 2007. Ethanol from lignocelluloses, *Handbook of Alternate Fuel Technologies*, New York: CRC Press, Ch. 11, p. 343.
145. Lee, S., Speight, J.G., and Loyalka, S.K., 2007. *Handbook of Alternate Fuel Technologies*, New York: CRC Press.

7

Mixed Feedstock

7.1 Introduction

As the world supply and demand picture for fossil energy changes and the environmental regulations for greenhouse gas (GHG) emissions become more stringent, more efforts are being made to find alternate sources of energy. One source is renewable bioenergy which addresses environmental regulations. Although bioenergy currently contributes to only a small percentage of worldwide energy production (it is about 5% of European Union energy supply and smaller in the United States), its worldwide usage is rapidly increasing. Furthermore, recent advances in sustainable waste management provide an additional opportunity for converting various cellulosic and polymer waste, rubber tires, MSW, and the like, to energy and products. Types of biomass feedstock used for energy purposes are described in Table 7.1 [1–3].

Although the use of biomass for power and fuel brings environmental benefits, its use involves high investment costs. Furthermore, the use of biomass raises concerns about the security of feedstock supply particularly for large power and fuel plants. The feedstock supply issue with biomass is caused by (a) the seasonal nature of biomass, (b) biomass resources are dispersed in many countries and an infrastructure for the biomass supply is not established, and (c) transportation of biomass can be very expensive because of its low mass and energy densities. Lower heating values and lower bulk densities compared to coal result in a much greater volume of biomass to be transported, handled, and stored and as a consequence, large biomass units (>300 MWe) are economically unattractive [4]. The argument of an inconsistent and unreliable feed supply to large plants also applies to the waste industry.

The biomass demand in stationary applications (heat and power) is the main driving force behind early expansion of bioenergy and this will remain the major demand source for bioenergy up to 2030. However, the need for more improved and efficient generation of stationary bioenergy, as well as the new efforts to use biomass and waste to generate biofuels and other products, will stimulate more and more use of biomass. These expansions

TABLE 7.1

Types of Biomass Feedstock Used for Energy Purposes [1-3]

Source	Types and Examples
Woody, forestry, and agricultural, and park and garden waste	1. Industrial waste wood from timber mills and sawmills (e.g., bark, sawdust, wood chips, slabs, off-cuts)
	2. Waste from paper and pulp industry including black liquor
	3. Forestry by products (e.g., wood blocks, logs, wood chips)
	4. Dry lignocellulosic agricultural residues (e.g., straw, sugar beet leaves and residue flows from bulb sector)
	5. Livestock waste (e.g., chicken, cattle, pig, and sheep manure)
	6. Herbaceous grass and woody pruning
Dedicated energy crops	1. Woody energy crops (e.g., willow, poplar and eucalyptus)
	2. Herbaceous energy crops (e.g., various types of reed grass, switch grass, miscanthus, Indian shrub and cynara cardunculus)
	3. Oil energy crops (e.g., rapeseed, sunflower seeds, soybean, olive-kernel, calotropis procera)
	4. Sugar energy crops (e.g., sugar beet, cane beet, sweet sorghum, sugar millet, Jerusalem artichoke)
	5. Starch energy crops (e.g., barley, wheat, potatoes, maize corn (cob), amaranth)
	6. Other energy crops (e.g., flax, hemp, tobacco stems, cotton stalks, kenaf, aquatic plants (lipids from algae))
Waste	1. Contaminated waste (e.g., biodegradable municipal waste, demolition wood, sewage sludge)
	2. Landfill gas
	3. Sewage gas
Miscellaneous	1. Roadside hay (e.g., grass)
	2. Husks/shells (e.g., olive, walnut, almond, palm pit, cacao)

Source: Maciejewska et al. 2006. *Co-Firing of Biomass with Coal: Constraints and Role of Biomass Pre-Treatment*, DG JRC Institute for Energy Report, EUR 22461 EN; Loo and Koppejan (Eds.) 2004. *Handbook of Biomass Combustion and Co-Firing*, Prepared by Task 32 of the implementing agreement on bioenergy under an auspices of the international energy agency, Twente University Press; VIEWLS. 2005. *Biofuel and Bio-Energy Implementation Scenarios*, Final report of VIEWLS WP5, modeling studies.

will also stimulate the establishment of a new and improved supply infrastructure for biomass in the long term.

At the present time, a mixed feedstock of coal and biomass offers a possible solution to the above-described problem. Such a mixture can help mitigate environmental concerns of plants running on coal alone. On the other hand, coal can mitigate the effects of variations in biomass feedstock quality and buffer the system when there is a lack of sufficient required biomass quantity [3,5]. A mixed feedstock can be used in large units that have better thermal and economical efficiencies compared to small-scale systems. Furthermore, it is possible to adapt existing coal power plants or coal-based fuel refineries for mixed feedstock at a relatively lower cost compared to building new and dedicated systems [4].

The use of mixed feedstock started in co-firing because its major purpose was to generate heat and electricity and the compositions of gas and solids were not important toward the end-use. However, now in addition to combustion, other thermochemical technologies such as gasification, plasma technology, pyrolysis, liquefaction, and supercritical technology have been further developed to generate heat, electricity, transportation fuel, chemicals, and materials. The use of mixed feedstock in these technologies is more complex because of the effects of mixed feedstock on the gas, liquid, and solid compositions and their subsequent use. As shown in Tables 7.2a to 7.2c [6, 7], the chemical compositions of various raw materials that can be used within a mixed feedstock vary substantially and these can significantly affect the gas, liquid, and solid product compositions.

Up till now at larger scales, co-utilization of waste and coal has received considerably more attention than co-utilization of coal and biomass. This is true for all thermochemical processes such as combustion, high severity pyrolysis, gasification (including supercritical), and plasma technology that are generating heat, electricity, or gaseous fuels and products. Investigated waste has been municipal solid waste (MSW) that has had minimal presorting or refuse-derived fuel (RDF) that has had significant pretreatment such as mechanical shredding and screening as well as shredded rubber tires, paper and pulp waste, and plastic waste.

In recent years, co-utilization of coal and biomass in combustion, gasification, pyrolysis, and plasma technology has been gaining significant acceptance. Recent reviews of cofiring literature identify over 100 successful field demonstrations in 16 countries using many types of biomass in combination

TABLE 7.2A

Selected Typical Properties of Several Coal and Biomass Fuels

Type	Coal	Peat	Olive Residue	Willow	Straw Corn	Stover	Cotton Gin	Rice Husk	Olive Husk
Ash (db)	9.7	5.5	4.5	2.55	5	3.25	14.5	20.61	1.6
Moisture	8.0	47.5	65	55	21	35	11.5	9.96	33
(wt%)									
C (%db)	81.5	54	49	49	46	42.5	42	34.94	47.8
H (%db)	4.25	5.75	6	6.25	5.9	5.04	5.4	5.46	5.1
O(%db)	7.05	35	34	43	43	42.6	35	38.86	45.4
N(%db)	1.15	2	1	0.5	0.5	0.75	1.4	0.11	0.1
S(%db)	1.8	<.17	0.12	0.06	0.125	0.18	0.5	–	

Source: Ratafia-Brown et al. 2007. *Assessment of Technologies for Co-converting Coal and Biomass to Clean Syngas-Task 2 Report (RDS)*, NETL report (May 10) and Sami, Annamalai, and Wooldridge, 2001. Co-firing of coal and biomass fuel blends, *Prog. Energy Combust. Sci.*, 27: 171–214.

TABLE 7.2B

Selected Typical Properties of Some Wood Products and Municipal Residue

Type	Sawdust	Hardwood	Softwood	Redwood	Switch Grass	Tan Oak	MSW	Tires Black	Locust
Ash (%db)	2.6	–	–	0.36	4.61	1.67	15.5	6.1	0.8
Moisture	7.3	–	–	–	11.99	–	27	0.5	–
(wt%)									
C(%db)	46.9	50.2	52.7	50.64	42.02	47.81	–	81.5	50.73
H(%db)	5.2	6.2	6.3	5.98	4.97	5.93	–	7.1	5.71
O(%db)	37.8	43.5	40.8	42.88	42.02	47.81	–	3.4	41.93
N(%db)	0.1	0.1	0.2	0.05	0.77	0.12	–	0.5	0.57
S(%db)	0.04	–	–	0.03	0.18	0.01	–	1.4	0.01

Source: Modified from Sami, Annamalai, and Wooldridge, 2001. Co-firing of coal and biomass fuel blends, *Prog. Energy Combust. Sci.*, 27: 171–214.

TABLE 7.2C

Selected Typical Composition of Energy Products and Food Processing Residue

Type	Poplar	Eucalyptus (Grandis)	Sugarcane Bagasse	Almond Shells	Olive Pits	Walnut Shells	Peach Pits
Ash(%db)	1.33	0.52	11.27	4.81	3.16	0.56	1.03
C(%db)	48.45	48.33	44.8	44.98	48.81	49.98	53.0
H(%db)	5.85	5.89	5.35	5.97	6.23	5.71	5.9
O(%db)	43.69	45.13	39.55	42.27	43.48	43.35	39.14
N(%db)	0.47	0.15	0.38	1.16	0.36	0.21	0.32
S(%db)	0.01	0.01	0.01	0.02	0.02	0.01	0.05

Source: Modified from Sami, Annamalai, and Wooldridge, 2001. Co-firing of coal and biomass fuel blends, *Prog. Energy Combust. Sci.*, 27: 171–214.

with various types of coals in boilers [8]. Significant new efforts at the laboratory level as well as at semi-commercial or commercial scales to generate energy or useful products have been carried out and these are reviewed in this chapter. A recent position paper indicates that co-firing represents among the lowest risk, least expensive, most efficient, and shortest term option for renewable-based electrical power generation [8]. Although in general co-firing is more expensive than the power plants based strictly on coal, CO_2 emission reduction, global climate change mitigation, and sustainable resource management for waste on a large scale are the main motivations behind the development of mixed feedstock strategy.

In this chapter, our discussions on various thermochemical technologies processing a mixed feedstock are divided in two parts: those that are largely generating gases such as combustion, gasification (including high temperature supercritical gasification and reforming technology), high severity (i.e., high temperature and high residence time) pyrolysis, and plasma technology;

and those that are focused in generating liquids such as low severity (i.e., low contact time or low temperature) pyrolysis, liquefaction (either by water or an organic solvent), and supercritical fluid extraction (either by water or other chemical substances). Because biochemical technologies are not extensively used for fossil fuels such as coal, tar sand, shale oil, and the like, their applications for a mixed feedstock are not addressed here. Gasification technologies are used for a mixed feedstock on a larger scale, and the development of liquefaction technologies for mixed feedstock is still at the laboratory and small scale stages.

Combustion, gasification, plasma technology, and high-severity pyrolysis are commercially proven technologies for a single feedstock, however, improvements are constantly being made to adapt these commercial operations for mixed feedstock. Major challenges in mixed feedstock processing are biomass fuel preparation, storage, and handling [8–10]. Other problems are associated with poor or incompatible fuel quality and these include fuel feeding, co-milling, deposit formation, increased corrosion and erosion, and need for new fly ash (and in general slag) utilization schemes [8, 10, 11]. These challenges are not significant in a mixed feedstock with low concentration of biomass but become more important as the concentration of biomass increases, particularly when low-quality biomass is used. In these cases an economy of the overall plant may be significantly affected. As for example, herbaceous biomass and coal are not as good a mixed feedstock as wood chips and coal. Similarly, stringlike biomass (e.g., straw or switchgrass) is not as good as RDF in the form of well-mixed pellets. As shown later, these drawbacks can to a certain extent be avoided by application of appropriate biomass pretreatments. The present chapter addresses these topics. Also, as shown later, several gasification technologies using mixed feedstock have been demonstrated commercially and these are briefly described. The development of supercritical gasification for a mixed feedstock is only at the laboratory and demonstration levels. Unlike gasification, direct liquefaction technologies are still at a laboratory or a small-scale developmental stage.

7.2 Advantages and Disadvantages of Mixed Feedstock

The advantages and disadvantages of the co-gasification of coal and biomass mixture are briefly illustrated in Table 7.3. [12–14]. As shown in this table, co-gasification provides many advantages to the production of the syngas as well as generation of power. The pure biomass gasification process is limited to the small scale, has high capital (fixed) cost, has lower thermal efficiency, and carries shutdown risk. All of these are alleviated by the use of coal. A mixture of coal and biomass provides a stable and reliable feed supply for large-scale operations. Coal can be considered as the "fly wheel" that allows

TABLE 7.3

Advantages and Disadvantages of Use of Coal and Biomass Mixture Feed for Gasification

Advantages	Disadvantages
Economy of scale – smaller biomass plants alone are inefficient and risky.	Expensive feed preparation.
Lower costs associated with fossil fuel consumption.	Complex feed systems for co-gasification.
Less CO_2 discharge.	Two separate feed injectors versus single feed injector may affect the gasifier performance.
More stable and reliable feed supply by the mixture.	Negative impact of the slagging behavior of the combined ash in the gasifier.
Provides more security, less risk.	Additional complication to gas cleaning system.
More positive public attitude toward use of fossil fuel, renewable fuel, and reliable multisource fuel supply.	More tar and oil formation in raw syngas can be a problem depending on the gasification technology.

Source: Prins, Ptasinski, and Janssen, 2006. Torrefaction of wood. Part 1. Weight loss kinetics, *J. Anal. Appl. Pyrolysis*, 77: 28–34; Shafizadeh, 1985. Pyrolytic reactions and products of biomass. In R.P. Overend, T.A. Mime, and L.K. Mudge (Eds.), *Fundamentals of Biomass Thermochemical Conversion*, London: Elsevier, pp. 183–217; and Shafizadeh, 1983. Thermal conversion of cellulosic materials to fuels and chemicals. In *Wood and Agricultural Residues*, New York: Academic Press, pp. 183–217.

a continuous plant operation when biomass fuel is not easily available. This concept in principle can be applied to any mixed feedstock. Co-gasification reduces the cost associated with fossil fuel consumption although some types of biomass can add significant cost to fuel production.

Co-gasification also reduces CO_2 discharge in the atmosphere. When emissions related to harvesting, transportation, and other elements of the biomass supply chain are not included, biomass is considered to be a CO_2-neutral fuel. A life-cycle assessment study shows that in comparison with coal-based systems, the use of biofuels for gasification results in environmental benefits. Unlike coal, the use of biomass fuel sources result in the generation of significantly lower quantities of anthropogenic CO_2 emissions during power or fuel productions. A 70/30 mixture of coal and biomass generally produces a carbon-neutral process [15]. These advantages provide more security and less risk for the project financiers than the use of pure biomass, and are likely to engender more positive public attitudes toward the use of a fossil fuel supply as a part of the mixed feedstock.

The mixed feedstock of biomass and coal also carries some disadvantages. As shown in Table 7.3, feed preparation and complex feed systems for mixed feedstock can be expensive. Two separate feed injectors, versus a single feed injector, may affect the gasifier performance. The gas cleaning system has the additional complications due to impurities in biomass. Although SO_x emissions during gasification with mixed feedstock generally decreases,

NO_x emission can increase, decrease, or remain the same depending on fuel type, firing, and operating conditions. Also, depending on the gasification technology, the presence of more tar or oil in the product gas may be problematic [1, 6]. Co-firing mixed feedstock with high chlorine content can increase corrosion in the system.

The ash coming from coal gasification processes is different from that coming from biomass gasification processes. When coal and biomass are gasified separately and ashes are kept separate, coal ash is generally used for the construction industry and biomass ash is recycled to the biomass origin or used as fertilizer, a building material, or as a fuel for power and heat generation. The last option is possible only for biomass ash with a high energy content such as that from a fluidized bed gasification process. Biomass ash is often sent to a landfill for disposal. When coal and biomass are gasified together, the combined ash may not be useful for concrete and other construction industry applications. The slagging behavior of the combined ash of coal and biomass in the gasifier may also have a negative impact [1, 6]. The hydrophilic character of biomass and alkaline metals in biomass ash can also create fouling of heat transfer surfaces within gasifiers.

The effects of various physical properties and chemical constituents of biomass on the various aspects of gasification process are summarized in Table 7.4. These and other constraints of mixed feedstock processing as well as possible solutions are further discussed in subsequent sections.

TABLE 7.4

Effects of Various Physical and Chemical Properties of Biomass on Gasification Process

Nature of Effects	Relevant Physical or Chemical Properties
Ash utilization/disposal	Ash content, K,Mg,Ca,P, heavy metals
Ash melting temperature	K,Na,Mg,Ca,P
Dust and particle emissions	Ash content, particle dimension and size distribution
Aerosol formation and emission of pollutants	F,K,Na,heavy metals, N,S
Corrosion	Cl,S,F,K,Na
GCV (positive or negative)	C(positive), H(positive), and O(negative)
Fuel logistics	Bulk density (for storage, transport, and handling), moisture content (for storage), particle dimension and size distribution (for feeding system and combustion technology)
Moisture content and drying	Particle dimension and size distribution, moisture content
NCV (low)	Moisture content
Self-ignition	Moisture content

Source: Modified from Maciejewska et al. 2006. *Co-Firing of Biomass with Coal: Constraints and Role of Biomass Pre-Treatment*, DG JRC Institute for Energy Report, EUR 22461 EN.

7.3 Transportation, Storage, and Pretreatment

In any mixed feedstock strategy, the transportation of raw materials to the plant is very important. Unlike coal, biomass is difficult and expensive to transport because of its low mass and energy densities (see Table 7.5). Generally, biomass is dispersed over a large area and will require to be transported to a central location from numerous places. Unless the plant is built where a large amount of biomass is easily available, a question of whether to prepare the biomass on site or transport it to a common location from various sites and then prepare at this common site requires a careful assessment. If biomass is transported a long distance under a natural environment, its biological decay may also occur. The on-site storage of unprepared biomass can lead to its biological decay releasing some heat. Low density of biomass necessitates the requirement of a relatively large storage space for biomass compare to coal.

The transportation and storage of waste materials is also an issue. MSW can be collected and delivered to a common location, but its amount depends on the size of the municipality. For densely populated areas this can work well for a large-size plant; however, for rural locations a supply of sufficient MSW feedstock may become an issue. The same principle applies to paper and pulp, plastic and polymer waste, rubber tires, and so on. In general, biomass and waste transportation and storage can be more expensive than that for coal.

7.3.1 Pretreatment

The goal of any gasification technology processing a mixed feedstock for specific applications such as IGCC, syngas generation, and the like is threefold. First, the gasifier operation and performance meet the desired

TABLE 7.5

Mass and Energy Densities of Various Feedstock

Feed	Bulk Density (Kg/M³)	Energy Density (GJ/M³)
Woodchips	400	8
Straw	100	2
Charcoal	300	9
Pyrolysis-oil	1200	30
Char-water slurry	1000 (50/50 mixture)	15 (50/50 mixture)
	1150 (20/80 mixture)	26 (20/80 mixture)

Source: Maciejewska et al. 2006. *Co-Firing of Biomass with Coal: Constraints and Role of Biomass Pre-Treatment*, DG JRC Institute for Energy Report, EUR 22461 EN, and Ratafia-Brown et al. 2007. *Assessment of Technologies for Co-Converting Coal and Biomass to Clean Syngas-Task 2 Report (RDS)*, NETL report (May 10).

objectives. Second, the products must meet the requirements of all the downstream operations such as gas turbine or FT catalysts, and so on, and third the gasifier must operate within its design limits. This goal is affected by the characteristics and properties as well as any synergistic effects of the mixed feedstock, composition of the total feed, the approach and equipment used to co-feed and the designs of the gasifier and the syngas clean-up equipment.

When possible, it is more economical to process a mixed feedstock in the process already being used for a single feedstock. For coal and biomass mixtures, it is more likely that coal processing plants will be retrofitted to accommodate the use of biomass in the feed. The pretreatment should allow this to happen as inexpensively as possible. Compared to coal, biomass has higher H/C and O/C contents (see Figure 7.1) which make biomass more active and easily degradable (more susceptible to fungal attack) when subject to the natural environment. Biomass also contains some inorganic elements (such as K, Na, Cl, Br, P, S, Ca, Mg, Si, Al, Fe, etc.) that if not removed can accumulate in the remaining solids of the various thermochemical processes causing slagging, reactor blockage, and other problems or end up with product gas requiring downstream cleaning operations. The ash from biomass gasification is different from that of coal gasification. Water removal from the biomass is also important for energy efficiency of the gasifier operation.

Pretreatment of biomass for its use in mixed feedstock is done for various reasons. A wide variety of biomass streams often does not match with narrow fuel specifications of feeding systems and the desired conversion processes. Other reasons are to reduce the plant's investment, maintenance, and personnel costs by using homogeneous fuel that is suitable for automatic fuel feeding and to reduce the need to invest in complex and novel gasification

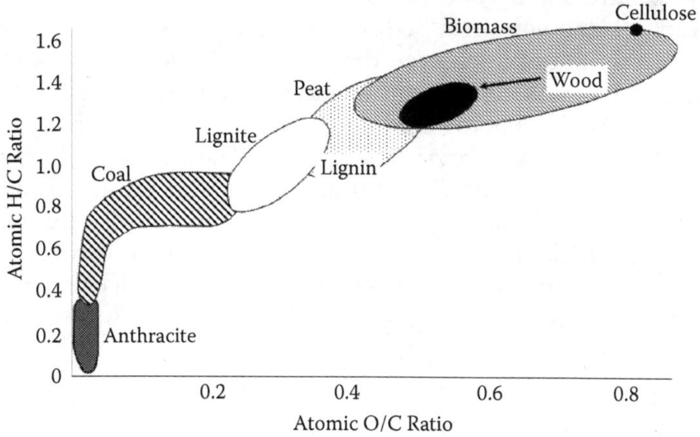

FIGURE 7.1
Van Krevelen plot illustrating the composition of various fuel sources. (Adapted from Hustad and Barrio. 2000. Biomass. *IFRF Online Combustion Handbook, Combustion File No. 23, Version No. 2, IFRF.* http://www.handbook.ifrf.net/handbook/cf.html?id=2).

systems. Pretreatment will also reduce the cost of transportation, storage, and handling of biomass. When coal and biomass are fed together as a mixture into a gasifier, the process of feeding must be uniform, consistent, and one that allows for easy particle fluidization in any type of gasifier. With popular entrained bed gasifiers, dry feeding is preferred because it allows maximum flexibility in allowable operating conditions and composition of the coal–biomass mixture feedstock. In addition to dry feeding a coal–biomass mixture successfully, the biomass needs to be prepared such that it forms a homogeneous mixture with coal.

The choice of pretreatment method used often depends on the downstream use of biomass as well as the local conditions and the need. As shown later, the choice of pretreatment method will also depend on the choice of the process configuration. The three most important parameters in feeding mixed feedstock are particle size and uniformity, level of inorganic impurities in the biomass, and moisture content of the biomass. The particle size affects the feeding and fluidization process within the reactor. A level of inorganics and other elements affects the quality of syngas and ash (and slag) formed within the reactor and the moisture content affects the energy efficiency of the gasifier. Various methods used to pretreat biomass for these purposes and their advantages and disadvantages are described in Table 7.6 [17]. The inorganic impurities can be partially removed by the leaching process mentioned in Table 7.6 [17]. Washing biomass with hot (around 60–80°C) water will largely remove ions of K, Na, Cl, and Br and somewhat remove the elements P, S, Ca, and Mg. Most other elements including Si, Fe, and Al will remain in the biomass. Process configuration and the downstream treatments are important for handling impurities in gas and solid product streams. The best method to handle volatile matter and particle size is torrefaction with or without pelletization. For an entrained bed reactor, a particle size of 1 mm. or less is required and this can be achieved through torrefaction. Sizing and milling of herbaceous biomass such as straw and switchgrass is very difficult and can be achieved via use of torrefaction. Pelletization of compressed biomass is possible, but it does not provide the additional benefits associated with the prevention of water absorption and fungal attack during storage and transportation of biomass that are provided by torrefaction. Furthermore, biomass pelletization alone does not easily allow the generation of fine particles on the order of several microns. In contrast, pyrolysis is a chemical process that significantly breaks down the chemical bonds within the biomass. It, however, requires biomass feeding in a slurry form.

7.3.1.1 Torrefaction

Traditionally, *torrefaction* is a thermal process for roasting biomass operated at 200° to 300° and for a relatively long residence time (30 to 60 minutes) under an inert atmospheric condition. The name torrefaction is adapted from the process used to roast coffee beans, which is performed at lower temperatures

TABLE 7.6

Summary of Advantages and Disadvantages of Various Biomass Pretreatments

Biomass Pre-treatment	Advantages	Disadvantages
Sizing (grinding, chipping, chunking, milling)	• Adjusts the feedstock to the size requirement of the downstream use.	• Nonbrittle character of biomass creates problems for sizing. • Should be done before transportation but storage of sized materials increase dry matter losses and microbiological activities leading to GHG (CH_4, N_2O) emissions.
Drying	• Reduces dry matter losses, decomposition, self-ignition, and fungi developments during storage. • Increases potential energy input for steam generation.	• Natural drying is weather dependent; drying in dryers requires sizing.
Bailing	• Better for storage and transportation; higher density and lower moisture content.	• Cannot be used without sizing for gasification.
Briquetting	• Higher energy density, possibility for more efficient transport and storage. • Possibility for utilization of coal infrastructure for storage, milling, and feeding; rate of combustion comparable with coal. • Reduces spontaneous combustion.	• Easy moisture uptake leading to biological degradation and losses of structure – require special storage conditions. Hydrophobic agents can be added to briquetting process, but increase their costs significantly.
Washing/leaching	• Reduction of corrosion, slagging, fouling, sintering, and agglomeration of the bed-washing is especially important in case of herbaceous feedstock. • Reduced wearing-out of equipment, and system shut down risks.	• Increased moisture content of biomass. • Addition of dolomite or kaolin, which increase ash melting point, can also reduce negative effects of alkali compounds.
Pelletizing	• Higher energy density leads to better transportation, storage, and grinding and reduced health risks. • Possible utilization of coal infrastructure for feeding and milling (permits automatic handling and feeding).	• Sensitive to mechanical damaging and can absorb moisture and swell, loose shape and consistency. • Demanding with regard to storage conditions.

(Continued)

TABLE 7.6 (CONTINUED)

Summary of Advantages and Disadvantages of Various Biomass Pretreatments

Biomass Pre-treatment	Advantages	Disadvantages
Torrefaction	• Possibility for utilization of coal infrastructure for feeding and milling. • Improved hydrophobic nature – easy and safe storage, biological degradation almost impossible. • Improved grinding properties resulting in reduction of power consumption during sizing. • Increased uniformity and durability.	• No commercial process. • Torrefied biomass has low volumetric energy density.
TOP process	• Combines the advantages of torrefaction and pelletizing. • Better volumetric energy density leading to better storage and cheaper transportation. • Desired production capacity can be established with smaller equipment. • Easy utilization of coal infrastructure for feeding and milling.	• No commercial process. • Does not address the problems related to biomass chemical propertied, that is, corrosion, slagging, fouling, sintering, or agglomeration.

Source: Modified from Shah and Gardner, in press. Biomass Torrefaction: Applications in Renewable Energy and Fuels. In *Encylopedia of Chemical Processes*, Boca Raton, FL: CRC Press.

and in the presence of air. Nevertheless, the important mechanical effect of torrefaction on biomass is similar to the effect on coffee beans. In the open literature, torrefaction is also referred to as roasting, slow-and-mild pyrolysis, wood cooking, and high-temperature drying [18–25]. The drying and grinding of biomass is not as easy as torrefaction and grinding due to the physical nature of biomass.

The process of torrefaction dehydrates and depolymerizes the long polysaccharide chains of biomass. This results in a product that is hydrophobic and has a higher energy density and improved grinding and combusting capabilities [26–32]. This process is best illustrated through the Van Krevelen plot shown in Figure 7.1 [18–24]. The figure illustrates that torrefaction results in the reduction of oxygen content and increased heating value of the biomass. Generally, during torrefaction an increase in both mass and energy density occurs because about 30% (by weight) of the biomass is transformed

TABLE 7.7

Aspects of Torrefied Biomass for Gasification and Other Applications

Torrefied Product
Has lower moisture content and higher heating value
Is easy to store and transport
Is hydrophobic, does not gain humidity in storage and transportation
Is less susceptible to fungal attack
Is easy to burn, forms less smoke and ignites faster
Significantly conserves the chemical energy in biomass
Has heating value (11,000 BTU/lb) that compares well with coal (12,000 BTU/lb)
Generates electricity with a similar efficiency to that of coal (35% fuel to electricity) and considerably higher than that of untreated biomass (23% fuel to electricity)
Has grindability similar to that of coal
Requires grinding energy 7.5 to 15 times less than that for untreated biomass for the same particle size
Has mill capacity 2 to 6.5 times higher compare to untreated biomass
Possesses better fluidization properties in the gasifiers
Is suitable for various applications in heating, fuel, steel and new materials manufacturing industries

Source: Bergman and Kiel, 2005. Torrefaction for biomass upgrading. *Proceedings of the Fourteenth European Biomass Conference and Exhibition,"* Paris, October, pp. 17–21. Bergman al. 2004. Torrefaction for entrained flow gasification of biomass. In: W.P.M. Van Swaaij, T. Fjällstrom, P.T. Helm, and P. Grassi (Eds.), *Proceedings of the Second World Biomass Conference on Biomass for Energy, Industry, and Climate Protection,* Rome, Italy, May 10-14, pp. 679–682, Energy Research Centre of the Netherlands (ECN), Petten, The Netherlands, Report No. ECN-RX--04-046; Bergman. 2005. *Combined Torrefaction and Pelletization: The TOP Process.* Energy Research Centre of the Netherlands (ECN), Petten, The Netherlands, Report No. ECN-C-05-073; and Bergman et al., 2004. *Torrefaction for Entrained Flow Gasification of Biomass.* Energy Research Centre of the Netherlands (ECN), Petten, the Netherlands, Report No. ECN-C-05-067.

into volatile gases. These gases carry only 10% of original biomass energy content [29, 30, 33–35]. This implies that during torrefaction, a substantial amount of chemical energy is transformed from the raw material to the product resulting in the enhanced fuel properties of the torrefied biomass. Mild pyrolysis of biomass results in gases such as H_2,CO,CO_2, CH_4,C_xH_y; the liquids such as toluene, benzene, H_2O, sugars, polysugars, acids, alcohol, furans, ketones, terpenes, phenols, fatty acids, waxes, and tannins and solids that contain char and ash.

As shown in Table 7.7, torrefied biomass possesses very valuable properties. It has lower moisture content and therefore higher heating value compared to untreated biomass. The storage and transportation capabilities of torrefied biomass are superior to those of untreated or only dried biomass. Torrefied biomass is hydrophobic and does not gain humidity in storage and transportation. It shows little water uptake on immersion (7–20% of mass) and is more stable and more resistant to fungal attack compared to charcoal and

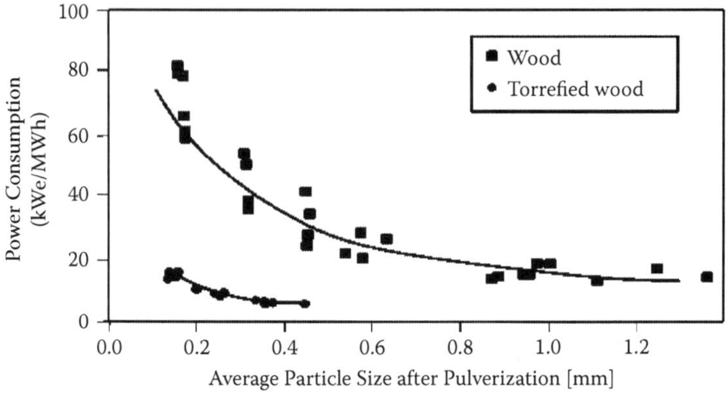

FIGURE 7.2
Power consumption for size reduction: untreated versus torrefied wood. (Modified from Van der Drift et al. 2004. *Entrained Flow Gasification of Biomass – Ash Behavior, Feeding Issues, and System Analysis,* Energy Research Center of the Netherlands (ECN), The Netherlands Report No. ECN-C-04-039, p. 58, April.)

an untreated biomass. Pelletization, by itself, produces biomass with higher mass density; however, the pellets are not hydrophobic and are susceptible to fungal attack. Torrefied biomass significantly conserves the chemical energy present in the biomass. The heating value of torrefied wood is approximately 11,000 BTU/lb and is nearly equal to that of a high volatile bituminous coal which is 12,000 BTU/lb. It generates electricity with an efficiency comparable to that of coal of approximately 35%, on a fuel to electricity basis [36, 37], and much higher than that of untreated biomass which has an efficiency of 23%, on a fuel to electricity basis [36, 37]. Bergmann et al. showed that torrefied biomass has better fluidization properties than that of untreated biomass, but similar to that of coal [38–41].

Untreated biomass requires many times the grinding energy (by a factor of 7.5 to 15) to achieve a similar particle size compared to torrefied biomass. This energy difference is significantly larger than the energy loss of biomass and energy supplied during torrefaction. The grindability of torrefied biomass versus that of untreated biomass is compared in Figure 7.2. The mill capacity of the torrefied biomass can also be as high as 6.5 times that of the untreated one. Finally, torrefied biomass is suitable for various applications such as working fuel, residential heating, new materials for the manufacture of fuel pellets, reducer in the steel smelting industry, the manufacture of charcoal and active carbon and gasification, and co-firing with other fuels in gasifiers, boilers, and so on [38–41]. Such wide usefulness makes torrefied biomass a valuable and marketable product.

Ratafia-Brown et al. [6] identified various feed preparation techniques needed for an entrained bed gasifier based on the nature of biomass and the nature of the feeding mechanism. For wet feeding, they proposed a pyrolysis process that produces a bioslurry which can be fed into the entrained bed gasifier by a feeder or an injector. This technique is analogous to the coal slurry feed used in the GE gasifier, and it is most appropriate for strawlike crops. For dry feeding they proposed three alternatives. For woody biomass, the feed can be milled (and cut) and broken down to size of 1 mm particles and fed to the entrained bed gasifier by either a screw feeder or a piston compressor. The process of milling can also be preceded by torrefaction. The torrefied biomass in this case can be fed to the gasifier by a pneumatic feeder, a screw feeder, or a piston compressor. For all other types of biomass they presented two options. One option is to follow the path of torrefaction, milling, and feeding just like the one for woody biomass. The other is to gasify the biomass in a pressurized fluidized bed to generate gas product and char and feed these materials to the entrained bed gasifier for the production of biosyngas.

7.4 Gasification Technologies

Over the last several decades most investigations on mixed feedstock were carried out for gasification technologies. This is because gasification technology in its different formats (i.e., combustion, pyrolysis, gasification, plasma technology, supercritical gasification, etc.) is very versatile and capable of generating heat and electricity as well as fuels, chemicals, and materials.

7.4.1 Literature Studies

Numerous outstanding reviews [43–63] have addressed the laboratory studies carried out on combustion, gasification, high-severity pyrolysis, plasma technology, and supercritical gasification for mixed feedstock. Sami, Annamalai, and Wooldridge [7] and Maciejewska et al. [1] have outlined a detailed review of combustion of mixed feedstock carried out at laboratory and demonstration scales. Similarly, Ricketts et al. [5] and Davidson [63] have presented outstanding reviews of the gasification, plasma technology, and pyrolysis of mixed feedstock. Some of recent gasification studies for mixed feedstock are summarized in Table 7.8. A summary of the technology developers in western Europe and the United States for gasification, pyrolysis, and plasma is given by Ricketts et al. [5].

The literature studies for mixed feedstock indicate that there seems to be very little or no synergistic effects on gasification kinetics between biomass and coal. The situation is different when polymeric waste is added into the mix. The compositions of the gases produced from this mixture are not simple additions of individual behaviors. More systematic work is needed to understand the nature of the synergy that may exist in polymeric and other

TABLE 7.8

Typical Studies of Gasification of Mixed Feedstock

Mixture	Authors	Comments
Coal and methane	Wu et al. 2011 [64]	Syngas production
Coal, biomass and iron ore	Kaupilla 2007 [65]	Syngas and steel production
Coal and biomass	Kim and Mitchell 2011 [66]	Supercritical water gasification
Coal and biomass	Kajitani et al. 2010 [67]	High temperature gasification
Coal and biomass	Seo et al. 2010 [68]	CFB reactor
Coal, biomass and plastic	Aznar et al. 2006 [69]	Catalysis of dolomite
coal and biomass	Williams et al. 2006 [70]	Environmental effects
Vacuum residue, plastics, coal, and petrocrop	Ahmaruzzaman and Sharma 2005 [71]	Kinetic studies
Waste plastics, HDPE and coal	Liu and Meuzelaar 1996 [72]	Catalytic kinetic studies
Coal and cattle manure	Sweeten et al. 2003 [73]	Co-firing to measure fuel quality
Coal and biomass	Chmielniak and Sciazko 2003 [74]	Methanol synthesis
Coal and biomass	Demirbas 2003 [75]	Sustainable co-firing
Residual biomass and poor coal	Pan et al. 2000 [76]	Fluidized bed gasification
Coal and biomass	Sjostrom et al. 1999 [77]	Promoted char reactivity
Coal and switchgrass	Brown et al. 2000 [78]	Catalytic effect on gasification
Biomass and waste filter	Sun et al. 2011 [79]	Steam gasification
Coal, activated carbon, coke and biomass	Gregg et al. 1980 [80]	Solar gasification
Coal and biomass	Prins et al. 2007 [81]	Thermodynamic efficiency
Coal and biomass	McLendon et al. 2004 [82]	High pressure gasification
Coal and woody biomass	Kumabe et al. 2007 [83]	Air and steam gasification
Coal and biomass	Andries et al. 1998 [84]	Pressurized fluidized bed
Coal and biomass	Sjostrom et al. 1992–1994 [85]	Synergistic effects in gasification
Coal and straw	Olsen et al. 1995 [86]	Pressurized gasification
Coal and industrial oil waste	Andre et al. 2005 [87]	Fluidized bed gasification
Coal and biomass	Uson et al. 2004 [88]	IGCC power plant operation
Coal and organic waste	Dayan et al. 2003 [89]	Simulation of two-chamber catalytic gasifier
Coal and biomass	Hirs 2000 [90]	Basic gasification study
Coal, biomass and plastics	Pinto et al. 2003 [91]	Parametric study of gasification
Coal and biomass chars	Backreedy et al. 2002 [92]	Modeling study of gasification
Biomass and fossil fuel	De Jong et al. 2003 [93]	Pressurized fluidized bed gasifier

(Continued)

TABLE 7.8 (CONTINUED)

Typical Studies of Gasification of Mixed Feedstock

Mixture	Authors	Comments
Biomass and plastics	Pinto et al. 2002 [94]	Parametric study of gasification
Biomass, charcoal and coal	Arcate 1998 [95]	Co-firing assessment
Coal and biomass	Collot et al. 1998 [96]	Pressurized fixed bed gasification
Coal and biomass	De Jong et al. 1998 [97]	Pressurized fluidized bed gasifier
Coal and biomass	Uson and Valero 2006 [98]	IGCC power plants study
Coal and biomass	Yan et al. 2005 [99]	Supercritical water gasification
Coal and biomass	Brage et al. 1995 [100]	Characteristics of tar

kinds of feedstock. The basic issues, as described in this section, for mixed feedstock gasification are the effects of mixture on the nature of the gases and solids produced as well as on the reactor and the downstream operations. These issues are extensively addressed in the literature, and they are detailed in the remaining part of this section.

7.4.2 Reactor Technology

For a mixed feedstock, the success of the gasification process depends on the reactor technology that carries out the desired objective with variable feed properties.

7.4.2.1 Combustion

For coal and biomass mixture combustion, in general three types of combustion systems can be identified: fixed bed combustion (i.e., stoker or grate combustion), fluidized bed combustion (bubbling and circulating), and pulverized fuel combustion (dust combustion). The rotary kiln outlined in Chapter 6 is mostly used for the hazardous waste.

A fixed bed combustor can handle both small- and medium-size coal and biomass particles. The grate-fired furnace and underfeed stokers can also handle shredded tires, plastic and polymer waste, and shredded MSW. In this type of reactor, drying, gasification, and charcoal combustion take place in the primary combustion chamber followed by incineration (usually in the separate combustion zone) of the gases produced after the addition of extra air [101]. Different types of grate furnace; fixed, moving, traveling, rotating, and vibrating are available with a maximum capacity of 20 MWe. Underfeed stokers are used in small- and medium-scale systems up to boiler capacity of 6 MWe [101].

The fixed bed boilers are normally used for parallel and indirect co-firing [1]. For a capacity of less than 20 MWe, they have low investment and operating costs. They can be used for varying particle sizes (except fines) and for any kind of wood with moisture content up to 10–60 wt% and high ash content. Although fixed bed boilers can be used for a mixture of woods, they cannot be used for a blend of wood and straw or grass which have different combustion behaviors and ash melting point. For a nonhomogeneous mixture, an increase in temperature may cause ash melting and resulting corrosion [102].

The underfeed stoker has low investment cost for a small boiler (<6 MWe) [1] and low emissions at partial load operations due to good fuel dosing [101]. This kind of combustor is only suitable for fuels with low ash content and high ash melting point. It also has less flexibility with respect to particle size (<50 mm).

In fluidized bed combustion, the bed contains a mixture of inert material (about 90–98% silica sand and dolomite) and fuel (about 2–10% of coal and biomass) particles, and they are fluidized by flowing air that is passed through a perforated plate at the bottom of the bed. The bed can either be operated as a one pass-through bubbling fluidized bed or as a circulating fluidized bed where the unconverted solids are recycled back in the reactor. The bed material provides high thermal inertia and stability in the combustion process. Good mixing and heat transfer by materials give good temperature control. The particle size for a bubbling fluidized bed (BFB) is generally less than 80 mm and that for circulating fluidized bed (CFB) is less than 40 mm [101, 103]. Low combustion temperature (800–900°C) prevents ash sintering and resulting defluidization [101, 103]. The bed materials remain at the bottom of the reactor in BFB, whereas in CFB, the material is carried upwards with flue gas and separated in a hot cyclone or U-beam separator, and fed back into the combustion chamber [101]. In CFB, light soot (unburned carbon) may leave cyclone along with ash. This contaminated ash may restrict its further use [104].

Fluidized bed combustors can be converted from coal into biomass/coal co-combustion with a very small investment, and they have large flexibility in calorific value and moisture and ash contents of the feedstock. The combustion temperature in these reactors is low, and they emit low CO (<50 mg/Nm^3) and NO_x (<70 mg/MJ) which can be further reduced to 10 mg/MJ after using SCR [1]. The combustors can have efficiency up to 90% even with low-grade fuels. Sulfur can be removed by a direct injection of limestone in the bed. For a mixed feedstock, if a common feeder for the mixture is not possible, the installment of separate feeders for different feedstock may be expensive. Slagging and fouling on the combustor walls and tubes can occur when the feedstock has high alkali content. High alkaline and aluminum contents can also cause bed agglomeration. Similarly, high chlorine content can cause corrosion on heat transfer surfaces [1].

Fluidized bed combustors are only cost-effective for capacity >20 MWe for bubbling fluidized bed and >39 MWe for the circulating fluidized bed [1].

These types of combustors have a low flexibility for particle size and bed materials can be lost with ash. They also have a high dust load in flue gas. Low temperature makes fuel combustion incomplete and in CFB unburned carbon content can appear in the ash [1, 101, 103, 104].

Pulverized fuel or dust combustion systems are fed pneumatically with fuels such as sawdust and fine shavings [104]. Fuel quality needs to be rather constant with low moisture content (<20 wt%) and particle size of 10–20 mm. This type of combustor gives increased efficiency due to low oxygen excess and when appropriate burners are used, it gives high NO_x reduction [1].

7.4.2.2 Gasification

Although there are various types of gasification reactors with different designs and operating characteristics that can be used for a coal/biomass mixture, they can be divided into three main categories: moving (sometimes called fixed) bed, fluidized bed, and entrained bed. Among other factors, these three types of gasifier are mainly distinguished by the reactor temperature, particle size, and the gas flow rate (or residence time). The entrained bed requires the finest particle size, highest gas flow rate, and temperature; the fluidized bed requires moderate particle size, temperature, and gas flow rate; and the moving bed can process large particle sizes and requires a low gas flow rate and temperature. These reactors can be operated with either air or oxygen, however, the oxygen-blown gasifier minimizes downstream syngas volume and vessel sizes. It also allows CO_2 to be more easily and cheaply separated. Because of these advantages, for all future IGCC plants, the oxygen-blown gasifier is preferred. In general, the oxidant requirement follows the order of: entrained bed > fluidized bed > moving bed. The important characteristics of these three basic types of gasifiers were well assessed by Ratafia-Brown et al. [6] and their assessment is briefly summarized below.

7.4.2.2.1 Entrained Bed Gasifier

Currently, large-scale commercial coal gasification technology is dominated by high-temperature, high-pressure entrained flow gasifiers. In fact, about 85% of existing commercial coal gasification reactors are entrained bed reactors. As shown in the previous section, recent successful tests with crops and biomass [6] indicate that this type of gasifier is reasonably well suited to the gasification of mixed feedstock. They eliminate tar formation and are not greatly affected by ash content differences with coal/coke; however, the Buggenum plant [6] noted a tendency to foul the syngas cooler when using sewage sludge. The primary drawback of this type of gasifier for the processing of mixed feedstock is the handling of feed materials that are difficult to pulverize such as switchgrass and straw, among others, as well as some woody biomass and high-pressure feed into the gasifier using either a slurry or dry feed system. The dry feed system requires pulverized particles of about 100 mesh which may be difficult to achieve with certain types of biomass. The reactor operates at temperatures in excess of about 1250–1300°C. Such a high temperature results in the production of syngas (CO and H_2) with no methane.

The entrained flow gasifier (a) is able to gasify all coal regardless of coal rank, caking characteristics, or amount of coal fines; (b) has uniform temperature and short residence time; and (c) slagging operation with some entrainment of molten slag in the raw gas. Although all new IGCC plant reactors will be entrained flow reactors, the design of various entrained flow reactors may vary depending on whether feed is dry or slurry, the internal design to handle the hot reaction mixture, and heat recovery configuration. One of the key technical issues is the method for cooling the product gases. Recent studies have shown [105] that this type of reactor is very suitable for mixed feedstock up to about 30% of biomass. Due to the high oxygen content of the biomass, the mixed feedstock will produce more water and carbon dioxide and less carbon monoxide, lower syngas heat content, and in general lower sulfur and ash content. In processing some mixed feedstock that contains biomass such as straw and other herbaceous biomass which contains chlorine can cause corrosion and biomass that contains calcium and sodium can alter ash melting point.

7.4.2.2.2 Fluidized Bed Gasifier

Fluidized bed gasifiers have a lower gas flowrate, higher residence time, uniform temperature, and good mixing which avoids clinker formation and possible defluidization. They are well suited for active reactants such as biomass and low-rank coal. If char particles are entrained, they can be recycled back by a cyclone. Ash particles that are removed at the bottom can exchange heat with incoming recycled gas and steam. Fluidized bed reactors require moderate oxygen and steam requirements and extensive char recycling. They can accept a wide variety of feedstock (e.g., solid waste, wood, and high ash coals) with a larger particle size then what is normally required for the entrained flow gasifiers. Commercial applications of fluidized bed reactors are high-temperature Winkler (HTW) and KRW designs, the latter gasifier is a part of the Pinon Pine Coal gasification plant [6].

Different fluidized bed gasifiers can vary in ash conditions (dry ash or agglomerated) and design configurations to improve char use. Both dry ash and agglomerating fluidized beds are operated with crushed feed (about ¼ in. size). The acceptability of fines is good for the dry ash bed and better for the agglomerating bed. The reactor is generally operated at 925–1025°C. The caking coal may be processed in a dry ash fluidized bed, but it can certainly be processed in an agglomerating bed. The dry ash bed is preferred for low-rank coal and agglomerating bed can process any rank of coal. The main technical issue of the fluidized bed is the carbon conversion per pass. For this reason, for a less reactive coal mixture, a circulating fluidized bed is preferred.

Mixed feedstock has been tested in fluidized bed reactors at the Royal Institute of Technology at Stockholm, Sweden. They found that for a coal–biomass mixture, the char from woody biomass is very sensitive to the thermal annealing effect which occurred at low (650°C) temperature and short soak time (less than 8 min). A mixture of birch and coal gasification [106] showed synergies by an enhanced gasification rate in the presence of oxygen and reduction of char formation. Also both tar and ammonia formations were lowered in the mixture gasification. Currently, there are very few large fluidized bed gasifiers in operation.

7.4.2.2.3 Moving Bed Gasifier

The British Gas Lurgi gasifier (BGL), which is a moving bed design type, is currently the only large-scale (>50 MWe) gasification technology with significant operational experience in the processing of mixed feedstock of coal and waste materials. BGL has the most operational experience in processing fuels of widely differing fuel properties. This technology is now owned by Advantica, a company leader in the large-scale gasification of variable property feedstock. Global Energy (part of ConocoPhillips since 2003) is the U.S. leader in mixed feedstock gasification. The BGL technology creating new plants in the United States and Europe for moving bed gasification of the materials which include coal, municipal refuse, and sewage sludge feedstock. Global Energy also operates the SVZ plant in Germany and a plant in Scotland alongside its natural gas-fired power plant for co-gasification using the BGL gasifier [6].

In a moving bed gasifier (often called fixed bed gasification), large particles (about 2 in.) of feedstock move down while gases move upward (or sometimes downward). The acceptability of fines is limited for a dry ash moving bed, although it is better for a slagging moving bed reactor. At the top of the bed, solids get heated by hot gases which get cooled before leaving the reactor. This is considered a drying zone. The solids then go through carbonization, gasification, and combustion zones as they move downward [6]. At the bottom of the reactor, oxygen reacts with the remaining char. In a dry-ash version (i.e., Lurgi dry ash gasifier) the temperature is moderated to below ash slagging temperature by reaction of char with steam. The ash below the combustion zone is cooled by the entering steam and oxidant. In a slagging version, less steam is used which maintains the temperature above the ash slagging temperature. The temperature in the reactor may vary from 300–650°C depending on the nature of the feedstock. The dry ash moving bed reactor prefers low-rank coal and slagging moving bed reactor prefers high-rank coal [6]. Both the dry ash (with some modifications) and slagging moving bed accept caking coals. The moving bed reactor technology has (a) a low oxidant requirement which produces hydrocarbon liquids such as oils and tars, (b) high "cold gas" thermal efficiency, when the heating values of the hydrocarbon liquids are included, and (c) a high steam requirement for the dry ash moving bed and low steam requirement for the slagging moving bed [6]. The key technical issue with moving bed technology is the utilization of fines and hydrocarbon liquids coming from the product gas. Two existing commercial processes using a moving bed reactor are the Lurgi dry-ash gasifier, which is used for town gas production and chemicals from coal in South Africa, and the BG Lurgi (BGL) slagging gasifier, which is currently used to process solid waste and a mixture of coal and waste.

In summary, although updraft and downdraft gasifiers have been used for small-scale applications, fluid bed, circulating fluidized bed, and entrained bed gasifiers are extensively used for large-scale operations. Updraft generates a high amount of tar, whereas the fluid bed and CFB generate a medium amount of tar and downdraft and the entrained bed generates a low amount of tar. Syngas can be produced by fluid bed, CFB, and entrained bed operations. Fluid bed and CFB have higher particle loading and can use larger particle sizes whereas the entrained bed requires a large amount of carrier gas and it has a particle size limit [6].

7.4.2.3 Plasma Gasifier

Plasma gasification uses an external electric source for heat and this results in the conversion of feedstock to fuel gas and the elimination of all tars, chars, and dioxins due to high temperatures. Unlike the reactors described above, plasma reactors are applied more to MSW and as shown in an earlier section they can be easily adapted to process a mixed feedstock including other types of waste such as rubber tires, polymer and plastic waste, and coal waste. The amount of electricity and temperature in the plasma reactor depends on the reactor configuration, energy content of the feedstock, and amount of air (or oxygen) allowed in the reactor. A plasma reactor can produce a variety of products by careful control of reaction conditions.

Westinghouse Plasma manufactures and supplies plasma torches to the industry. The Westinghouse Plasma gasifier is essentially a classic downflow moving bed system operating at 1 atm and about 2,300°F. Various types of plasma reactors described in the previous chapters can also be used for processing the mixed feedstock.

In addition to Westinghouse, the Solena group uses the Solena gasifier which locates the torch at the bottom of the gasifier to vitrify inorganics in the feed, forming glass aggregate. This reactor uses less energy, and it also uses a carbon-based catalyst to enhance gasification. This group is planning to develop a 10 MWe plant in Malaysia using Padi husks and a number of 130 MWe integrated plasma gasification combined cycle plants in the eastern United States to use waste coal and coal fines in partnership with Stone and Webster [107, 108]. Startech Environmental Corporation (SEC) is also developing a plasma gasification unit that feeds shredded materials using a pump, screw, or ram, depending on the consistency of the feed. In this reactor, the plasma torch is located at the top of the reactor and is then directed to dissociate organics and to melt inorganics in the feed, forming clean gas and glass aggregate. Acid gases, volatile metals, and particulate matter are removed from the cooled gasifier effluent. Starcell^{TM} technology can also produce hydrogen. SEC is also planning a 200 T/D MSW plant in Panama jointly with Hydro-Chem (a division of Linde) and a 10T/D plant in China [6].

7.4.3 Handling of Product Streams

The exit gases and solids coming out of any type of gasification unit (combustion, gasification, high severity pyrolysis, plasma technology, etc.) should not only meet EPA standards but also facilitate subsequent processing of gases (such as for Fischer–Tropsch) and solids (for the construction and fertilizer industries).

7.4.3.1 Syngas Treatment

In a combustion process, gases do not have a significant fuel value (they mostly contain CO_2, O_2, H_2O, and N_2) and as shown earlier, the main objective of gas purification is to remove any impurities and particulate materials. Earlier sections outline the steps taken by commercial processes for this purpose. The extent to which impurities are present in syngas (by gasification, high severity pyrolysis, and plasma technology) will be a function of the nature of the process and the nature of the feedstock. The presence of minor impurities such as sulfur, chlorine, ammonia, and soot will, however, be inevitable. Because concentrations of these compounds normally exceed specification of a gas turbine or a catalytic synthesis reactor such as Fischer–Tropsch, methanol, and the like, which process the syngas (see Table 7.9), the gas cleaning is necessary.

The impurities in syngas can be poisonous to the catalysts used in FT and methanol syntheses. The required minimum for these impurities in syngas is outlined in Table 7.9. The definition of gas cleaning is therefore based on the economic consideration of investment in cleaning versus accepting a lower catalyst life. Generally, an investment in cleaning is less expensive than replacing expensive catalyst materials. Co-gasification adds another complexity because different products coming out of coal and biomass need to be handled. Additional compounds will include chlorides, sulfur compounds, and very toxic carcinogenic tars (unless tar is recycled back in the reactor such as in the slagging entrained flow reactor). For mixed feedstock of coal and biomass, higher hydrogen chloride, and more toxic organic compounds need to be treated compared to those found in coal gasification alone. The gas cleaning associated with the mixed feedstock is not the same as that for a single feedstock. As long as there is a single syngas purification system, it needs to handle impurities coming from all components of the mixed feed stock whether the components of mixed feedstock are gasified together or separately.

TABLE 7.9

Allowable Concentrations of Contaminants in Syngas for Catalytic Systems

Syngas Contaminants	Maximum Allowable Concentrations
Sulfur, nitrogen and oxygen compounds as well as other hetero-organic compounds (e.g., $H_2S + COS + CS_2$; $NH_3 + HCN$; etc.)	Total in each category <1 ppmv
Halogenated compounds and alkali metals (e.g., HCl + HBr + HF; Na + K)	Total in each category <10 ppmv
Solid particles such as ash and soot	Almost completely removed
Other compounds such as CO_2, N_2, and higher hydrocarbons	Lower the better; however, "loose maximum" CH_4 of total up to 15 vol % is acceptable
Tar (complex liquid and solid hydrocarbons)	Should be at the level that no condensation occurs when syngas is pressurized to the required pressure of approx. 25–60 atm for Fischer–Tropsch synthesis

Source: Ratafia-Brown et al. 2007. *Assessment of Technologies for Co-converting Coal and Biomass to Clean Syngas-Task 2 Report (RDS)*, NETL report (May 10); Boerrigter, H. et al. 2005. *OLGA Tar Removal Technology-Proof of Concept for Application in Integrated Biomass Gasification Combined Heat and Power (CHP) Systems*, ECN-C-05-009 (January); and Boerrigter et al. 2004. *Gas Cleaning for Integrated Biomass Gasification (BG) and Fischer-Tropsch (FT) Systems*, ECN, Petten,The Netherlans, ECN- report number ECN-C-04-056, 59 pp.

Sulfur compounds generally result in the production of sulfur dioxide, and this can be removed by several existing processes. More development is, however, being pursued to remove sulfur at high syngas temperature to improve the energy efficiency of an integrated FT plant. Co-gasification of coal with low sulfur substances such as biomass or waste reduces hydrogen sulfide partial pressure in an FT reactor and hence makes the gas cleaning system larger. The tar from wood gasification and hydrochloric acid from waste gasification may further complicate the issue, as will the presence of liquid products (e.g., tar) in syngas.

One problem for syngas treatment for gasification of mixed feedstock is the production of a relatively large amount of tar and the possibility of this condensing as the gas is cooled. For biomass, tar condensation occurs in the temperature range of 200–500°C and this can rapidly blind the filters. The tar generation is not a problem for an entrained bed gasifier because the temperature in such a gasifier is greater than about 2,300°F and tar cracks at such a high temperature. In processing a mixed feedstock, the composition and rates of feedstock should be designed such that the tar formation is not an issue [111]. For an entrained bed gasifier, the biosyngas is cleaned with standard techniques used for fossil syngas: dust filters, wet scrubbing techniques for the removal of NH_3 and HCl, and zinc oxide (ZnO) filters for the removal of H_2S. After adjusting the required H_2/CO ratio and CO_2 removal (which may require a steam or a tri-reforming step), the gas is compressed to the required FT synthesis pressure (approximately 40 atm) and fed to the FT reactor.

In other types of gasifiers where generally the temperature in the gasifier is low, tar in the exit gas stream needs to be handled. For example, in a CFB gasifier, tar is removed by either installing a tar cracking unit operated at 1,300°C or by using OLGA tar removal technology. In a tar cracker all organic compounds in the product gas (i.e., tar, BTX, CH_4, and C_2-hydrocarbons) are destroyed to produce additional syngas. In the OLGA technique, tar and BTX are separated and recycled back into the gasifier until they are extinct. Lower hydrocarbons are not removed by OLGA technology. Hence, in this case wet cleaning and the filter process are followed by a reforming step to convert lower hydrocarbons into additional syngas. There are reports [109, 112] that the tar problem can also be reduced by an addition of dolomite in the gasification process. In both cases, the gas purification step is followed by a gas conditioning step before gas is introduced in the FT reactor.

In co-gasification, gas is used for further conversion to products via FT and methanol synthesis, and solids are directly used for industrial purposes. The exact species formed from the ash constituents when biomass is gasified depends on the reactor temperature and oxygen partial pressure. For example, in the IGCC system, the sulfur species will be present as sulfides rather than as sulfates (alkalis and alkaline earths as sulfides; Fe as FeS_x), whereas Al^-- and Si^--containing species probably would present as oxides. This means the gas coming from the gasifier may encounter contaminants in their reduced form. To some extent this also depends upon the interaction between solids and gas as they travel through the reactor. It is therefore important to monitor gas and solids at several locations within the various process streams.

7.4.3.2 Solids Handling

As mentioned before, the co-firing of coal and biomass creates problems for the use of mixed ash that contains fly ash as well as inorganic materials. Coal ash is used in the construction industries and biomass ash is used in the fertilizer industry. The ash content of a feedstock (biomass) has a major impact on gasifier operation. This type of impact depends on the gasifier type: slagging or nonslagging. For a moving bed (or fixed bed) nonslagging operation, ash below its fusion temperature forms clinker which stops the flow of feedstock within the reactor. Although ash fusion temperature depends on the amount of sodium present in the feedstock, clinkers can become a significant problem for reactor operation. The ash can also affect the fuel's reaction response. For high sodium content woody biomass such as birchwood, the formation of sticky sodium silicates by the interaction between sodium and silica bed materials used in a circulating fluidized bed can cause agglomeration and potential interruption in the gasification operation.

For an entrained flow gasifier, ash from biomass does not melt even at temperatures of 1,300–1,500°C because ash is rich in CaO, and alkali metals are removed by the gas phase. Despite the high melting temperature of ash, the slagging entrained flow reactor is preferred because melt can never be avoided and the slagging entrained flow gasifier is more fuel flexible. Slagging co-gasification may require fluxing materials such as silica or clay in order to obtain proper slag properties at reasonable temperatures. By adding flux material to biomass, coal-based slag (generally coal ash with added limestone) is mimicked and slag properties become comparable. Solids handling is one of the important reasons why new gasifiers tend to be high-pressure, high-temperature entrained bed gasifiers. Such gasifiers will also handle solids produced with the mixed feedstock well.

7.4.4 Process Configurations for Gasification Technologies

Mixed feedstock process options for gasification technology depend on the nature of the technology.

7.4.4.1 Combustion

There are basically three process configurations for co-combusting coal and biomass [1]. The most popular option is direct co-firing where biomass and coal are fed together in the same combustor. This is because with this method an existing coal power plant can be converted to a co-firing plant with a relatively low financial investment. In this method, both coal and biomass are either fed together or separately in the same combustor. When they are fed together, biomass feed can either be prepared jointly with coal or prepared separately and then injected in the combustor using the coal injection and burning system. When they are injected separately, a new and dedicated biomass feed preparation and sometimes also burning equipment is used. Thus there are three separate subconfigurations for feed preparation and burning within a co-firing system. Ultimately, the common combustor unit produces steam from both coal and biomass for power generation.

Many different types of biomass can be co-fired with coal. These include wood, residues from forestry and related industries, agricultural residues, and various biomass in refined forms such as pellets (RDF). Energy crops except oil, sugar, and starch can also be used for co-firing [1]. Constraints in the use of co-firing originate from feedstock properties. Raw biomass has high moisture content and low bulk and energy densities compared to coal, a low ash melting point, higher chlorine and oxygen content, and a hydrophilic and nonfriable character. The constraints related to co-firing include fuel preparation, handling and storage, possible decrease in overall efficiency, deposit formation (slagging and fouling) agglomeration, corrosion or erosion, and ash utilization. The degree of these constraints depends on the quality and percentage of biomass in the fuel blend, type of combustor used, co-firing configuration of the system (as mentioned above), and the properties of coal. The importance of these problems increases with increased biomass/coal ratios and the use of poor quality biomass without a dedicated biomass preparation infrastructure.

Biomass pretreatment can help alleviate several concerns. As mentioned earlier, leaching, pelletizing, and torrefaction are preferred pretreatment methods; however, they can be expensive. Another interesting pretreatment option is fast pyrolysis to produce pyrolysis oil of high energy density. This oil can be mixed with coal and the slurry can be injected in the combustor. This method, however, requires a dedicated infrastructure for transport, storage, and feeding as well as a separate conversion unit. Herbaceous biomass is presently not considered a suitable biomass for co-firing. It can probably be used with pretreatment by torrefaction and pelletization.

The second method is parallel co-firing (hybrid system) where coal and biomass are fed into separate combustors (with a separate set of operating conditions) producing steam that can be combined in a common header before being sent to the turbine for power generation. The advantage of this method is that ash coming from coal and biomass is kept separate and can be used separately, thus allowing the use of biomass with high chlorine and alkali content. The disadvantage of this method is that it is more expensive, and it may require a larger capacity for the steam turbine. This method is also particularly popular in the paper and pulp industries.

The third and final method is indirect co-firing. In this method biomass is gasified in a separate gasifier or combustor and the gases are injected to the coal combustor thus providing additional heat for the combustion process. The method once again keeps the ash from coal and biomass separate, but the overall process is more expensive than methods one and two. This method with pregasification is currently practiced in a number of demonstration plants in Austria, Finland, and the Netherlands [1].

7.4.4.2 Gasification and Pyrolysis

Process configuration is more important and more complex for gasification and pyrolysis than combustion because the products are used for downstream upgrading as well as power generation. Process configuration has many options [6]: (a) gasify both coal and biomass together in the same gasifier and have a unified downstream operation, (b) gasify both coal and biomass in separate gasifiers and then combine the product gas for a unified downstream operation, (c) have separate syngas cleanup steps but combine syngas coming from coal and biomass at some downstream step, (d) use different forms of feed (dry or liquid) for biomass, or (e) share a common or separate air separation unit to produce oxygen for the biomass and coal gasifiers. All these options lead to numerous process configurations. Ratafia-Brown et al. [6] considered six possible configurations and their advantages and disadvantages. The analysis of Ratafia-Brown et al. [6] is briefly summarized below. This analysis will allow others to consider other possible configurations.

Configuration 1: Co-feeding coal and biomass to the gasifier as a mixture, either in dry or slurry form.

Configuration 2: Co-feeding biomass and coal to the gasifier using separate gasifier feed systems, either in dry or slurry form.

Configuration 3: Pyrolyzing as-received biomass followed by co-feeding pyrolysis char and coal to the gasifier and separately feeding pyrolysis gas to the syngas cleanup system.

Configuration 4: Biomass and coal are co-processed in separate gasifiers followed by combined syngas cleanup.

Configuration 5: Biomass and coal are co-processed in separate gasifiers followed by separate syngas cleanup trains, and the syngas feeds are combined prior to sulfur and CO_2 removal unit operations.

Configuration 6: Same as 4 and 5 but share common air separation unit (ASU) for oxygen feed to the separate gasifiers.

Ratafia-Brown et al. [6] have given an extensive assessment of these six process options. The following paragraphs briefly summarize their assessments of each option.

According to Ratafia-Brown et al. [6], the advantages of the first two configurations are: (a) they take advantage of economy of scale; gasification can proceed with or without biomass; (b) in Configuration 1, a separate biomass feed system is not required and it reduces gasifier complexity via the use of a single feedstock injection; (c) in Configuration 2, a separate biomass system can be designed without affecting the coal system; and (d) both Configurations 1 and 2 depend on a single syngas processing system. The disadvantages of the first two configurations are: (a) gasifier design should avoid tar production or the syngas system should treat tars, (b) the common gasifier will need to be larger than a single coal processing gasifier because of the low energy content of biomass, (c) control of the syngas composition (H_2/CO ratio) for subsequent FT processes may become an issue with a single gasifier processing both coal and biomass, and (d) the value of biomass ash which contains phosphate and potash for fertilizer purposes will be reduced due to the addition of coal ash.

As shown by Ratafia-Brown et al. [6],the advantages of Configuration 3 are: (a) it is thermodynamically more efficient than off-site biomass processing, (b) the pyrolysis gas can be used as a feed to syngas, (c) and the pyrolysis process can handle biomass with different properties and char can be directly injected to the gasifier with coal, although this will depend on the gasifier type and design and properties of char. The disadvantage of this configuration is that it may not be cost-effective depending on the scale of the gasification facility and level of biomass consumed.

According to Ratafia-Brown et al. [6], the advantages of Configurations 4, 5, and 6 are: (a) they allow maximum flexibility in coal and biomass properties and an independent control of biomass gasification for syngas production, (b) they take advantage of the economy of scale of syngas processing, particularly in Configuration 4, (c) depending on the type of gasifier employed, they allow separate recovery of biomass ash, and (d) for Configurations 5 and 6, independent control of two gasification processes but taking advantage of economy of scale for the air separation unit.

The disadvantages of Configurations 4, 5, and 6 are: (a) they do not take care of coal gasification economy of scale; (b) parallel biomass gasification increases investment cost, design, and operational complexity; (c) a tar and particulate removal system may be required depending on the type of biomass gasifier used; and (d) more land area and parallel ash handling and storage systems for coal and biomass are required.

Although the above discussion is mainly focused on a mixed coal and biomass feedstock, the same thought process can be applied to the different types of mixed feedstock (including different types of biomass) with other variable properties. Earlier in Section 7.3, various process configurations and feed preparation options for biomass with different types of properties as identified by Ratafia-Brown et al. [6] were briefly outlined. These options are also valid for the mixed (coal and biomass, coal and waste, etc.) feedstock.

7.4.4.3 Plasma Technology

In principle, plasma technology process options are the same as those described above for gasification. Because plasma technology is used either for heat and electricity generation as well as for fuels and chemical production, it is more analogous to the gasification or pyrolysis process than a combustion process. The use of a mixed feedstock is very prevalent in plasma technology. A challenging feedstock is automobile rubber tires. They can be pyrolyzed (with or without the use of plasma technology) as is, however, they are often shredded and then converted to useful chemicals by high-temperature pyrolysis. However, just as with switchgrass, the feed preparation for rubber tires can be expensive.

7.4.5 Industrial Processes

There are numerous gasification processes with mixed feedstock currently in operation [1, 6, 114]. We briefly describe three such processes. The descriptions of two gasification processes that use an entrained bed gasifier follow the ones outlined in the report by Ratafia-Brown [6].

7.4.5.1 Nuon Power Buggenum BV-Willem-Alexander Centrale (WAC)—250 MWe IGCC Plant

A commercial-scale gasification plant using mixed feedstock of biomass and coal has been demonstrated at the 253 MWe Nuon Power Plant in Buggenum, the Netherlands [6]. The plant was built in 1993, and it uses biomass to reduce CO_2 emissions based on dry feed Shell gasification technology. The Shell gasifier is an oxygen-blown continuous slagging, entrained flow reactor. The plant is capable of using different types of coal and contains several advanced design features that are different from U.S. IGCC plants. The air separation unit and gas turbine are very closely coupled, with the gas turbine compressor supplying all the air to the ASU. Although this improves efficiency, it makes the plant more complex and harder to start.

Pulverized coal is fed into the gasifier with transport gas via dense phase conveying. Either product gas or nitrogen can be used as a carrier gas. The reactor is operated at 1,500–1,600°C and pressure ranging from 350 to 650 psig to produce syngas principally composed of H_2 and CO and very little CO_2. Operation at the elevated temperatures eliminates the production of hydrocarbon gases and liquids in the product gas. The molten slag at high temperature runs down the refractory-lined water wall of the gasifier into a water bath, where it is solidified and is removed through a lock hopper as slurry in the water. The hot gas leaving the reactor is first cooled by the recycling product gas and then by a waste heat recovery (syngas cooler) unit. The syngas is further cooled before particle removal in a wet scrubber. The syngas is then treated to remove carbonyl sulfide and hydrogen sulfide before going to a gas turbine.

The plant processed mixed feedstock of coal and biomass first from 2001–2004 with about 18% by weight of pure and mixed biomass. More recently, biomass concentration has increased up to 30 wt%. In addition to gasification of demolition wood, tests were also conducted with chicken litter and sewage sludge. The test program evaluated the effects of biomass on product gas and ash quality. The NUON/IGCC plant uses the coal and biomass composition shown in Table 7.10. As shown, the plant takes about 30% by weight biomass, most of which is waste wood to provide about 17% of energy input to the gasifier.

7.4.5.2 250 MWe IGCC Plant of Tampa Electric's Polk Power Station

This integrated gasification combined cycle power plant (see Figure 7.3) is operated by Tampa Electric [6]. In 2001/2002, the plant used about 1.5 wt% woody biomass harvested from a five-year-old eucalyptus grove along with coal to test whether biomass can be converted to fuel gas and whether a fuel handling system accommodates this change. The original system was not designed to handle softer fibrous biomass.

TABLE 7.10
Coal and Biomass Compositions of NUON/IGCC Plant

Feedstock Type	Lower Heating Value (MJ/kg)	Feedstock Input (1000 Metric Tons/Yr)	Feedstock (% by Weight)	Feedstock (% by Energy Input)
Waste Wood	15.4	130	22.5	14
Dried Sewage Sludge	8.2	40	7	2.3
Other Biomass	10.2	10	0.5	0.7
Total Biomass	**13.6**	**185**	**30**	**17**
Coal	29	400	70	83
TOTAL FEED	**24.4**	**577**	**100**	**100**

Source: Ratafia-Brown, et al., 2007. "Assessment of Technologies for Co-converting Coal and Biomass to Clean Syngas-Task 2 Report (RDS)", NETL report (May 10).

The Polk Power Station used old ChevronTexaco IGCC technology that is now owned by General Electric. In this process, 60–70% coal–water slurry is fed to the gasifier at the rate of 2,200 tons (on dry basis) of coal per day. The normal feed is a blend of coal and petroleum coke, the solid residue from crude oil refining. The fresh feed is mixed with unconverted recycled solids and finely ground in rod mills until 98% of the particles are less than 12 mesh in size. The slurry passes through a series of screens before being pumped into the gasifier. The slurry and oxygen are mixed in the gasifier process injector. The gasifier is designed to convert 95% of carbon per pass, and it produces syngas of 250 BTU/Scf heat content.

A schematic of the Polk Power Station plant is shown in Figure 7.3. As shown in the diagram, the syngas coming out of gasifier is cooled in a series of steps, each recovering heat in the form of saturated high-pressure steam. The first syngas cooler, called the "radiant syngas cooler" (RSC), produces 1,650 psig saturated steam. The gas from RSC is split into two streams and they are sent to parallel convective syngas coolers (CSC) where the process of cooling and generating additional high-pressure steam (at lower

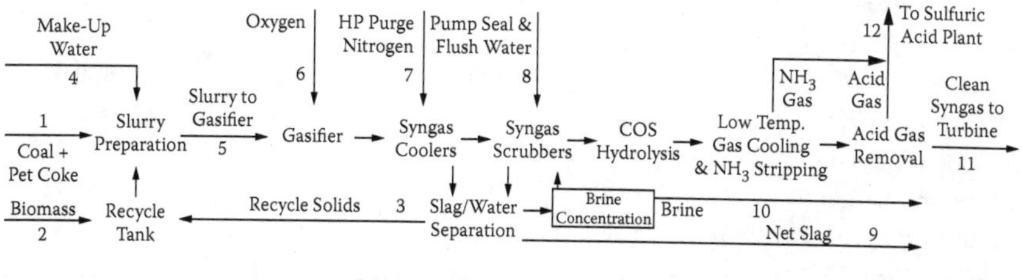

FIGURE 7.3
Polk Power Station co-gasification configuration. (From Ratafia-Brown et al. 2007. *Assessment of Technologies for Co-converting Coal and Biomass to Clean Syngas-Task 2 Report (RDS)*, NETL report (May 10); and McDaniel, 2002. *Biomass Gasification at Polk Power Station-Final Technical Report*, DOE award DE-FG26-01NT41365, May.)

temperature) is repeated. The gases then further go through a simultaneous cooling and impurity removal (particulates, hydrogen chloride) process. A final trim cooler reduces the syngas temperature to about 100°F for the cold gas clean-up (CGCU). The CGCU system is a traditional amine scrubber system, and it removes sulfur which is then converted to sulfuric acid and sold to the local phosphate industry.

The eucalyptus feedstock used in this power plant contained about 1/3 of heating value per pound at about half the bulk density of coal. The characteristics of the mixed feedstock for the Polk Power Plant are shown in Table 7.11a. These numbers indicate that even a modest concentration of this biomass will require a massive and expensive feed system. Although the combined characteristics of the mixed feedstock are not significantly different from the baseline, it increases hydrogen, oxygen, and ash content by 4.6, 11, and 3.4%, respectively. The CO_2 discharge is reduced by 0.87%. Biomass used in the Polk plant did not lend itself to size separation and screening, and it caused minor plugging of the suction to one of the pumps [6]. The results indicate that for a slurry system, feed preparation must be tailored to the nature of the biomass in order to prevent any malfunction by the slurry pump as well as downstream gas cleaning and turbine operation. Typical test results for the Polk Power Station are described in Tables 7.11a and b. The experience of the Polk Power Station can be extended to coal and other materials.

TABLE 7.11A

Polk IGCC Plant Coal/Coke and Biomass Combined Feedstock

Feed Composition (Wt%)	Coke + Coal	Biomass	Combined Feed	Recycle Solids to Gasifier
C	82.88	49.18	82.02	66.26
H	4.50	5.78	4.71	0.29
N	1.85	0.24	1.81	0.95
S	2.99	0.06	3.13	2.31
O	3.53	39.42	3.92	0.00
ASH	4.25	5.32	4.4	30.19
TOTAL	100.00	100.00	100.00	100.00
HHV, BTU/lb dry	14,491	8,419	14,470	9,698
% of Original Feed Recycled	–	–	–	48.6
Lb Carbon/Million BTU	**57.2**	**58.41**	**57.21**	68.32
Effective Lb Carbon/Million BTU[a]	**57.2**	**1**	**56.71**	

Source: Ratafia-Brown, et al., 2007. "Assessment of Technologies for Co-converting Coal and Biomass to Clean Syngas-Task 2 Report (RDS)", NETL report (May 10); and McDaniel, J., 2002. "Biomass gasification at Polk Power Station-Final Technical Report," DOE award DE-FG26-01NT41365 (May).

[a] Accounts for biomass carbon recycle and carbon released during biomass preparation.

TABLE 7.11B

Polk Power Station Biomass Co-Gasification Test Results

Property	Base Fuel	Biomass	Total/Average
Fuel Feed Rate (Lb/Hr, As-Received)	164,840	1,945	166,786
Moisture Content (Wt%)	7.82%	46.8%	8.27%
Higher Heating Value (BTU/lb, As-Received)	13,322	4,424	13,218
Higher Heating Value (MMBTU/Hr)	2196	8.6	2,205
Net Power Production (kW)	219,640	860	220,500

Source: Ratafia-Brown, et al., 2007. "Assessment of Technologies for Co-converting Coal and Biomass to Clean Syngas-Task 2 Report (RDS)", NETL report (May 10); and McDaniel, J., 2002. "Biomass gasification at Polk Power Station-Final Technical Report," DOE award DE-FG26-01NT41365 (May).

7.4.5.3 WPC Plasma Process

Alter NRG's WPC two-stage plasma process [114] (described in the previous chapter) has been used to build plants of various sizes in the United States, Canada, and Japan. Table 7.12 depicts basic descriptions of some of these plants. This technology provides clean fuel (toxin free) from a variety of mixed feedstock. The basic description of the process is described in Figure 7.4. The plasma technology is most suitable for a mixed feedstock. It uses a moving bed reactor. The temperatures are high enough so that ash is melted and molten ash and slag are collected, vitrified, and discarded into landfill. The composition of the fuel gas produced by this technology depends on the nature of the feedstock, but high temperature and lack of oxygen can lead to the production of clean syngas containing hydrogen and

TABLE 7.12

Plant Locations and Their Capacities for Various WPC Operations

Location/Start-Up	Feed Characteristics	Plant Capacity	Treatment Capacity (Tons per Year)
General Motors, Defiance, Ohio, U.S. (1989)	Iron and steel scrap (Plasma melting)	50–80 tons per hour	–
ALCAN, Jonquiere, Canada (1992)	Aluminum dross (Plasma melting)	–	36,000
JHI, Kinuura, Japan (1995)	MSW incinerator ash (Plasma melting)	60–80 tons per hour	18,000–24,000
Yoshi, Japan (1999)	MSW (Plasma gasification)	24 tons per day	7,200
Utashinai, Japan (2003)	MSW+ASR (50/50) Plasma gasification	180 tons per day	49,500
Mihama-Mikata, Japan (2003)	MSW/dried sewage sludge (80/20)	22 tons per day	6,600

Source: The Alter NRG/Westinghouse Plasma Gasification Process. 2008. *Independent Waste Technology* report by Juniper Consulting Services Limited, Bathurst House, Bisley UK, November.

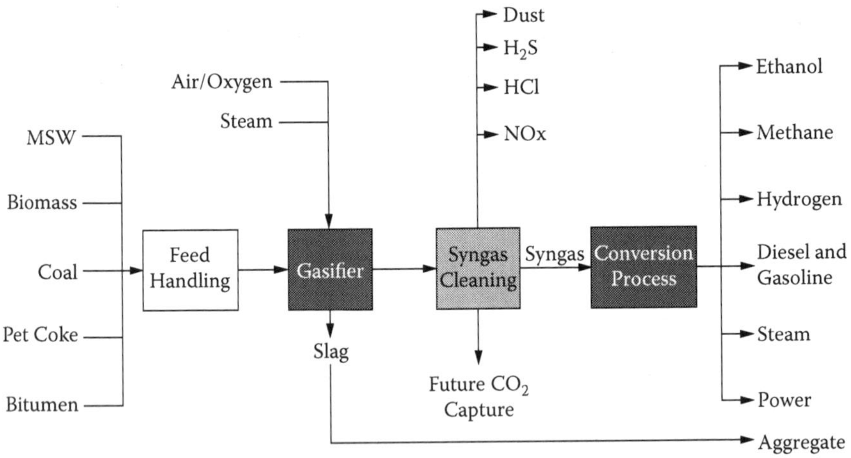

FIGURE 7.4
Basic flow process for WPC process. (From The Alter NRG/Westinghouse Plasma Gasification Process. 2008. *Independent Waste Technology* report by Juniper Consulting Services Limited, Bathurst House, Bisley UK, November.)

carbon monoxide. The product gas composition also depends on the operating conditions. The syngas is purified and conditioned and then converted to numerous different products via FT, methanol, or other syntheses. As shown in Figure 7.4, the process is divided into a number of different steps such that each can be optimized individually or collectively. The feed preparation and handling are a very important part of all new plants. In the second step, the feed can be gasified in one or multiple reactors. Generally one reactor is preferred, because unlike in gasification the slag is not used for further applications.

7.5 Liquefaction Technologies

Unlike gasification technologies, liquefaction technologies are more at the developmental stage. For both coal and biomass there are direct and indirect methods for producing liquid products. The indirect methods are based on the use of a syngas platform from which liquid fuels and methanol can be produced using Fischer–Tropsch syntheses employing different types of catalyst. Both biosyngas and petroleum- (or coal-) derived syngas can be obtained from coal–biomass mixture gasification. This subject has already been discussed in detail. The gas-to-liquid conversion process via FT synthesis depends strictly on purity and syngas composition irrespective of its generation source.

In this section we address direct methods for producing liquids from biomass and coal. There are three basic methods for liquid production: (a) direct liquefaction, (b) low severity pyrolysis, and (c) supercritical extraction and liquefaction. Methods (a) and (c) can be carried out either by an organic solvent or by water. We briefly evaluate both options.

7.5.1 Direct Liquefaction

Direct liquefaction of coal has been carried out in a hydrogen donor solvent at a temperature between 350–450°C, pressure between 2,000–5,000 psig, and in the presence of a hydrogenation catalyst along with hydrogen or another suitable reducing agent (such as carbon monoxide). During the 1970s and 1980s, significant research on direct coal liquefaction was carried out and this was summarized by Shah [115]. Although liquefaction is technically feasible, high hydrogen consumption, high pressure, short catalyst life, and poor liquid composition (without upgrading) made the process economically unattractive. The oil produced was stable, insoluble with water but not of the same quality as that derived from crude oil. Although direct liquefaction can be applied to bituminous, subbituminous, and lignite coals, the liquid production per ton of coal was higher for bituminous and subbituminous coals compared to lignite coals. Lignite coals are more active due to the high amount of hetero atoms and oxygen, but they also contain a large amount of water which reduces the oil production per ton of coal.

To some extent biomass is similar to lignite (see Figure 7.1) in that it contains a high amount of oxygen, a large amount of volatile matter, and its H/C and O/C ratios are not too far from the ones for biomass (see Figure 7.1). This is to be expected considering the fact that lignite is the lowest rank and youngest coal. The process of biomass liquefaction in the presence of a hydrogen donor solvent, catalyst, and hydrogen is likely to be similar to that for lignite coal liquefaction. Biomass liquefaction generally requires a temperature between 250–450°C and pressure between 700 to 3,000 psig [116]. A number of different hydrogen donor solvents have been used including creosote oil, ethylene glycol, tetralin, methanol, phenol, and recycled oil. The catalysts used for biomass liquefaction are alkaline oxides, carbonates and bicarbonates, metals such as zinc, copper, and nickel, formate, iodine, cobalt sulfide, zinc chloride, ferric hydroxide, and so on. Some of these are very similar to those used in coal liquefaction, for example, Fe, Co, Mo, Zn, Cu, and their derivatives. Both biomass and coal can be dissolved by solvolysis using a reactive liquid solvent. A review of biomass liquefaction research from 1920–1980 is presented by Moffatt and Overend [117].

In recent years, studies [118–123] have been reported to co-liquefy a mixture of coal and biomass. These studies have used lignite coal and various types of cellulosic waste materials. These studies indicated that the conversion is not significantly affected by the particle size of coal and biomass. Although all biomass gave good liquefaction results, waste paper gave the most desirable product distribution under both catalytic and noncatalytic conditions. The optimum reaction conditions were a solvent/solid ratio of 3, temperature of 400°C, and a reaction time of 90 min. The most suitable catalyst for the mixture was Fe_2O_3, although $Cr(CO)_6$, $Mo(CO)_6$, and MoO_3 were also effective catalysts.

The studies on direct liquefaction of mixed feedstock carried out thus far have come to the same conclusion that was arrived at for direct coal liquefaction. High pressure, hydrogen, and low catalyst life make the co-liquefaction economically unattractive at the present time. A cheap catalyst or a catalyst with long life, low pressure, use of recycled solvent, and low hydrogen partial pressure will make this process more attractive. The quality of bio-oil is better than that produced from fast pyrolysis, however, it can only be used as boiler fuel and the like without further upgrading.

7.5.1.1 Hydroliquefaction

Water, particularly at high temperature and pressure is a good solvent for both biomass and coal [124–131]. For biomass, as described in the previous chapter, between temperatures of about 180–280°C, hydrothermal carbonization occurs which produces a heavy biocrude. For temperatures close to the critical temperature, hydrothermal liquefaction occurs producing a higher quality crude and this can be further upgraded with the use of hydrogen by a hydrothermal upgrading process. These processes are described in Chapter 6 and are not repeated here. Similarly finely pulverized coal and high-temperature water can also depolymerize coal producing asphaltene and preasphaltene types of materials. Initial tests done at Auburn University indicate that liquefaction of a mixed feedstock in water at high temperatures results in a product similar to biocrude, bio-oil, and asphaltene types of materials. There was no synergistic effect between liquefaction of biomass and coal. The tests were done for different types of coal and biomass. Because water is not a hydrogen donor, further upgrading of these crudes (or heavy materials) will require the use of hydrogen. Hydroliquefaction is a good method to produce heavy crude from coal and biomass mixture which may have some practical applications in the construction (e.g., cement) or fertilizer industries.

7.5.2 Pyrolysis

Just as with direct co-liquefaction, in recent years significant attention to co-pyrolysis of a variety of mixed feedstock has been given. Some of these studies have been focused on low-severity (low residence time as with fast pyrolysis) conditions to produce more oil and these are summarized in Table 7.13. Although the studies have examined a wide range of feedstock and reaction conditions, the major conclusion that emerges is that the pyrolytic reactivity of biomass and coal are different because biomass in general is more reactive than coal and requires a lower temperature to devolatilize

TABLE 7.13
Some Typical Studies of Co-Pyrolysis

Mixed Feedstock	Author	Comment
Coal/biomass	Li and Xu, 2006 [132]	Coal/biomass pyrolysis review
Wood/synthetic polymer, also biomass/plastic	Sharypov et al., 2002 [133]	For biomass/plastic <1, non-additive effects were observed
Coal/biomass	Li and Xu, 2007 [134]	Free fall reactor
Coal/woody biomass	Moghtaderi et al., 2004 [135]	No synergy; additive effects
Biomass/plastic	Zhou et al., 2006 [136]	Significant synergy
Biomass/poor coal	Pan et al., 1996 [137]	No synergy
Capsicum stalks/coal	Niu et al., 2011 [138]	No synergy
Biomass/coal	Rodjeen et al., 2006 [139]	Catalytic pyrolysis in CFB
Coal/plastic	Huffman et al., 1995 [140]	High liquefaction of plastic with zeolite catalyst
Coal/biomass	Collot et al., 1999 [141]	No synergy
Sugarcane bagasse/ Argentinean subbituminous coal	Bonelli et al., 2007 [142]	Results similar to other agro-Industrial byproducts
Coal/biomass	Vuthaluru, 2004 [143]	No synergy
Coal/biomass, coal/sewage	Storm et al., 1999 [144]	No synergy
Coal/biomass, biomass/ polymer	Ma et al., 2006 [145]	Additive effects for biomass and polymer; no synergy for coal and biomass

and depolymerize compared to coal. However, there seem to be no or very little reactions or interactions between species of biomass and coal. Hydrogen release from biomass does not appear to interact with the species from coal. Thus, in general there is no synergy between coal and biomass pyrolysis. The final results for co-pyrolysis are additives of individual pyrolysis. Each species within the mixture produces its own liquids and gases. The situation is, however, different for the mixtures of biomass and a variety of plastics. In these cases, there appears to be a significant synergy between pyrolysis of these two substances. Not many studies using catalyst and hydrogen (i.e., hydropyrolysis) for a mixed feedstock have been reported. It is possible that under those conditions more synergies between coal and biomass pyrolysis would be observed. More work in this area is needed.

7.5.3 Supercritical Extraction

As reported in the previous chapter, supercritical extraction of coal, biomass, and other organic wastes by CO_2, water, and other organic solvents have been extensively reported [146–160]. These studies have clearly indicated the value of this process. Outstanding reviews of supercritical extraction by water both at high and low temperatures are given by Guo, Cao, and Liu [161] and Peterson et al. [124]. Only limited studies [161–163] for supercritical extraction of a mixed feedstock by water have been reported. Once again, there seems be no synergy for the high-temperature supercritical extraction of coal and biomass in the mixture environment by water. The work of Veski, Palu, and Kruusement [163], however, shows an interesting interaction between liquefaction reactions of oil shale and wood. As shown in Chapter 6 and the previous section, syngas and hydrogen productions by supercritical water gasification of pure and mixed feedstock have been the subjects of significant recent interest [66, 124, 161].

7.6 Summary

The use of a mixed feedstock in combustion, pyrolysis, gasification, plasma technology, liquefaction, and supercritical technology is becoming more and more prevalent and popular. Going forward, there are, however, several issues that need to be addressed to make the use of mixed feedstock in each of these technologies more economical, environmentally acceptable, and technologically feasible. Some of these issues are briefly outlined below.

1. Although there are several successful commercial co-firing installations, the effects of increased biomass/coal ratio and the use of low-quality biomass on the effectiveness of co-firing need to be further examined. Some of the issues include transportation, handling, storage, milling and feeding problems, slagging and fouling due to deposit formation, agglomeration, corrosion or erosion, and ash utilization. There are several methods to address these issues. One method is the cleaning of deposits by soot blowing or exchange of agglomerated materials. It is also possible to avoid the effects of agglomeration and deposit formation by adding chemicals to reduce corrosion and increase the ash melting point. It is also possible to create an independent infrastructure for feeding, milling, storage, and conveying biomass just like one for coal. More expensive parallel and indirect firing modes where both feed treatments and gasification infrastructure can be separate at different levels are also options that need to be further examined for specific types of mixed feedstock. For each combustion process and mixed feedstock, an optimum logistic detail for biomass transportation, storage, and pretreatment needs to be examined.

2. Unlike combustion, co-gasification is more complex because the process of gasification can be used to generate fuels, chemicals, and materials along with heat and electricity. Syngas generated via gasification has many downstream applications, but it needs to be pure and of the right composition. It is generally accepted that a future gasification reactor, even for a mixed feedstock, is most likely to be a pressurized oxygen-blowing entrained bed reactor [6]. For this type of reactor, an optimum burner design for solid biomass feeding and optimum gasification conditions with respect to biomass particle size (1 mm or less), maximum efficiency, maximum heat recovery, minimum flux use, minimum inert gas consumption, complete conversion, production of biosyngas with low CH_4, and no tars will need to be obtained for each type of mixed feedstock [6]. In a slagging gasifier the ash and flux are present as a molten slag that protects the inner wall against high temperature. The slag must have the right viscosity and flow behavior at the temperature in the gasifier and its behavior as a function of gasification temperature, biomass ash properties, and selected flux should be known. The reactor should be operated at the lowest temperature possible to get high cold gas efficiency and oxygen consumption [6]. In a gas cooling and purification system, maximum energy efficiency is desirable. Multiple cooling steps should be further optimized. The reactor should also be capable of processing feedstock of a wide variety of properties and produce no waste. The bottom ash should be recoverable as non-leachable slag with a value as construction material whereas fly ash needs to be used for mineral recycling [6]. This goal will require a continuous and steady flow of mixed feedstock for safe and steady operation. An appropriate pretreatment of feedstock, which may include leaching, torrefaction, and pelletizing, and an optimum process configuration are critically important to achieve the desired process objectives.

As shown in Section 7.5 [6], there are numerous process options possible. Milling of woody biomass and pressurization in a piston compressor with negligible inert consumption and feeding to the gasifier with a screw feed system (or a stamet type of system) is one option. The use of flash pyrolysis for grassy and straw biomass materials to produce bioslurry that can be fed to the gasifier is another option. The torrefaction can allow the generation of fine and homogeneous particles (with low energy consumption). This can then be homogeneously mixed with coal and pressurized by a piston compressor and fed to the reactor by a screw or a pneumatic feeding system. Finally, all feed materials can be gasified in a fluidized bed gasifier and the product gas then be processed in an entrained bed gasifier. These are some of the options when using coal and biomass of different properties. For other mixtures, which include materials such as oil shale, plastics, rubber tires, tar sand, and the like, additional modifications of the above-described options may be required.

3. Just as with combustion and gasification, the use of a plasma reactor for a variety of feedstock needs to be optimized. Plasma reactors often generate valuable chemicals along with fuel gas. The reactor design and operating conditions will need to be such that maximum production of valuable chemicals and fuel gas is possible. More experience with a larger scale plasma process is needed.
4. For high severity pyrolysis, the above-described statements for gasification apply. For low severity pyrolysis, a focus needs to be on the liquid production rate and its quality. As shown earlier, there is no synergy between pyrolysis of coal and biomass although there is a significant synergy between pyrolysis of biomass and plastics. Thus, an appropriate optimization of the operating conditions for the pyrolysis process will need to be examined based on the nature of the feedstock mixture. More experience with a larger scale pyrolysis operation using mixed feedstock is needed.
5. An optimization of liquefaction of mixed feedstock is challenging and interesting. There seems to be very little synergy between liquefaction of coal and biomass. There is, however, synergy between liquefaction of biomass and polymeric products. This is true for both organic (hydrogen donor) solvents as well as water. More data on synergy for other types of feed mixture are needed. At the present time, just as in direct coal liquefaction, the liquefaction process for the mixture is, in general, economically unattractive for the production of high-quality fuels. More research is needed to test the liquefaction process which uses less pressure, less gaseous hydrogen, and recycled solvent. More markets for lower quality liquefaction products need to be found.
6. Supercritical technology, both for gasification and liquefaction, is becoming more popular. The application of this technology using water, carbon dioxide, and some alcohols for mixed feedstock needs to be fully explored. The technology may be very attractive for selected feedstock mixtures and products.

Until now at larger scales, co-utilization of waste and coal has received considerably more attention than co-utilization of coal and biomass. This has been true for all thermochemical processes. The pricing factor for the raw material and government subsidies often make the use of waste for energy and products more attractive. New concepts of sustainable resource management and enhanced landfill management will make thermochemical conversion of mixed waste more popular. Investigated waste has been (a) municipal solid waste that has had minimal presorting, (b) refuse derived fuel that has had significant pretreatment such as mechanical shredding and screening, and (c) shredded rubber tires, paper and pulp waste, and plastic waste.

7.7 Future of Mixed Feedstock

Currently, the entire energy industry is striving for more alternate, economical, and environmentally friendly options. The industry is divided into three parts. Old fossil energy relies on the energy sources coming from the ground. The use of this energy source releases carbon to the atmosphere which is becoming more and more unacceptable due to its impact on the environment. Although over the next decade, oil and coal may be replaced more and more by a cleaner natural gas, this form of energy will continue to receive some environmental objections unless the problems of carbon dioxide and volatile hydrocarbons released to the environment are appropriately handled. Over the next several decades, fossil energy will still be the dominant force in the energy industry.

Renewable bioenergy comes from materials that grow on the ground and in its life cycle, it is carbon neutral to the environment. Thus it is more acceptable to the public on environmentally friendly grounds. Bioenergy coming from biomass is, however, of limited supply and carries some serious logistic problems of transportation, storage, and pretreatment. Waste (which is predominantly a form of bio or cellulosic energy) is an interesting source for energy and products, and it will play an increasingly important role in the energy industry.

The third and final source of energy is green and clean energy such as wind, solar, hydroelectric, hydrothermal, hydrogen, nuclear, and so on which do not release any carbon into the atmosphere. Wind, solar, hydroelectric and hydrothermal energy sources are also renewable, but they are location-dependent and more work is needed to make them economical. At the present time they are largely used for stationary sources of energy. Hydrogen is the purest and most abundant form of energy. Public acceptance of nuclear energy at this time is uncertain.

The resources available for these three types of energy industries vary greatly among countries and in the future each country will optimize its own local situation. It is clear that the use of mixed feedstock for producing energy has to be an important part of the strategies for future energy industry development. It will allow more flexibility and a greater acceptance by the public on environmental grounds. As summarized in the previous section, more work is needed to make this strategy more practical and attractive. The future for mixed feedstock is, however, bright, and novel ideas will further accelerate its use in the energy industry.

There are those who believe that in the long term, electricity and hydrogen are the solutions to our energy needs. Even if this is true, a mixed feedstock should play an important role in achieving this goal.

References

1. Maciejewska, A., Veringa, H., Sanders, J., and Peteves, S.D. 2006. *Co-firing of Biomass with Coal: Constraints and Role of Biomass Pre-Treatment*, DG JRC Institute for Energy report, EUR 22461 EN.
2. Loo van, S. and Koppejan, J. (Eds.). 2004. *Handbook of Biomass Combustion and Co-Firing*, Prepared by Task 32 of the implementing agreement on bioenergy under an auspices of the international energy agency, Twente University Press.
3. VIEWLS. 2005. *Biofuel and Bio-energy Implementation Scenarios*, Final report of VIEWLS WP5, modeling studies.
4. IEA Clean Coal Center, 2005. *Fuels for Biomass Cofiring*.
5. Ricketts, B., Hotchkiss, R., Livingston, B., and Hall, M. 2002. Technology status review of waste/biomass co-gasification with coal, *Chem. Fifth European Gasification Conference*, Noordwijk, The Netherlands (April 8–10).
6. Ratafia-Brown, J., Haslbeck, J., Skone, T.J., and Rutkowski, M. 2007. *Assessment of Technologies for Co-converting Coal and Biomass to Clean Syngas-Task 2 Report (RDS)*, NETL report (May 10).
7. Sami, M., Annamalai, K., and Wooldridge, M. 2001. Co-firing of coal and biomass fuel blends, *Prog. Energy Combust. Sci.*, 27: 171–214.
8. Baxter, L. 2005. Biomass-coal co-combustion opportunity for affordable renewable enrgy, *Fuel*, 84: 1295–1302 .
9. EUBION (European Bioenergy Networks). ALTENER, Biomass co-firing-an efficient way to reduce greenhouse gas emissions, http://europa.eu.int/comm/energy/res/sectors/doc/bioenergy/cofiring_ eu_bionet.pdf.
10. Jarvinen, T. and Alakangas, E. 2001. (VTT Energy), Co-firing of biomass-evaluation of fuel procurement and handling in selected existing plants and exchange of information (COFIRING), Altener Programme (January).
11. Karki et al., 2005. The performance and operation economics of co-fired biomass boilers, *Bioenergy in Wood Industry-Book of Proceedings*, pp. 373–380.
12. Prins, M.J., Ptasinski, K.J., and Janssen, F.J.J.G. 2006. Torrefaction of Wood. Part 1. Weight loss kinetics, *J. Anal. Appl. Pyrolysis*, 77: 28–34.
13. Shafizadeh, F. 1985. Pyrolytic reactions and products of biomass. In R.P. Overend, T.A. Mime, and L.K. Mudge (Eds.), *Fundamentals of Biomass Thermochemical Conversion*, London: Elsevier, pp. 183–217.
14. Shafizadeh, F. 1983. Thermal conversion of cellulosic materials to fuels and chemicals, *Wood and Agricultural Residues*, New York: Academic Press, p. 183–217.
15. Tarka, T.J. 2008. Carbon Capture and Storage for CTL/CBTL, *Seventh Annual Conference on Carbon Capture and Sequestration*, U.S. Department of Energy, National Energy Technology Laboratory, Office of Systems, Analysis and Planning, May.

16. Hustad, J. and Barrio, M. 2000. Biomass. *IFRF Online Combustion Handbook, Combustion File No. 23, Version No. 2, IFRF*. http://www.handbook.ifrf.net/handbook/cf.html?id=2.
17. Shah, Y.T. and Gardner, T. In press. Biomass Torrefaction: Applications in Renewable Energy and Fuels. In *Encylopedia of Chemical Processes*, Boca Raton, FL: CRC Press.
18. Korbee, R., Eenkhoorn, S., Heere, P.G.T., and Kiel, J.H.A. 1998. *Co-gasification of Coal and Biomass Waste in Entrained-Flow Gasifiers, Phase 2: Exploratory Lab-Scale Experimentation*, Energy Research Centre of the Netherlands (ECN), Petten, The Netherlands, Report No. ECN, ECN-C-98-056, December.
19. Hustad, J. and Barrio, M. 2000. Biomass, *IFRF Online Combustion Handbook*, Combustion File No. 23, Version No. 2, IFRF, http://www.handbook.ifrf.net/handbook/cf.html?id=2.
20. Li, J. and Gifford, J. 2001. Evaluation of woody biomass torrefaction, *Forest Res. Rotorua*, New Zealand, September.
21. Lipinsky, E.S., Arcate, J.R., and Reed, T.B. 2002. Enhanced wood fuels via torrefaction, Prepr. Pap. *Am. Chem. Soc., Div. Fuel Chem.*, 47: 408–410.
22. Livingston, W.R. 2005. A review of the recent experience in Britain with the co-firing of biomass with coal in large pulverized coal-fired boilers, Mitsui Babcock, Renfrew, Scotland, *Proceedings of the IEA Exco Workshop on Biomass Co-firing*, Copenhagen.
23. Maciejewska, A., Veringa, H., Sanders, J., and Peteves, S.D. 2006. *Co-Firing of Biomass with Coal: Constraints and Role of Biomass Pretreatment*, Report No. EUR 22461 EN, DG JRC, Institute for Energy, pp. 1–113.
24. Nimlos, M.N., Brooking, E., Looker, M.I., and Evans, R.J. 2003. Biomass torrefaction studies with a molecular beam mass spectrometer, Prepr. Pap. *Am. Chem. Soc., Div. Fuel Chem.*, 48: 590–591.
25. Obernberger, I. and Thek, G. 2004. Physical characterization and chemical composition of densified biomass fuels with regard to their combustion behavior, *Biomass Bioenergy*, 27: 653–669.
26. Olah, G.A., Geoppart, A., and Surya Prakash, G.K. 2006. *Beyond Oil and Gas: The Methanol Economy*, Weinheim: Wiley-VCH , pp. 1–50.
27. Orfo, J.J.M., Antunes, F.J.A., and Figueiredo, J.L. 1999. Pyrolysis kinetics of lignocellulosic materials - Three independent reactions model, *Fuel*, 78: 349–358.
28. Pach, M., Zanzi. R., and Bjømbom, E. 2002. Torrefied biomass a substitute for wood and charcoal, *Proceedings of the 6th Asia-Pacific International Symposium on Combustion and Energy Utilization*, Kuala Lumpur, Malaysia, May 20–22.
29. Panshin, A.J. and de Zeeuw, C. 1980. *Textbook of Wood Technology*, Fourth Edition. New York: McGraw-Hill, p. 736.
30. Pentanunt, R., Mizanur Rahman, A.N.M., and Bhattacharya, S.C. 1990. Updating of biomass by means of torrefaction, *Energy*, 15: 1175–1179.
31. Pétrissans, M., Gérardin, P., El Bakali, I., and Serraj, M. 2003. Wettability of heat-treated wood, *Holzforschung*, 57: 301–307.
32. Prins, M.J., Ptasinski, K.J., and Janssen, F.J.J.G. 2006. Torrefaction of Wood. Part 1. Weight loss kinetics, *J. Anal. Appl. Pyrolysis*, 77: 28–34.

33. Prins, M.J., Ptasinski, I.G., and Janssen, F.J.J.G. 2004. More efficient biomass gasification via torrefaction. In R. Rivero, L. Monroy, R. Pulido, and G. Tsatsaronis (Eds.), *Proceedings of 17th Conference on Efficiency, Costs, Optimization, Simulation, and Environmental Impact of Energy Systems (ECOS 2004)*, Guanajuato, Mexico, July 7–9.
34. Prins, M.J. 2005. Thermodynamic analysis of biomass gasification and torrefaction. Ph.D. Thesis, Technische Universiteit Emdhoven: Eindhoven, The Netherlands.
35. Ramiah, M.V. 1970. Thermogravimetric and differential thermal analysis of cellulose, hemicelluloses, and lignin, *J. Appl. Polymer Sci.*, 14: 1323–1337.
36. Arcate, J.R. 2002. Torrefied wood, an enhanced wood fuel, *Bioenergy 2002*, Boise, Idaho, September 22–26, Paper #207.
37. Arias, B., Pevida, C., Fermoso, J., Plaza, M.G., Reubiern, F., and Pis, J.J. 2008. Influence of torrefaction on the grindability and reactivity of wood biomass, *Fuel Process. Technol.* 89: 169–175.
38. Bergman, P.C.A. and Kiel, J.H.A. 2005. Torrefaction for biomass upgrading. *Proceedings of the 14th European Biomass Conference and Exhibition*, Paris, October, pp. 17–21.
39. Bergman, P.C.A., Boersma, A.R., Kiel, J.H.A., Prins, M.J., Ptasinski, K.J., and Janssen, F.J.J.G., 2004. Torrefaction for entrained flow gasification of biomass. In W.P.M. Van Swaaij, T. Fjällstrom, P.T. Helm, and P. Grassi (Eds.), *Proceedings of the Second World Biomass Conference on Biomass for Energy, Industry, and Climate Protection*, Rome, May 10–14, pp. 679-682; Petten, The Netherlands: Energy Research Centre of the Netherlands (ECN), Report No. ECN-RX--04-046 .
40. Bergman, P.C.A. 2005. *Combined Torrefaction and Pelletization: The TOP Process.* Petten, The Netherlands: Energy Research Centre of the Netherlands (ECN), Report No. ECN-C-05-073.
41. Bergman, P.C.A., Boersma, A.R., Kiel, J.H.A., Prins, M.J., Ptasinski, K.J., and Janssen, F.J.J.G. 2004. *Torrefaction for Entrained Flow Gasification of Biomass*, Petten, the Netherlands: Energy Research Centre of the Netherlands (ECN), Report No. ECN-C-05-067.
42. Van der Drift, A., Boerrigter, H., Coda, B., Cieplik, M.K., and Hemmes, K. 2004. *Entrained Flow Gasification of Biomass – Ash Behavior, Feeding Issues, and System Analysis*, The Netherlands: Energy Research Center of the Netherlands (ECN), Report No. ECN-C-04-039, p. 58, April.
43. Brouwer, J., Owens, W.D., Harding, S., Heap, M.P., and Pershing, D.W., Cofiring waste biofuels and for emission reduction. In Proceedings of the Second Biomass Conference of Americas, Portland, OR, p.390-9(1995).
44. Hansen, L.A., Michelsen, H.P., and Dam-Johansen, K. 1995. Alkai metals in a coal and biosolid fired CFBC-measurements and thermodynamic modeling. In *Proceedings of the 13th International Conference on Fluidized Bed Combustion*, vol. 1, May 7–10, p. 39–48.
45. Dhanaplan, S., Annamalai, K., and Daripa, P. 1997. Turbulent combustion modeling coal: Biosolid blends in a swirl burner, *Energy Week*, IV, ETCE, ASME: 415-423.
46. Spliethoff, H. and Hein, K.R.G. 1998. Effects of co-combustion of biomass on emissions in pulverized fuel furnaces, *Fuel Process. Technol.* 54: 189–205.
47. Frazzitta, S., Annamalai, K., and Sweeten, J. 1999. Performance of a burner with coal and coal manure blends, *J. Propulsion Power*, 15(2): 181–186.

48. Abbas, T., Costen, P., Kandamby, N.H., Lockwood, F.C., and Ou, J.J. 1994. The influence of burner injection mode on pulverized coal and biosolid co-fired flames, *Combustion Flame*, 99: 617–625.
49. Aerts, D.J., Bryden, K.M., Hoerning, J.M., and Ragland, K.W. 1997. Co-firing switchgrass in a 50 MW pulverized co-boiler. *Proceedings of the 1997 59th Annual American Power Conference*, Chicago Illinois, vol 59(2), p.1180–1185.
50. Sampson, G.R., Richmond, A.P., Brewster, G.A., and Gasbarro, A.F. 1991. Co-firing of wood chips with coal in interior Alaska, *For. Prod. J.*, 41(5): 53–56.
51. Siegel, V., Schweitzer, B., Spliethoff, H., and Hein, K.R.G. 1996. Preparation and co-combustion of cereals with hard coal in a 500kW pulverized fuel test unit. Biomass for energy and environment. *Proceedings of the Ninth European Bioenergy Conference*, Copenhagen, Denmark, June 24–27, vol. 2, pp. 1027–1032.
52. Ohlsson, O. 1994. *Results of Combustion and Emission Testing When Co-Firing Blends of Binder-Enhance Densified Refuse Derived Fuel (B-Drdf) Pellets and Coal in A 440 Mwe Cyclone Fired Combustor, Vol. 1, Test Methodology and Results*, Subcontract report No. DE94000283. Argonne, Il: Argonne National Laboratory, p.60.
53. Skodras, G., Grammelis, P., Kakaras, E., and Sakellaropoulos, G.P. 2004. Evaluation of the environmental impact of waste wood co-utilization for energy production, *Energy*, 29: 2181–2193.
54. Hartmann, D. and Kaltschmitt, M. 1999. Electricity generation from solid biomass via co-combustion with coal energy and emission balances from a German case study, *Biomass BioEnergy*, 16: 397–406.
55. Malkki, H. and Virtanen, Y. 2003. Selected emissions and efficiencies of energy systems based on logging and sawmills residues, *Biomass Bioenergy*, 24: 321–337.
56. Benetto, E., Popovici, E.-C., Rousseaux, P., and Blondin, J. 2004. Life cycle assessment of fossil CO2 emissions reduction scenerios in coal-biomass based electricity production, *Energy Conversion Manage.*, 45: 3053–3074.
57. Ney, R.A. and Schnoor, J.L. 2002. Incremental life cycle analysis: Using uncertainty analysis to frame greenhouse gas balances from bioenergy systems for emission trading, *Biomass Bioenergy*, 22: 257–269.
58. Heller, M.C., Kaoleian, G.A., and Volk T. 2003. A life cycle assessment of a willow bioenergy cropping system, *Biomass Bioenergy*, 25: 147–165.
59. Heller, M.C., Kaoleian, G.A., Mannb, M.K., and Volk, T.A. 2004. Life cycle energy and environmental benefits of generating electricity from willow biomass, *Renewable Energy*, 29: 1023–1042.
60. Benetto, E., Rousseaux, P., and Blondin, J. 2004. Life cycle assessment of coal by-products based electric power production scenerios, *Fuel*, 83: 957–970.
61. van der Horst, D. 2002. "Sustainability of biomass electricity systems. An assessment of costs, environmental and macro-economic impacts in Nicaragua, Ireland and Netherlands, " *Energy Policy*, 30(2): 167–169.
62. Siemons, R.V. 2002. "A development perspective for biomass fuelled electricity generation technologies. Economic technology assessment in view of sustainability" PhD thesis. University of Amsterdam, the Netherlands.
63. Davidson, R. 1997. Co-processing waste with coal, IEA Coal Research, *IEAPER/36*.
64. Wu, J., Fang, Y., and Wang, Y. 2011. Production of syngas by methane and coal co-conversion in fluidized bed reactor (personal communication).

65. Kaupilla, R. 2007. Concurrent production of syngas and steel from coal, biomass, and iron ore, U.P. Steel, in response to solicitation: DE-PS02-07ER07-36, Topic 57a, November 26.
66. Kim, B.J. and Mitchell, R.E. 2011. Coal and Biomass Gasification under Supercritical Water Conditions, Global Climate and Energy Project, Stanford University, U.S. (Personal communication).
67. Kajitani, S., Zhang, Y., Umemoto, S., Ashizawa, M., and Hara, S. 2010. Co-gasifiaction reactivity of coal and woody biomass in high temperature gasification, *Energy and Fuels*, 24(1): 145–151.
68. Seo, M.W., Goo, J.H., Kim, S.D., Lee, S.H., and Choi, Y.C. 2010. Gasification characteristics of coal/biomass blend in a dual circulating fluidized bed reactor, *Energy Fuels*, 24(5): 3108–3118.
69. Aznar, M.A., Caballero, M.A., Sancho, J.A., and Frances, S.E. 2006. Plastic waste elimination by co-gasification with coal and biomass in fluidized bed with air in pilot plant, *Fuel Process. Technol.*, 87: 409–420.
70. Williams, R.H., Larson, E.D., and Jin, H. 2006. Comparing climate-change mitigating potentials of alternative synthetic liquid fuel technologies using biomass and coal, *Fifth Annual Conference on Carbon Capture and Sequestration*, May 8–11.
71. Ahmaruzzaman, M. and Sharma, D.K. 2005. Non-isothermal kinetic studies on co-processing of vacuum residue, plastics, coal, and petrocrop, *J. Anal. Appl. Pyrolysis*, 73(2, June): 263–275.
72. Liu, K. and Meuzelaar, H.L.C. 1996. Catalytic reactions in waste plastics, HDPE and coal studied by high-pressure thermogravimetry with on-line GC/MS, *Fuel Processing Technology*, 49(1–3, Oct.–Dec.): 1–15.
73. Sweeten, J.M., Annamalai, K., Thien, B., and McDonald, L.A. 2003. Co-firing of coal and cattle feedlot biomass (FB) fuels, Part I, Feedlot biomass (cattle manure) fuel quality and characteristics, *Fuel*, 82(10): 1167–1182.
74. Chmielniak, T. and Sciazko, M. 2003. Co-gasification of biomass and coal for methanol synthesis, *Appl. Energy*, 74: 393–403.
75. Demirbas, A. 2003. Sustainable Cofiring of biomass and coal, *Energy Conversion Manage.*, 44(9, June): 1465–1479.
76. Pan, Y.G., Velo, E., Roca, X., Manya, J.J., and Puigjaner, L. 2000. Fluidized bed co-gasification of residual biomass/poor coal blends for fuel gas production, *Fuel*, 79: 1317–1326.
77. Sjostrom, K., Chen, G., Yu, Q., Brage, C., Rosen, C., 1999. Promoted reactivity of char in co-gasification of biomass and coal: synergies in the thermochemical process, *Fuel*, 78: 1189–1194.
78. Brown, R.C., Liu, Q., and Norton, G. 2000. Catalytic effects observed during the co-gasification of coal and switchgrass, *Biomass Bioenergy*, 18: 499–506.
79. Sun, H., Song, B.H., Jang, Y.W., and Kim, S.D. 2011. The characteristics of steam gasification of biomass and waste filter carbon (personal communication).
80. Gregg, D.W., Taylor, R.W., Campbell, J.H., Taylor, J.R., and Cotton, A. 1980. Solar gasification of coal, activated carbon, coke and coal and biomass mixtures, *Solar Energy*, 25 (4): 353-364.
81. Prins, M.J., Ptasinski, K.J., and Janssen, F.J.J.G. 2007. From coal to biomass gasification: Comparison of thermodynamic efficiency, *Energy*, 32: 1248–1259.

82. McLendon,T.R., Lui, A.P., Pineault, R.L., Beer, S.K., and Richardson, S.W. 2004. High-pressure co-gasifiaction of coal and biomass in a fluidized bed, *Biomass Bioenergy*, 26: 377–388.
83. Kumabe, K., Hanaoka, T., Fujimoto, S., Minowa, T., and Sakanishi, K. 2007. Co-gasification of woody biomass and coal with air and steam, *Fuel*, 86: 684–689.
84. Andries, J., de Jong, W., and Hein, K.R.G. 1998. Gasification of biomass and coal in a pressurized fluidized bed gasifier, DGMK Tagugsbericht, 9802 Beitraege zur DGMK-FaChbereichstatung Energetische und Stoffliche Nutzung von Ab faellen und Nachwachsenden Rohstoffen, pp. 319–326.
85. Sjostrom, K., Bjornbom, E., Chen, G., Brage, C., Rosen, C., and Yu, Q. 1992–1994. Synergistic effects in coal gasification of coal and biomass. *APAS Clean Coal Technology Program*, vol. 3, p. c3.
86. Olsen, A., Rathmann, O., Gjernes, E., Fjellerup, J., Liierup, J.B., Hald, P. Hansen, L.K., and Kirkegaard, M. 1995. Combustion and gasification of coal and straw under pressurized conditions, Riso-R-808 (EN).
87. Andre, R.N., Pinto, F., Franco, C., Dias, M., Gulyurtiu, I., and Matos, M.A.A. 2005. Fluidized bed co-gasification of coal and olive oil industry wastes, *Fuel*, 84(12-13, Sept.): 1635–1644.
88. Uson, S., Valero, A., Correas, L. and Martinez, A. 2004. Co-gasification of coal and biomass in an IGCC power plant: Gasifier modeling, *Int. J. Thermodynamics*, 7(4, Dec.): 165–172.
89. Dayan, J., Brion, V., Shavit, A., Zimmels, Y., and Zvegilsky, D. 2003. Simulation of a two chamber catalytic organic-waste/coal gasifier. *22nd International Conference on Modelling Identification and Control*, Innsbruck, Austria.
90. Hirs, G.G. 2000. ECOS 2000: Biomass and coal gasification and acoustics of combustion, *Eurotherm Seminars 66 and 67 Biomass, Coal and Oil Gasification and Acoustics of combustion*, ABB, Essent, EZH, Gemeente Enschede, November.
91. Pinto, F., Franco, C., Andre, R.N., Tavares, C., Dias, M., Gulyurtlu, I., and Matos, M. 2003. Effect of experimental conditions on co-gasification of coal, biomass, and plastic waste, *INETI DEECA Edificio J.*, 1649-038, Lisboa Codex, Portugal; also *Fuel*, 82 (15-17 Oct./Dec.): 1967–1976.
92. Backreedy, R.I., Jones, J.M., Pourkashanian, M., and Williams, A. 2002. Modelling of reaction of oxygen with coal and biomass chars, *Proceedings of Combustion Institute*, 29 (1): 415–420.
93. De Jong, W., Unal, O., Andries, J., Hein, K.R.G., and Spliethoff, H. 2003. Biomass and fossil fuel conversion by pressurized fluidized bed gasification using char, *Biomass Bioenergy*, 25(1): 59–83.
94. Pinto, F., Franco, C., Andre, R.N., Miranda, M., Gulyurtlu, I., and Cabrita, I. 2002. Co-gasification study of biomass mixed with plastic wastes, *Fuel*, 81 (3, Feb.): 291–207.
95. Arcate, J.R. 1998. Biomass charcoal cofiring with coal, *Proceedings of International Gas Turbine and Aeroengine Congress Conference*, Stockholm.
96. Collot, A., Megaritis, A., Herod, A.A., Dugwell, D., and Kandiyoti, K. 1998. *Proceedings of International Gas Turbine and Aeroengine Congress Conference*, Stockholm.
97. De Jong, W., Andries, J., and Hein, K.R.G. 1998. Coal-biomass gasification in a pressurized fluidized bed gasifier In *ASME International GT and Aerospace Congress*, Stockholm, SE, June 2–5, pp. 1–7.

98. Uson, S. and Valero, A. 2006. Oxy-co-gasification of coal and biomass in an integrated gasification combined cycle, *Energy*, 31(10-11): 1643–1655.
99. Yan, Q., Guo, L., Xing, L., and Zhang, X. 2005. Hydrogen production from co-gasification of coal and biomass in supercritical water, *J. Xian Jiaotong Univ.*, 39(5, May): 454–457.
100. Brage, C., Yu, Q.Z., and Sjostrom, K. 1995. Characterization of tars from coal-biomass gasification, *Proceedings of Third International Symposium on Coal Combustion Science and Technology*, Beijing, pp. 45–52.
101. Loo van, S. and Koppejan, J. (Eds.) 2004. *Handbook of Biomass Combustion and Co-Firing*, prepared by Task 32 of the Implementing Agreement on Bioenergy under the auspices of the International Energy Agency, Twente University Press.
102. European Commission, 2000. Addressing the Constraints for Successful Replication of Demonstration Technologies for Co-Combustion of Biomass/Waste, booklet DIS 1743/98-NL.
103. EUBION (European Bioenergy Networks), ALTENER, Biomass co-firing-an efficient way to reduce greenhouse gas emissions.
104. Veringa, H. 2005. ECN Biomass, Petten, The Netherlands (Dec.). Personal communication.
105. Valero, A. and Uson, S. 2005. Oxy-co-gasification of coal and biomass in an integrated gasification combined cycle (IGCC) power plant, University of Zaragoza: Center for Research and Energy Resources and Consumptions (CIRCE), April.
106. Mudulodu, S. 2002. *A Systems Analysis of Pyrolysis Systematics to Serve Energy Requirements*, Graduate School thesis, Univ. of Florida.
107. UC Davis, http://biomass.ucdavis.edu/pages/reports/UC_CIMWB_Appendices.pdf.
108. Cobb, J. 2007. Production of Synthesis gas by Biomass Gasification.
109. Boerrigter, H. et al. 2005. OLGA tar removal technology-proof of concept for application in integrated biomass gasification combined heat and power (CHP) systems, ECN-C-05-009, January.
110. Boerrigter, H., Calis, H., Slort, D., Bodenstaff, H., Kaandro, A., den Uil, H., and Rabou, L. 2004. Gas cleaning for integrated biomass gasification (BG) and Fischer-Tropsch FT systems, ECN, Petten, The Netherlands, ECN- report number ECN-C-04-056, 59.
111. Argonne Premium Fuel Sample, Analytical Data, Ultimate and Proximate Analyses, http://www.anl.gov/PCS/report/part2.html.
112. Kurkela, E., Stahlberg, P., and Laatikainen, J. 1993. *Pressurized Fluidized-Bed Gasification Experiments with Wood, Peat, and Coal at VTT in 1991–1992*, VTT No. 161.
113. McDaniel, J. 2002. *Biomass Gasification at Polk Power Station-Final Technical Report*, DOE award DE-FG26-01NT41365 (May).
114. The Alter NRG/Westinghouse plasma gasification process. 2008. *Independent Waste Technology* report by Juniper Consulting Services Limited, Bathurst House, Bisley UK (Nov.).
115. Shah, Y.T. 1981. *Reaction Engineering in Direct Coal Liquefaction*, Reading, MA: Addison-Wesley.
116. Huber G.W., Iborra, S., and Corma, A. 2006. Synthesis of transportation fuels from biomass: Chemistry, catalysts, and engineering, *Chem. Rev. Amer. Chem. Soc.*
117. Moffatt, J.M. and Overend, R.P. 1985. *Biomass*, 7: 99.

118. Karaca, H. and Koyunoglu, C. 2010. Co-liquefaction of Elbistan Lignite and Biomass. Part I: The effect of process parameters on the conversion of liquefaction products, *Energy Sources, A.*, 32: 495–511.
119. Karaca, F. and Bolat, E. 2000 Coprocessing of a Turkish lignite with a cellulosic waste material: 1. The effect of coprocessing on liquefaction yields at different reaction temperatures, *Fuel Process. Technol.*, 64 (1–3, May), 47–55.
120. Karaca, F. and Bolat, E. 2002. Coprocessing of a Turkish lignite with a cellulosic waste material: 2. The effect of coprocessing on liquefaction yields at different reaction pressures and sawdust/lignite ratios, *Fuel Process. Technol.*, 75(2, Feb.): 109–116.
121. Karaca, F., Bolat, E., and Dincer, S. 2002. Coprocessing of a Turkish lignite with a cellulosic waste material: 3. A statistical study on product yields and total conversion, *Fuel Process. Technol.*, 75(2, Feb.): 1117–1127.
122. Stiller, A.H., Dadyburjor, D.B., Wann, J.P., Tian, D., and Zondlo, J.W. 1996. Co-processing of agricultural and biomass waste with coal, *Fuel Process. Technol.*, 49 (1–3, Oct.-Dec.): 167–175.
123. Ikenaga, N., Ueda, C., Matsui, T., Ohtsuki, M., and Suzuki, T. 2001. Co-liquefaction of micro algae with coal using coal liquefaction catalysts, *Energy Fuels*, 15: 350–355.
124. Peterson, A.A., Vogel, F., Lachance, R.P., Froling, M., Antal, M.J., and Tester, J.W. 2008. *Energy Environ. Sci.*, 1: 32–65.
125. Kumar, S. and Gupta, R.B. 2009. *Energy Fuels*, 23: 5151–5159.
126. Byrd, A.J., Kumar, S., Kong, L., Ramsurn, H. and Gupta, R.B. 2011. *Int. J. Hydrogen Energy*, 36: 3426–3433.
127. Akhtar, J. and Amin, N.A.S. 2011. "A review on process conditions for optimum bio-oil yield in hydrothermal liquefaction of biomass," Renewable and Sustainable Energy Reviews, 15: 1615–1824.
128. Sevilla, M. and Fuertes, A.B. 2009. "The production of carbon materials by hydrothermal carbonization of cellulose," *Carbon*, 47, 2281–2289.
129. Funke, A. and Ziegler, F. 2010. "Hydrothermal carbonization of biomass: A summary and discussion of chemical mechanisms for process engineering," *Biofuels, Bioproducts, Biorefining*, 4: 160–177.
130. Demirbas, A. 2010. Green energy and technology, *Biorefineries for Biomass Upgrading Facilities*, London: Springer-Verlag.
131. Kranich, W.L. 1984. *Conversion of Sewage Sludge to Oil by Hydroliquefaction*, EPA-600/2 84-010. Report for the U.S. EPA, Cincinnati, Ohio.
132. Li, S. and Xu, S. 2006. *Copyrolysis of Coal and Biomass*, Beijing: Tsinghua Tongfang Knowledge Network Technology.
133. Sharypov, V.I., Marin, N., Beregovtsova, N.G., Baryshnikov, S.V., Kuznetsov, B.N., Cebolla, V.L., and Weber, J.V. 2002. Copyrolysis of wood biomass and synthetic polymer mixtures. Part I: Influence of experimental conditions on the evolution of solids, liquids, and gases, *J. Anal. Appl. Pyrolysis*, 64(1, July): 15–28.
134. Li, Z. and Xu, S. 2007. Co-pyrolysis of biomass and coal in a free fall reactor, *Fuel*, 86(3, Feb.): 353–359.
135. Moghtaderi, B., Meesri, C., and Wall, T. 2004. Pyrolytic characteristics of blended coal and biomass, *Fuel*, 83: 745–750.
136. Zhou, L., Wang, Y., Huang, Q., and Cai, J. 2006. Thermogravimetric characteristics and kinetic of plastic and biomass blends co-pyrolysis, *Fuel Process. Technol.*, 87(11, Nov.): 963–969.

137. Pan, Y.G., Velo, E., and Puigjaner, L. 1996. Pyrolysis of blends of biomass with poor coals, *Fuel*, 75(4): 412–418.
138. Niu, Y., Tan, H., Wang, X., and Xu, T. 2011. Synergistic effect on co-pyrolysis of capsicum stalks and coal, *African J. Biotechnol.*, 10(2, Jan.): 174–179.
139. Rodjeen, S., Mekasut, L., Kuchontara, P., and Piumsomboon, P. 2006. Parametric studies on catalytic pyrolysis of coal-biomass mixture in a circulating fluidized bed, *Korean J. Chem. Engrg.*, 23(2)L 216–223.
140. Huffman, G.P., Zhen, F., and Mahajan, V. 1995. Direct liquefaction of plastics and coliquefaction of coal-plastic mixtures, *Amer. Chem. Soc. Div. Fuel Chem.*, pp. 34–37 (Dec.).
141. Collot, A.G., Zhuo, Y., Dugwell, D.R., and Kandiyoti, R. 1999. Co-pyrolysis and co-gasification of coal and biomass in bench-scale fixed bed and fluidized bed reactors, *Fuel*, 78: 667–679.
142. Bonelli, P.R., Buonomo, E.L., and Cukierman, A.L. 2007. Pyrolysis of sugarcane bagasse and co-pyrolysis with Argentinean Subbituminous coal, *Energy Sources, A*, 29: 731–740.
143. Vuthaluru, H.S. 2004. Thermal behavior of coal/biomass blends during co-pyrolysis, *Fuel Process. Technol.*, 85(2–3, Feb.): 141–155.
144. Storm, C., Rudiger, H., Spliethoff, H., and Hein, K.R.G. 1999. Co-pyrolysis of coal/biomass and coal/sewage sludge mixtures, *J. Eng. Gas Turbines Power*, 121(1, Jan.): 55.
145. Ma, G.L., Liu, G., and Cao, Q. 2006. *Research Progress of Co-Pyrolysis of Biomass with Polymers and Coal*, Beijing: Tsinghua Tongfang Knowledge Network Technology.
146. Xu, X.D. and Antal, M.J. 1998. Gasification of sewage sludge and other biomass for hydrogen production in supercritical water, *Environ. Progress*, 17(4): 215–220.
147. Sinag, A., Kruse, A., and Schwarzkopf, V. 2003. *Ind. Eng. Chem. Res.*, 42: 3516–3522.
148. Demirbas, A., Ozturk, T., and Demirbas, M.F. 2006. *Energy Sources, A*, 28: 1473–1482.
149. Demirbas, A. 2000. Recent advances in biomass conversion technologies, *Energy Edu. Sci. Technol.*, 6, 19–40.
150. Valenzuela, M.B., Jones, C.W., and Agarwal, P.K. 2006. *Energy Fuels*, 20: 1744–1752.
151. Markovic, Z., Markovic, S., Engelbrecht, J.P., and Visser, F.D. 2000. *S. African J. Chem.*, 53(3): 2.
152. Bimakr, M., Rahman, R., Taip, F., Chuan, L., Ganjloo, A., Selamat, J., and Hamid, A. 2009. *European J. Sci. Res.*, 33(4): 679–690.
153. Sajfrtova, M., Lickova, I., Wimmmerova, M., Sevova, H., and Wimmer, Z. 2010. *Int. J. Mol. Sci.*, 11: 1842–1850.
154. Stahl, E., Schultz, E., and Mangold, H.K. 1980. *J. Agric. Food Chem.*, 28: 1153–1157.
155. Xiu, S., Shabhazi, A., Wang, L., and Wallace, C.W. 2010. *Am. J. Eng. Appl. Sci.*, 3(2): 494–500.
156. Wang, T. and Zhu, X. 2004. Conversion and kinetics of the oxidation of coal in supercritical water, *Energy Fuels*, 18: 1569–1572.
157. Erzengin, M. and Kuchuk, M.M. 1998. Liquefaction of sunflower stalk by using supercritical extraction, *Energy Conversion Manage.*, 39(11, Aug.): 1203–1206.
158. Sener, U., Genel, Y., Saka, C., Kilicel, F., and Kucuk, M.M. 2010. Supercritical fluid extraction of cotton stalks, *Energy Sources A*, 32: 20–25.

159. Demirbas, A. 2000. Liquefaction of olive husk by supercritical fluid extraction, *Energy Conversion Manage.*, 41(17, Nov.): 1875–1883.
160. Aresat, M., Dibenedetto, A., Carone, M., Colinna, T., and Fragale, C., 2005. Production of biodiesel from microalgae by supercritical CO2 extraction and thermochemical liquefaction, *Environ. Chem. Lett.*, 3: 136–139.
161. Guo, L., Cao, C., and Liu, Y., 2010. *Biomass*. M. Momba and F. Bux (Eds.) pp. 202 (September).
162. Matsumura, Y., Nonaka, H., Yokura, H., Tsutsumi, A., and Yoshida, K., 1999. Co-liquefaction of coal and cellulose in supercritical water, *Fuel*, 78(9, July): 1049–1056.
163. Veski, R., Palu, V., and Kruusement, K., 2006 Co-liquefaction of Kukersite oil shale and pine wood in supercritical water, *Oil Shale Publisher, Estonian Academy*, 23 (3, Sept.): 1–8.

Index

Page references in **bold** refer to tables.

D

E

Q

R

S